D1748724

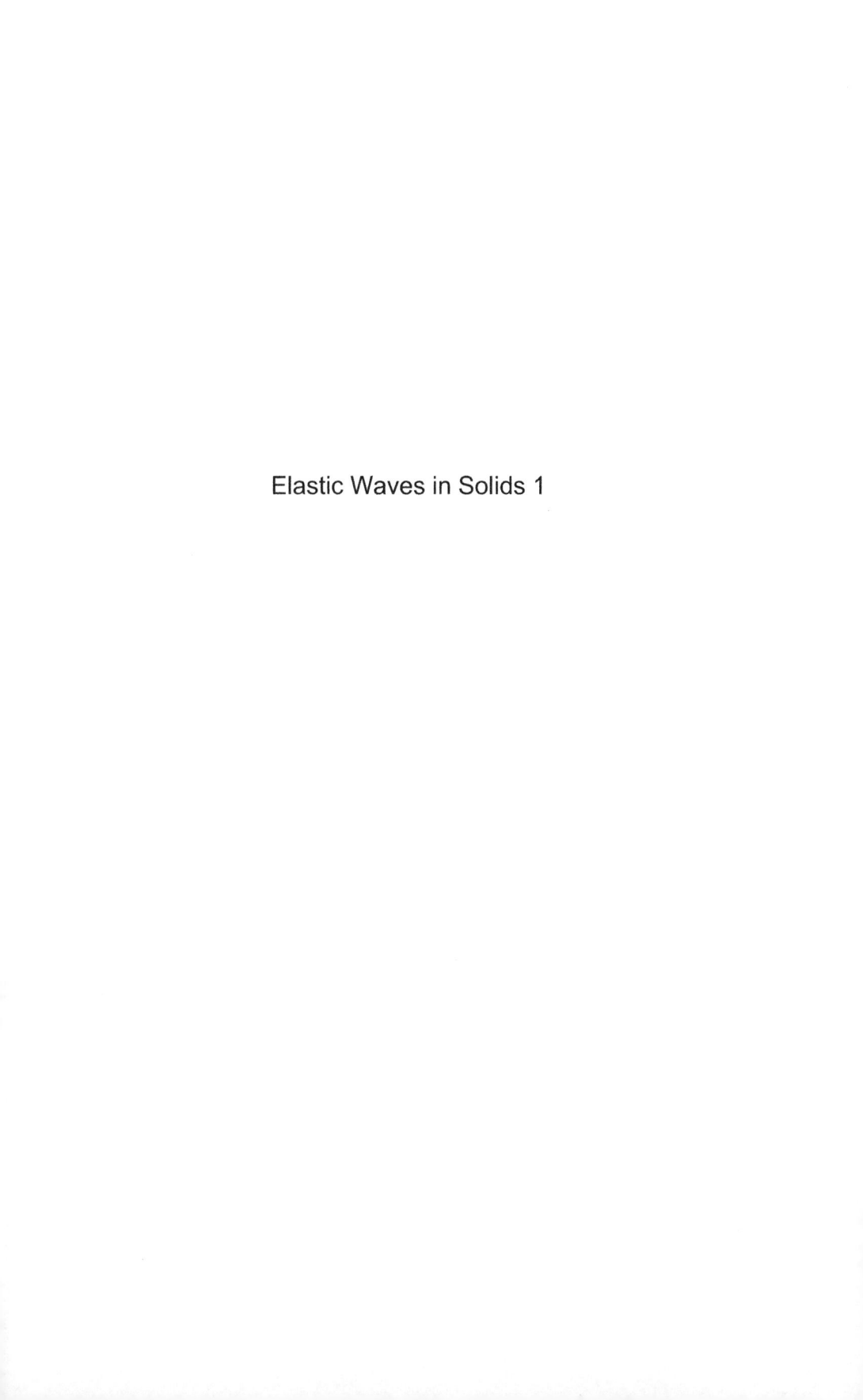

Elastic Waves in Solids 1

*Series Editors
Pierre-Noël Favennec† and Frédérique de Fornel*

Elastic Waves in Solids 1

Propagation

Daniel Royer
Tony Valier-Brasier

iSTE   WILEY

First published 2022 in Great Britain and the United States by ISTE Ltd and John Wiley & Sons, Inc.

Apart from any fair dealing for the purposes of research or private study, or criticism or review, as permitted under the Copyright, Designs and Patents Act 1988, this publication may only be reproduced, stored or transmitted, in any form or by any means, with the prior permission in writing of the publishers, or in the case of reprographic reproduction in accordance with the terms and licenses issued by the CLA. Enquiries concerning reproduction outside these terms should be sent to the publishers at the undermentioned address:

ISTE Ltd
27-37 St George's Road
London SW19 4EU
UK

www.iste.co.uk

John Wiley & Sons, Inc.
111 River Street
Hoboken, NJ 07030
USA

www.wiley.com

© ISTE Ltd 2022

The rights of Daniel Royer and Tony Valier-Brasier to be identified as the authors of this work have been asserted by them in accordance with the Copyright, Designs and Patents Act 1988.

Any opinions, findings, and conclusions or recommendations expressed in this material are those of the author(s), contributor(s) or editor(s) and do not necessarily reflect the views of ISTE Group.

Library of Congress Control Number: 2021951482

British Library Cataloguing-in-Publication Data
A CIP record for this book is available from the British Library
ISBN 978-1-78630-814-6

Contents

Preface ix

List of Main Symbols xiii

Chapter 1. Propagation in an Unbounded Solid 1
   1.1. Reviewing the mechanics of continuous media 2
      1.1.1. Conservation equations 2
      1.1.2. Kinematics of continuous media 9
      1.1.3. Poynting's theorem: energy balance 10
      1.1.4. Stress–strain relationship: Maxwell relations 12
   1.2. Isotropic solid 14
      1.2.1. Constitutive equations 14
      1.2.2. Equations of propagation, wave decoupling 16
      1.2.3. Traveling, plane, sinusoidal waves 21
      1.2.4. Polarization 25
      1.2.5. Acoustic intensity 26
      1.2.6. Cylindrical and spherical waves 27
   1.3. Anisotropic solid 32
      1.3.1. Symmetry and elasticity tensor 32
      1.3.2. Propagation equation, phase velocity, polarization 41
      1.3.3. Propagation in an orthotropic material 43
      1.3.4. Group velocity and energy velocity 45
      1.3.5. Slowness surface and wave surface 48
   1.4. Piezoelectric solid 54
      1.4.1. Constitutive equations 54
      1.4.2. Reduction in the number of independent piezoelectric constants . . 59
      1.4.3. Plane waves in a piezoelectric crystal 61
   1.5. Viscoelastic media 70
      1.5.1. Constitutive equation of linear viscoelasticity 71

1.5.2. Simple rheological models 72
1.5.3. Velocity and attenuation in a viscoelastic medium 74
1.5.4. Time–temperature superposition principle 77
1.5.5. Newtonian fluid 78

Chapter 2. Reflection and Transmission at an Interface 81

2.1. Boundary conditions 82
2.2. Direction and polarization of reflected and transmitted waves 85
    2.2.1. Graphical construction 86
    2.2.2. Wave decoupling 87
    2.2.3. Critical angle, evanescent wave and total reflection 89
    2.2.4. Conservation of energy 91
2.3. Isotropic solid: transverse horizontal wave 93
    2.3.1. Reflection and transmission between two solids 93
    2.3.2. Plate between two solids, impedance matching 96
2.4. Isotropic media: longitudinal and transverse vertical waves 100
    2.4.1. Reflection on a free surface 100
    2.4.2. Solid–fluid interface 105
2.5. Anisotropic medium: diffraction matrix 116
    2.5.1. Analytical resolution 117
    2.5.2. Expression for the stresses 119
    2.5.3. Sorting the solutions 120
    2.5.4. Considerations of symmetry 121
    2.5.5. Reflection and transmission coefficients, interface waves 124
    2.5.6. Interface between an orthotropic solid and an isotropic solid 127

Chapter 3. Surface Waves and Interface Waves 131

3.1. Surface waves 132
    3.1.1. Isotropic solid: Rayleigh wave 132
    3.1.2. Anisotropic solid 141
    3.1.3. Piezoelectric crystal 151
3.2. Interface waves 164
    3.2.1. Isotropic solid-perfect fluid interface 164
    3.2.2. Interface between two isotropic solids 169
3.3. Bleustein–Gulyaev wave 173

Chapter 4. Guided Elastic Waves 179

4.1. Waveguide, group velocity 180
    4.1.1. Elementary planar waveguide 181
    4.1.2. Velocity of a wave packet 184
    4.1.3. Propagation of a Gaussian pulse 187
4.2. Transverse horizontal waves 189

  4.2.1. Guided TH modes 190
  4.2.2. Love wave 190
  4.2.3. Love wave in an inhomogeneous medium 192
 4.3. Lamb waves 196
  4.3.1. Free isotropic plate 196
  4.3.2. Isotropic plate immersed in a fluid 221
  4.3.3. Free anisotropic plate 226
 4.4. Cylindrical guides 235
  4.4.1. Compressional modes 239
  4.4.2. Flexural modes 243
  4.4.3. Torsional modes 244
  4.4.4. Tubular waveguide 246

Appendix 1. Differential Operators in Cylindrical and Spherical Coordinates 247

Appendix 2. Symmetry and Tensors 253

Appendix 3. Transport of Energy 279

References 287

Index 295

Preface

This book follows two books co-authored with Eugène Dieulesaint devoted to *Elastic Waves in Solids*; the first book is subtitled *Free and Guided Propagation*, and the second, *Generation, Acousto-optic Interaction, Applications*.

This book is also divided into two volumes. It is designed for students who are pursuing their masters in physics, mechanics or geophysics, as well as for other graduate students, PhD students, engineers, researchers and professors. The objective is to analyze the propagation, interactions and generation of elastic waves in a large variety of solid media and structures. Wherever possible, a common formalism has been used that is applicable to both bulk and surface waves, as well as to guided waves.

Elastic waves are vibrations that propagate in any medium: gaseous, liquid or solid. The term "elastic" is used to describe the mechanical behavior of the propagation medium. When the frequency of these waves is in the audible range (approximately between 20 Hz and 20 kHz), they are commonly called "acoustic waves" or "sound waves"; they are called infrasound or ultrasonic waves if their frequency is below or above this range. The term "acoustics" is often broadly used for anything related to matter waves, regardless of their frequency. Given the earlier specifications, this is not the most appropriate term; however, it has the advantage of defining a discipline, such as mechanics, optics, thermodynamics, and so on. Acoustics is often considered as the oldest of the physical sciences. A brief review of the historical evolution of this field and a summary of the applications of elastic waves are used to explain the contents of this book.

It was known since Poisson's memoir, published in 1829, that longitudinal or transverse matter waves can propagate in the bulk of an isotropic, elastic solid. At the end of the 19th century, on the earliest seismic recordings, P wave trains (arriving first) and S (or secondary) wave trains were identified with the arrivals of bulk longitudinal waves (the fastest) and bulk transverse waves. A third, late echo was attributed to surface waves, discovered in 1885 by Lord Rayleigh. In the early 20th century, seismic waves were used to study the interior of the Earth and to determine its structure. Therefore, it is not surprising that most elastic waves were discovered by geophysicists and carry their names: Lamb, Love, Stoneley and Scholte waves.

Until 1915 and the research carried out by Paul Langevin and Constantin Chilowsky, earthquakes were the only means for generating these elastic waves, and this phenomenon was hard to reproduce and was quite destructive. However, experiments carried out at the École Supérieure de Physique et de Chimie Industrielles (ESPCI) in Paris, then in the Seine, and finally in Toulon demonstrated that the piezoelectric effect (discovered in 1880 by Pierre and Jacques Curie) could generate ultrasound waves in water and detect the echo reflected by a target. For many decades, quartz was the only piezoelectric crystal used. Given its exceptional mechanical properties and thermal stability, it began to be used in the emerging field of telecommunications to stabilize and filter the frequency of broadcasting transmitters and receivers. The usage of quartz resonators spread after K.S. Van Dyke and D.W. Dye independently developed an equivalent circuit of a piezoelectric resonator.

Following intense research carried out after the Second World War, new piezoelectric materials were developed. These included notably lead zirconate titanate (PZT) ceramics, which could be made piezoelectric by applying an electric field. These materials, whose electromechanical coupling coefficient is much higher than that of quartz, greatly facilitated the generation and detection of elastic waves of frequency lower than 10 MHz. As they could be used to create sensitive transducers of large dimensions and of varied shapes, these ceramics constitute the active elements of transmitters and receivers in sonars and of medical and metallurgical ultrasonic equipment.

In 1965, R.M. White and F.W. Voltmer implemented the idea of generating and detecting surface acoustic waves using two comb-shaped electrodes deposited on a piezoelectric material. The exploitation of this technique led to a series of spectacular applications of Rayleigh waves. These surface acoustic wave filters, which are very compact, can be mass-produced owing to microelectronic technologies and operate in a wide range of frequencies (50 MHz–5 GHz). They have invaded the consumer electronics market, in products such as televisions and mobile phones. The continuous improvement in their performances required the elaboration of new piezoelectric crystals of large dimensions, as well as a finer and finer analysis of the propagation, interaction and generation of surface waves.

From the 1990s onward, we have witnessed a large-scale development of elastic waves in varied domains, such as medical imaging and therapy, non-destructive evaluation of materials, sensors of physical or biological quantities, acoustic detection and localization systems for submarines applications, and for locating and exploring oil fields. These advances have different origins, sometimes involving a combination of factors: the elaboration of composite materials and thin piezoelectric layers, the refinement of high-performance instruments (acoustic microscopes, multi-element transducers), innovative methods (time-reversal acoustics), as well as progress in computer sciences, whether for simulations or running processes. Given the scope of our current text, it is not possible to fully describe these applications, which evolve so fast.

Our objective is to provide the reader with the theoretical bases required to understand these developments, to explore them in detail and to elaborate on other advances. In order to move away from abstraction, the phenomena are illustrated by many figures corresponding to commonly used structures and materials. Furthermore, the text regularly specifies the order of magnitude of physical parameters.

The first volume is divided into four chapters and three appendices. Chapter 1 deals with the propagation of elastic waves in an unbounded solid. In the absence of any boundary, the only problem is to establish the equations for the propagation of a mechanical disturbance and to solve them. However, in order to do this, the constitutive laws for the solid must be known. This solid may be isotropic, anisotropic, piezoelectric, purely elastic or viscoelastic. Chapter 2 focuses on the phenomena of reflection and transmission between two semi-infinite media separated by a planar interface. The propagation equations in each medium must be supplemented by limiting conditions that ensure the connection of the physical quantities on the boundary. Several examples are treated: solid–vacuum interfaces, solid-perfect fluid and isotropic solid-anisotropic crystal. For each configuration, the analysis shows modes that exist in the absence of any incident waves; these surface and interface waves are studied in Chapter 3. A very important example is that of the semi-infinite solid, along whose surface a pure or generalized Rayleigh wave is always propagated, regardless of the material. Other waves appear at the interface between a solid and a fluid (Scholte waves), while the conditions for existence are very restrictive at the interface between two solids (Stoneley waves). When the solid is limited by several parallel planes, the successive reflections of the bulk elastic waves give rise to guided waves. The properties of these guided waves are analyzed in Chapter 4. The simplest are transverse horizontal waves, discovered by Love, which are partly propagated in a layer and partly in its substrate. Given their practical importance, the Lamb waves, with two components, progressing in a plate are treated in detail both in the case of isotropic and anisotropic solids. The propagation of these plate modes is dispersive, like that of guided waves in cylinders, studied at the end of this chapter.

The second volume investigates the properties of acoustic fields emitted by various sources of finite dimensions, the interaction of elastic waves with cylindrical and spherical targets immersed in a fluid or buried in a solid, the means of generating and detecting the waves studied in the first volume.

The authors wish to extend their gratitude to Claire Prada, Director of Research at CNRS, institut Langevin Ondes et Images, for her help during the writing of this first volume.

October 2021

List of Main Symbols

a, b : amplitude of the waves.

c_{ijkl} (C_{IJ}) : stiffnesses (Voigt notation).

D : directivity factor.

$\underline{D}$: electric induction (displacement) vector.

E : Young modulus.

$\underline{E}$: electric field vector.

e_c (e_p) : kinetic (potential) energy per unit volume.

e_{ijk} (ε_{ij}) : piezoelectric (dielectric) constants.

$\underline{F}$: force density per unit mass.

$H(t)$: Heaviside step function.

I : acoustic intensity.

$\underline{J}$: Poynting vector.

K : bulk modulus.

$\underline{k}$ (k) : wave vector (number).

K_M : electromechanical coupling coefficient (mode M).

$\underline{l}$: unit vector normal to a surface.

$\underline{n}$: unit vector of the propagation direction.

P : transported power.

p_a : acoustic pressure.

$p\,(p_0)$: thermodynamic pressure (at rest).

$\underline{p},\underline{q}$: polarization vectors.

$P_s\,(p_s)$: power supplied by the source (per unit volume).

$r\,(t)$: amplitude reflection (transmission) coefficients.

$R\,(T)$: intensity reflection (transmission) coefficients.

s : entropy per unit volume.

$\underline{s}$: phase slowness vector.

T : absolute temperature.

$\underline{T}$: mechanical traction, stress vector.

$\underline{u}$: mechanical displacement vector.

U : internal energy per unit volume.

$\underline{v}$: particle velocity vector.

$\underline{V}^e$: energy velocity vector.

$V\,(V_g)$: phase (group) velocity.

$V_L\,(V_T)$: velocity of longitudinal (transverse) bulk waves.

$W\,(w)$: work (per unit volume).

$\underline{x}$: position vector.

X, Y, Z : crystallographic axes.

Z_e (Y) : electrical impedance (admittance).

Z ($\mathcal{Z}$) : acoustic (mechanical) impedance.

α : attenuation coefficient.

$\underline{\underline{\Gamma}}$ ($\underline{\underline{\gamma}}$) : elastic (piezoelectric) Christoffel tensor.

δ_{ij} : Kronecker's symbol.

$\delta(t)$: Dirac function.

$\underline{\underline{\varepsilon}}$ ($\underline{\underline{S}}$) : linearized strain tensor (piezoelectric solid).

η : viscosity coefficient.

$\theta_{i,r,t}$: angle of incidence, reflection, transmission.

Θ : dilatation.

κ : bulk wave velocity ratio (V_T/V_L).

λ, μ : Lamé constants.

ν : Poisson's ratio.

ρ : mass density.

ρ_e (σ_e) : electric charge density per unit volume (area).

σ (σ_d) : effective (differential) scattering cross-section.

$\underline{\underline{\sigma}}$: Cauchy stress tensor.

ϕ ($\underline{\psi}$) : scalar (vector) potential.

Φ : electric potential.

χ : decay factor.

ω : angular frequency.

1

Propagation in an Unbounded Solid

The objective of this chapter – divided into five sections – is to describe the propagation of mechanical vibrations in an unbounded solid. As elastic wavelengths are very long, compared to interatomic distances, the medium is considered to be continuous. The phenomena studied are macroscopic: we do not consider the individual motion of the molecules that constitute the medium, but that of a solid particle. This term designates an element of infinitesimal volume at the scale of the physical dimensions of the medium, which nonetheless contains a large number of molecules. The propagation equations for an elastic wave are, thus, deduced from the *conservation equations* of matter and momentum, complemented by the constitutive equation of the propagation medium.

In a *deformable solid*, two tensors play a fundamental role. The *mechanical stress*, whose tensor character comes from the fact that a force has three components and that the surface element on which it is exerted is defined by the components of its normal. The *strain*, which expresses the relative difference between the displacements of two neighboring material points, that is, the extremities of an element of infinitesimal length (section 1.1). A solid is elastic if it returns to its initial state when the external forces that deformed it are removed. This return to the state of rest is the work of internal stresses, which vanish with the strains. When these strains are small, the linear relationship between stresses and strains, which generalizes Hooke's law, defines a fourth rank tensor: the stiffness tensor.

The equation of propagation of elastic waves can be obtained by writing the generalized Hooke's law in the fundamental relation of dynamics. In an unbounded isotropic solid, a longitudinal displacement and a transverse displacement propagate independently, at two different velocities (section 1.2). In an anisotropic medium, for a given direction, three waves can propagate and the direction of propagation of energy is, in general, not parallel to the wave vector: it is given by the direction of the

Poynting vector (section 1.3). The solutions are represented by a slowness surface, analogous to the index surface in optics. This surface, composed of two sheets in optics, is formed here of three sheets. This representation reveals the importance of the symmetry axes, along which the propagation modes are pure, with the wave vector and energy vector generally being carried by this axis. In a piezoelectric crystal, at least one of the elastic waves is accompanied by an electric field (section 1.4). The importance of this electromechanical coupling, depending on the direction, can also be deduced from the slowness surface. A dimensionless parameter that expresses the ability of piezoelectric materials to generate or detect elastic waves is defined.

In practice, elastic waves propagate in any material medium: gaseous, liquid, homogeneous or inhomogeneous solid, isotropic or anisotropic solid. However, their amplitude decreases during the propagation because the bonds between atoms or molecules are not purely elastic (section 1.5). The attenuation of the waves is smaller when the medium is more ordered. Thus, losses are larger in a liquid than in a solid, while losses are larger in an amorphous or polycrystalline solid, compared to a single crystal. On the other hand, these losses increase rapidly with frequency, so that liquids are rarely used beyond 50 MHz and only single crystals are used at frequencies in the GHz domain.

1.1. Reviewing the mechanics of continuous media

In this section, we will establish the equations for the stress and the mechanical displacement fields that are independent of the medium (linear or nonlinear, isotropic or anisotropic, elastic or viscoelastic), supporting the elastic waves. The passage of an elastic wave in a solid is accompanied by a transport of energy without any permanent displacement of the matter. As in electromagnetism, the acoustic power crossing a surface element is equal to the flux of a vector called the Poynting vector.

1.1.1. *Conservation equations*

The fundamental equations of the mechanics of continuous media may be derived from the four laws of conservation of classical physics: the conservation of mass, the conservation of linear momentum, the conservation of angular momentum and the conservation of energy:

– the first law states that in the absence of a flow source, the mass of a given set of particles is invariant;

– Newton's second law indicates that in an inertial frame of reference, the change per unit time of linear momentum $m\underline{v}$ is equal to the external force applied to the particle with mass m:

$$\frac{\mathrm{d}(m\underline{v})}{\mathrm{d}t} = \underline{F}_{ext} \qquad [1.1]$$

– according to the theorem of angular momentum, the change per unit time of the angular momentum $\underline{x} \wedge m\underline{v}$, with respect to a fixed point, is equal to the moment of the external forces:

$$\frac{\mathrm{d}(\underline{x} \wedge m\underline{v})}{\mathrm{d}t} = \underline{M}_{ext} \qquad [1.2]$$

– the law of conservation of energy expresses, for an isolated mechanical system, the integral conversion of kinetic energy into potential energy and *vice versa*, so that their sum remains constant. However, a complete energy balance must include the power supplied by external sources and the energy dissipated by internal friction processes, through which the mechanical energy linked to the overall movement[1] progressively disappears, giving rise to the thermal excitation of molecules. This excitation is manifested through a (local) increase in the temperature of the medium. To extend the law of conservation of energy to non-conservative systems, it is necessary to introduce the notion of internal energy: the dissipation of part of the mechanical energy increases the internal energy of the system by a quantity δq, equal to the amount of heat released. It must be noted that other transformations are also possible, for example, structural changes.

When writing conservation equations, we have the choice between two equivalent descriptions of motion (Salençon 1988). In the Lagrangian description, each particle is identified by its initial position in a configuration taken as the reference. The motion is then described by the position of each particle over time (trajectory). The Eulerian description (chosen here) defines the motion by the knowledge, at each time t, of the velocity field of the particles $\underline{v}(\underline{x}, t)$.

Similarly, all physical quantities are represented by functions of position $\underline{x}$ and time t in the reference frame, which is assumed to be Galilean. To establish the integral forms of the conservation laws, let us consider any fixed volume V, inside the medium, bounded by a closed surface S of unit normal $\underline{l}$ oriented toward the exterior. The variation per unit time of the considered quantity consists of two terms: the derivative of the quantity contained in the volume V and the flux across the surface S arising from the transport of matter. The first term is expressed by a volume integral, the second one by a surface integral involving the scalar product $\underline{v}.\underline{l}$, which is written as:

$$\underline{v}.\underline{l} = v_1 l_1 + v_2 l_2 + v_3 l_3 = v_i l_i \qquad [1.3]$$

---

1 For example, the passage of an elastic wave.

taking into account the summation rule over dummy indices (Einstein convention). To convert a surface integral into a volume integral and *vice versa*, we will use Green's theorem:

$$\iint_S g_j l_j \, dS = \iiint_V \frac{\partial g_j}{\partial x_j} \, dV \qquad [1.4]$$

1.1.1.1. *Integral equations*

The mass density ρ of the medium varies according to the *law of conservation of matter*. The material flow at the point $\underline{x}$, with coordinates x_i ($i = 1, 2, 3$) and at time t, is equal to the product of the density $\rho(\underline{x}, t)$ and the velocity vector of the particles $\underline{v}(\underline{x}, t)$. The mass that crosses per unit time the surface element dS of unit normal $\underline{l}$ is equal to $\rho \underline{v} \cdot \underline{l} \, dS$. The integral of this quantity[2] on the surface S represents the decrease per unit time of the mass contained in the fixed volume V:

$$\frac{d}{dt} \iiint_V \rho \, dV + \iint_S \rho v_j l_j \, dS = 0 \qquad [1.5]$$

According to Newton's second law, the time derivative of the *linear momentum* is equal to the resultant force acting on the fixed volume V. These forces are exerted directly in the volume, like gravity, with a density $\underline{F}(\underline{x}, t)$ per unit mass, or through the intermediary of the surface S delimiting the volume V with a surface density $\underline{T}$. This is the case with actions exerted by the matter situated outside V, which are progressively transmitted by the bond strengths between molecules. Since their radius of action is very small on a macroscopic scale, their resultant is expressed by the integral on the surface S of the elementary forces $\underline{T}(\underline{l}) \, dS$. The vector $\underline{T}(\underline{l})$, called *mechanical traction* or *stress vector*, depends on $\underline{x}$ and t, but also on the orientation of the surface element dS, defined by the unit vector $\underline{l}$ normal to S (Figure 1.1).

In the case of a perfect fluid, this force is normal to the surface element and directed toward the interior of the volume V. It is expressed as a function of the hydrostatic pressure p:

$$T_i = -p l_i \qquad [1.6]$$

In a solid or in a viscous fluid, tangential forces appear. Let us write the condition for equilibrium of the volume V, taking into account the forces in the volume. The resultant of the forces must be zero for each component ($i = 1, 2, 3$):

$$\iint_S T_i(\underline{l}) \, dS + \iiint_V \rho F_i \, dV = 0 \qquad [1.7]$$

---

[2] Contrary to the case of a fluid, it is only in special cases that we need to introduce a flow source for a solid.

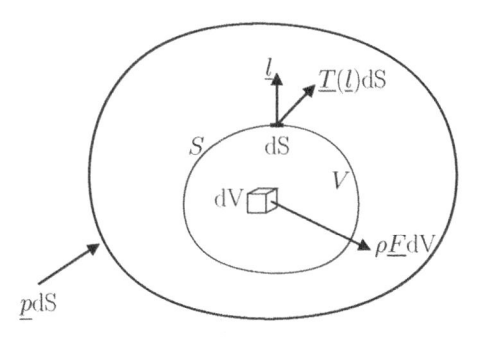

Figure 1.1. *Equilibrium of a fixed volume V inside the solid*

Due to equality between the volume integral and the surface integral, ρF_i must be the divergence of a second-rank tensor:

$$\rho F_i = -\frac{\partial \sigma_{ij}}{\partial x_j} \quad [1.8]$$

According to Green's theorem, the volume integral is transformed into a surface integral, leading to the relation:

$$\iint_S [T_i(\underline{l}) - \sigma_{ij} l_j] \, dS = 0 \quad [1.9]$$

Since this equality is satisfied for any surface S, the quantity within the square brackets is necessarily zero at each point. Each component of the mechanical traction $\underline{T}(\underline{l})$ is, therefore, a linear combination of the direction cosines of the normal:

$$T_i(\underline{l}) = \sigma_{ij} l_j \quad [1.10]$$

The nine quantities $\sigma_{ij}(\underline{x}, t)$, which only depend on the position, $\underline{x}$, and time t, form the *Cauchy stress tensor*. This tensor defines the mechanical state of the material: the force exerted on a unit surface by matter situated on the side of its normal is $T_i = \sigma_{ij} l_j$. If the surface element is perpendicular to one of the axes x_k ($l_j = \delta_{jk}$), then $T_i = \sigma_{ik}$, hence σ_{ik} is the ith component of the force acting per unit area on a surface perpendicular to the x_k axis (Figure 1.2). It appears that the Cauchy stress is the macroscopic quantity that associates to a surface element $d\underline{S}$ of the deformed configuration the force $d\underline{f}$ acting on this element:

$$d\underline{f} = \underline{\underline{\sigma}} \cdot d\underline{S} \quad \text{or} \quad df_i = \sigma_{ij} l_j \, dS = \sigma_{ij} \, dS_j \quad [1.11]$$

Let us write that the change in *linear momentum* over time is equal to the total force applied on the volume V:

$$\frac{d}{dt} \iiint_V \rho v_i \, dV + \iint_S (\rho v_i) v_j l_j \, dS = \iiint_V \rho F_i \, dV + \iint_S \sigma_{ij} l_j \, dS \quad [1.12]$$

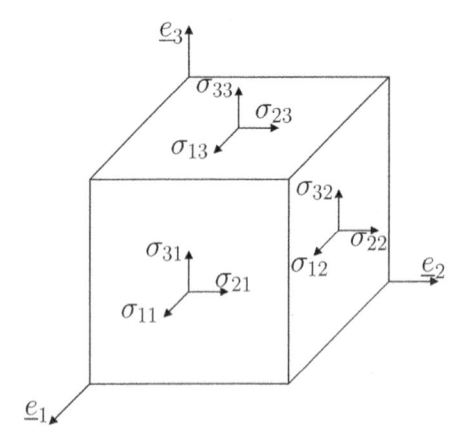

Figure 1.2. *Three stresses act on each of the three orthogonal faces. From the point of view of the macroscopic mechanics, these nine components express the physical interactions between the elementary constituents of the continuous medium*

On the left-hand side, the time derivative is the rate of change of the linear momentum ρv_i contained in the volume V and the surface integral represents the outgoing linear momentum per unit time. The right-hand side is the linear momentum created by the forces ρF_i exerted in the volume and by the mechanical tractions $T_i = \sigma_{ij} l_j$ exerted on the surface S.

1.1.1.2. *Local equations*

The integral forms [1.5] and [1.12] always apply, even in the presence of discontinuities. In a region where the quantities are continuous and differentiable, let us consider the volume V tend to zero in order to replace them by local differential equations.

We must first transform the time derivatives, given the fact that the volume V is fixed, that is, for any quantity g:

$$\frac{d}{dt} \iiint_V g(\underline{x},t) \, dV = \iiint_V \frac{\partial g(\underline{x},t)}{\partial t} \, dV \qquad [1.13]$$

For any volume V, however small, the equation for *the conservation of matter* or the continuity equation [1.5] is written as:

$$\iiint_V \left[\frac{\partial \rho}{\partial t} + \frac{\partial \rho v_i}{\partial x_i} \right] dV = 0 \qquad [1.14]$$

that is:

$$\frac{\partial \rho}{\partial t} + \frac{\partial \rho v_i}{\partial x_i} = 0 \qquad [1.15]$$

In the same way, the local form of the equation for the *conservation of linear momentum* [1.12] is written as:

$$\frac{\partial \rho v_i}{\partial t} + \frac{\partial}{\partial x_i}(\rho v_i v_j - \sigma_{ij}) = \rho F_i \qquad [1.16]$$

The differential equations [1.15] and [1.16] govern the dynamics of continuous media. They can be simplified by introducing the *Lagrangian derivative*:

$$\frac{d}{dt} = \frac{\partial}{\partial t} + v_i \frac{\partial}{\partial x_i} \qquad [1.17]$$

which represents the derivative for a particle followed in its motion. d/dt is also called the *particular derivative*. Indeed, during a time interval Δt, a particle P moves from $v_j \Delta t$ along the axis x_j and the value of a quantity g associated with this particle is, at time $t + \Delta t$:

$$g(\underline{x} + d\underline{x}, t + \Delta t) = g(x_j + v_j \Delta t, t + \Delta t) = g(\underline{x}, t) + \frac{\partial g}{\partial t}\Delta t + v_j \frac{\partial g}{\partial x_j}\Delta t \quad [1.18]$$

When Δt tends to zero, the limit of the ratio $\Delta g/\Delta t$ is the particular derivative:

$$\frac{dg}{dt} = \frac{\partial g}{\partial t} + v_j \frac{\partial g}{\partial x_j} = \frac{\partial g}{\partial t} + \underline{v} \cdot \text{grad } g \qquad [1.19]$$

The *Eulerian derivative* $\partial g/\partial t$, which represents the variation of g at the fixed point $\underline{x}$, is a local derivative. The nonlinear term expresses the variation of g due to the change in position of the particle (convection phenomenon). For the *conservation of matter*, it follows that:

$$\frac{d\rho}{dt} + \rho \frac{\partial v_i}{\partial x_i} = 0 \quad \text{or} \quad \frac{d\rho}{dt} + \rho \text{ div } \underline{v} = 0 \qquad [1.20]$$

By grouping the terms differently, equation [1.16] becomes:

$$\rho\left(\frac{\partial v_i}{\partial t} + v_j \frac{\partial v_i}{\partial x_j}\right) + v_i\left[\frac{\partial \rho}{\partial t} + \frac{\partial(\rho v_j)}{\partial x_j}\right] = \rho F_i + \frac{\partial \sigma_{ij}}{\partial x_j} \qquad [1.21]$$

and, taking into account the conservation of mass [1.15], the conservation of linear momentum can be written in the form of the *local equation of dynamics*:

$$\rho \frac{dv_i}{dt} = \rho F_i + \frac{\partial \sigma_{ij}}{\partial x_j} \quad \text{or} \quad \rho \frac{d\underline{v}}{dt} = \rho \underline{F} + \underline{\text{div }} \underline{\underline{\sigma}} \qquad [1.22]$$

which directly expresses Newton's second law: the product of the mass density and the acceleration of the solid particle followed in its motion are equal to the force per unit volume. In the absence of any external force in the volume, this force density arises only from the internal stresses, σ_{ij}, through which the external medium acts upon the particle.

The law of *conservation of angular momentum* indicates that relative to a fixed point, taken as the origin, the rate of change of the angular momentum is equal to the torque resulting from the volume forces ($\rho \underline{F} \, dV$) and surface forces ($\underline{T} \, dS = \underline{\underline{\sigma}} \cdot d\underline{S}$):

$$\iiint_V \left(\underline{x} \wedge \rho \frac{d\underline{v}}{dt}\right) dV = \iiint_V (\underline{x} \wedge \rho \underline{F}) \, dV + \iint_S (\underline{x} \wedge \underline{\underline{\sigma}}) \cdot d\underline{S} + \iiint_V \rho \underline{G} \, dV \quad [1.23]$$

where $\rho \underline{G}$ represents a torque applied in the volume V. According to the divergence theorem, the surface integral transforms into a volume integral:

$$\iint_S (\underline{x} \wedge \underline{\underline{\sigma}}) \cdot d\underline{S} = \iiint_V \underline{\text{div}} \, (\underline{x} \wedge \underline{\underline{\sigma}}) \, dV \quad [1.24]$$

Let us introduce the Levi–Civita tensor ε_{ijk} to express the divergence of the vector product:

$$\varepsilon_{ijk} = \begin{cases} +1 & \text{if the permutation } (i, j, k) \text{ is even} \\ -1 & \text{if the permutation } (i, j, k) \text{ is odd} \\ 0 & \text{if two or three indices are equal} \end{cases} \quad [1.25]$$

that is, for the component im of the tensor $\underline{x} \wedge \underline{\underline{\sigma}}$:

$$[\underline{x} \wedge \underline{\underline{\sigma}}]_{im} = \varepsilon_{ijk} x_j \sigma_{km} \quad [1.26]$$

Given the independence of the coordinates x_j and x_m, the ith component of the divergence is given by:

$$[\underline{\text{div}} \, (\underline{x} \wedge \underline{\underline{\sigma}})]_i = \frac{\partial}{\partial x_m} (\varepsilon_{ijk} x_j \sigma_{km}) = \varepsilon_{ijk} \delta_{jm} \sigma_{km} + \varepsilon_{ijk} x_j \frac{\partial \sigma_{km}}{\partial x_m} \quad [1.27]$$

and for the component i, equation [1.23] can be written as:

$$\iiint_V \left[\underline{x} \wedge \left(\rho \frac{d\underline{v}}{dt} - \rho \underline{F} - \underline{\text{div}} \, \underline{\underline{\sigma}}\right)\right]_i dV = \iiint_V (\rho G_i + \varepsilon_{ijk} \sigma_{kj}) \, dV \quad [1.28]$$

Taking into account the local equation of dynamics [1.22], the first integral is zero. Canceling the second integral for any volume V implies:

$$\varepsilon_{ijk} \sigma_{kj} = -\rho G_i \qquad i = 1, 2, 3 \quad [1.29]$$

In the absence of any torque applied in the volume ($G_i = 0$), the conservation of angular momentum leads to the equality:

$$\sigma_{ij} = \sigma_{ji} \quad \forall i,j = 1,2,3 \quad [1.30]$$

The *stress tensor is symmetric* and has six independent components: the normal stresses σ_{11}, σ_{22} and σ_{33}, and the tangential stresses $\sigma_{12} = \sigma_{21}$, $\sigma_{13} = \sigma_{31}$ and $\sigma_{23} = \sigma_{32}$.

1.1.2. Kinematics of continuous media

A solid is said to be elastic when it is deformed under the effect of any solicitation, and when it returns to its original form once this solicitation is removed. In the context of linear elasticity, the deformations induced are assumed to be small. Let $\underline{u}(\underline{x})$ be the displacement, in an orthonormal frame, of the material point M with coordinates $\underline{x}$ at rest. The components of displacement of the neighboring material point N, with initial coordinates $\underline{x} + \mathrm{d}\underline{x}$, are therefore written as:

$$u_i(\underline{x} + \mathrm{d}\underline{x}) = u_i(\underline{x}) + \frac{\partial u_i}{\partial x_j}\mathrm{d}x_j = u_i(\underline{x}) + \mathrm{d}u_i \quad [1.31]$$

The medium is only deformed if the displacement gradient $\partial u_i/\partial x_j$ is non-zero. However, this second-rank tensor is not appropriate to express the deformation because it is not canceled out in a simple overall rotation, which conserves the distances between the material points and therefore does not modify the internal state of the solid. Like any tensor, the displacement gradient is composed of a symmetric part ε_{ij} and an antisymmetric part Ω_{ij}:

$$\varepsilon_{ij}(\underline{x},t) = \frac{1}{2}\left[\frac{\partial u_i}{\partial x_j} + \frac{\partial u_j}{\partial x_i}\right] \quad \text{and} \quad \Omega_{ij}(\underline{x},t) = \frac{1}{2}\left[\frac{\partial u_i}{\partial x_j} - \frac{\partial u_j}{\partial x_i}\right] \quad [1.32]$$

Only the symmetric part ε_{ij} is canceled for any overall movement: translation and infinitesimal rotation. This is the appropriate quantity to define linearized deformations as a function of mechanical displacement:

$$\underline{\underline{\varepsilon}} = \frac{1}{2}\left(\operatorname{grad}\underline{u} + \operatorname{grad}^t\underline{u}\right) \quad [1.33]$$

When the solid is deformed, the volume V occupied by a fixed number of particles changes by the quantity:

$$\Delta V = \iint_S \underline{u}\cdot\underline{l}\,\mathrm{d}S = \iiint_V \operatorname{div}(\underline{u})\,\mathrm{d}V \quad [1.34]$$

where the second integral is deduced from the previous one by applying the Green's theorem. Since the variation ΔV of the volume V is the sum of variations of elementary volumes $\delta(\mathrm{d}V)$: $\Delta V = \iiint_V \delta(\mathrm{d}V)$, the local dilatation Θ of the medium is equal to the divergence of the mechanical displacement vector:

$$\Theta = \frac{\delta(\mathrm{d}V)}{\mathrm{d}V} = \operatorname{div} \underline{u} = \frac{\partial u_1}{\partial x_1} + \frac{\partial u_2}{\partial x_2} + \frac{\partial u_3}{\partial x_3} \qquad [1.35]$$

Θ can be expressed by the trace of the strain tensor at the considered point:

$$\Theta(\underline{x},t) = \varepsilon_{11} + \varepsilon_{22} + \varepsilon_{33} = \operatorname{Tr}\left[\underline{\varepsilon}\left(\underline{x},t\right)\right] \qquad [1.36]$$

1.1.3. *Poynting's theorem: energy balance*

Mechanical sources are needed to generate elastic waves. In this section, we establish the law of conservation that links, at each instant, the power supplied by sources contained in a volume V inside the solid to the rate of change of mechanical energy stored in this volume, and to the power radiated by the elastic waves through the surface S enclosing the volume V. The notion of Poynting vector is naturally introduced to express the acoustic power transported per unit area.

The instantaneous power p_s supplied per unit volume by the sources of volume density $\rho \underline{F}$ is deduced from the work of the mechanical forces $\mathrm{d}w_m = \rho F_i \, \mathrm{d}u_i$:

$$p_s(\underline{x},t) = \frac{\mathrm{d}w_m}{\mathrm{d}t} = \rho F_i \dot{u}_i \qquad [1.37]$$

Given the local equation of dynamics [1.22], this power can be put in the form:

$$p_s = \rho \ddot{u}_i \dot{u}_i - \frac{\partial \sigma_{ij}}{\partial x_j} \dot{u}_i \qquad [1.38]$$

or again:

$$p_s = \rho \ddot{u}_i \dot{u}_i + \sigma_{ij} \frac{\partial \dot{u}_i}{\partial x_j} - \frac{\partial (\sigma_{ij} \dot{u}_i)}{\partial x_j} \qquad [1.39]$$

The first term of the right-hand side corresponds to the time derivative of the kinetic energy density per unit volume:

$$e_k(\underline{x},t) = \frac{1}{2} \rho \dot{u}_i^2 \qquad [1.40]$$

and the second term corresponds to the time derivative of the elastic potential energy, per unit volume, defined by its differential:

$$\mathrm{d}e_p(\underline{x},t) = \sigma_{ij} \, \mathrm{d}\varepsilon_{ij} \qquad [1.41]$$

If the constitutive equation is linear then by integrating from the initial state ($\sigma_{ij} = 0$) to the final state, we get:

$$e_p(\underline{x},t) = \frac{1}{2}\sigma_{ij}\varepsilon_{ij} \qquad [1.42]$$

The last term in equation [1.39] is the divergence of the *Poynting vector* $\underline{J}$, defined by the relation:

$$\underline{J} = -\underline{\underline{\sigma}} \cdot \underline{\dot{u}} \quad \text{or} \quad J_i = -\sigma_{ij}\dot{u}_j \qquad [1.43]$$

Finally, the local expression for the law of energy conservation takes the form:

$$p_s = \frac{d(e_k + e_p)}{dt} + \text{div}\,\underline{J} \qquad [1.44]$$

Using:

$$W(t) = \iiint_V w\,dV, \quad E_k(t) = \iiint_V e_k\,dV \quad \text{and} \quad E_p(t) = \iiint_V e_p\,dV \qquad [1.45]$$

to denote the work of the sources, the kinetic energy and the potential energy contained in a volume V inside the solid, the integration of equation [1.44] shows that the (instantaneous) power supplied by the sources acting in this volume is given by:

$$P_s(t) = \frac{dW}{dt} = \frac{d(E_k + E_p)}{dt} + \iint_S \underline{J} \cdot \underline{l}\,dS \qquad [1.46]$$

At each time, this relation expresses the law of energy conservation: the power supplied by the source inside the volume V is either stored in this volume in the form of kinetic and potential energies or is radiated toward the exterior with a power equal to the flux of the Poynting vector across the surface S (Poynting's theorem). If the material is dissipative, it is necessary to add the power P_d dissipated in the volume V to the right-hand side of equation [1.46].

REMARK.–

– In the case of a perfect fluid, since $\sigma_{ij} = -p\delta_{ij}$, the Poynting vector defined by the relation [1.43]:

$$J_i = -\sigma_{ij}v_j = p\delta_{ij}v_j = pv_i, \quad \text{i.e.} \quad \underline{J} = p\underline{v} \qquad [1.47]$$

is the product of the acoustic pressure p and the particle velocity $\underline{v} = \underline{\dot{u}}$.

– Given the expression [1.10] for mechanical traction $T_i(\underline{l})$, the instantaneous power $\underline{J}\cdot\underline{l}$ per unit area takes the form:

$$\underline{J}\cdot\underline{l} = -\sigma_{ij}l_j\dot{u}_i = -T_i(\underline{l})\dot{u}_i \qquad [1.48]$$

which shows that the acoustic Poynting vector has a concrete physical significance as the power density per unit area. Indeed, $-T_i(\underline{l})\,\mathrm{d}S$ represents the force exerted on the surface element $\mathrm{d}S$ by the matter located inside the volume V acting on the matter located outside the volume: $-T_i \dot{u}_i \,\mathrm{d}S$ is, therefore, the power delivered from the inside to the outside across $\mathrm{d}S$. Thus, contrary to what happens in electromagnetism, the use of the acoustic Poynting vector is not limited to closed surfaces.

1.1.4. *Stress–strain relationship: Maxwell relations*

In an elastic solid, there exists a one-to-one relationship between stresses and strains. In the hypothesis of small strains, the behavior of the solid material is described by a linear relationship that generalizes the law of proportionality $\sigma = E\varepsilon$ stated in the 17th century by Hooke in the case of a thin bar of Young's modulus E subjected to a traction along its axis.

When the strains are smaller than 10^{-4}, the behavior of most elastic solids is accurately described by the first-order term in the Taylor expansion of the stress–strain relation:

$$\sigma_{ij}(\varepsilon_{kl}) = \sigma_{ij}(0) + \left.\frac{\partial \sigma_{ij}}{\partial \varepsilon_{kl}}\right|_{\varepsilon_{kl}=0} \varepsilon_{kl} + \frac{1}{2}\left.\frac{\partial^2 \sigma_{ij}}{\partial \varepsilon_{kl} \partial \varepsilon_{mn}}\right|_{\substack{\varepsilon_{kl}=0 \\ \varepsilon_{mn}=0}} \varepsilon_{kl}\varepsilon_{mn} + \ldots \quad [1.49]$$

At rest, that is, in the absence of a prestress: $\sigma_{ij}(0) = 0$, we have:

$$\sigma_{ij} = c_{ijkl}\varepsilon_{kl} \quad [1.50]$$

where the coefficients:

$$c_{ijkl} = \left.\frac{\partial \sigma_{ij}}{\partial \varepsilon_{kl}}\right|_{\varepsilon_{kl}=0} \quad [1.51]$$

that link the nine components of each of the tensors σ_{ij} and ε_{kl}, form a fourth-rank tensor, called the *elastic constant tensor* or *stiffness tensor*. The symmetry of the tensors σ_{ij} and ε_{kl} imposes relations of symmetry on the stiffness tensor:

$$c_{ijkl} = c_{jikl} = c_{ijlk} \quad [1.52]$$

which reduce the number of independent elastic constants from 81 to 36. According to thermodynamic considerations given below, the stiffness tensor c_{ijkl} has other symmetry properties.

The work done by the external forces during a deformation is stored within the solid in the form of elastic potential energy. When these forces are removed, this energy is returned by the internal stresses that restore the solid to its initial state. For a strain variation $\mathrm{d}\varepsilon_{ij}$, the density of elastic potential energy is given by relation [1.41].

If δq is the amount of heat received per unit volume, the variation in internal energy U per unit volume is:

$$dU = \delta q + \sigma_{ij}\, d\varepsilon_{ij} \qquad [1.53]$$

According to the second principle of thermodynamics, for a reversible transformation: $\delta q = T\, ds$, where T is the absolute temperature and s the entropy per unit volume. The expression

$$dU = T\, ds + \sigma_{ij}\, d\varepsilon_{ij} \qquad [1.54]$$

shows that the internal energy of a deformed body is a function of the entropy and the strains: $U = U(s, \varepsilon_{ij})$. It follows that the stresses are partial derivatives of the internal energy at constant entropy:

$$\sigma_{ij} = \left.\frac{\partial U}{\partial \varepsilon_{ij}}\right|_s \qquad [1.55]$$

that is, given the definition [1.51] for elastic constants:

$$c_{ijkl}^{(s)} = \left.\frac{\partial^2 U}{\partial \varepsilon_{ij}\partial \varepsilon_{kl}}\right|_s \qquad [1.56]$$

Since the order of the derivations does not change the second derivative, the isentropic stiffness tensor is invariant in a permutation of the first two indices with the last two:

$$c_{ijkl}^{(s)} = c_{klij}^{(s)} \qquad [1.57]$$

These are isentropic moduli that are involved in the propagation of elastic waves whose wavelength (except at very high frequencies) is much larger than the thermal diffusion length; consequently, it does not allow heat exchange to take place between a compressed region ($T > T_0$) and a dilated region ($T < T_0$) separated by half a wavelength. If the transformation is reversible, the entropy is constant during the propagation phenomenon. However, the symmetry relation [1.57] – *Maxwell relation* – is valid under other thermodynamic conditions. For example, for isothermal transformation ($T = $ Const), the appropriate state function to be considered is the free energy $F = U - Ts$, whose differential is:

$$dF = -s\, dT + \sigma_{ij}\, d\varepsilon_{ij} \qquad [1.58]$$

The free energy depends on the temperature and the strains: $F = F(T, \varepsilon_{ij})$; consequently:

$$\sigma_{ij} = \left.\frac{\partial F}{\partial \varepsilon_{ij}}\right|_T \quad \text{and} \quad c_{ijkl}^{(T)} = \left.\frac{\partial^2 F}{\partial \varepsilon_{ij}\partial \varepsilon_{kl}}\right|_T = c_{klij}^{(T)} \qquad [1.59]$$

for the isothermal stiffness constants.

In the domain of validity of the generalized Hooke's law [1.50] and after the permutation of pairs of dummy indices (ij) and (kl), the variation of internal energy is given by the expression [1.54]:

$$dU = T\,ds + c^{(s)}_{ijkl}\varepsilon_{kl}\,d\varepsilon_{ij} = T\,ds + \frac{1}{2}\left(c^{(s)}_{ijkl}\varepsilon_{kl}\,d\varepsilon_{ij} + c^{(s)}_{klij}\varepsilon_{ij}\,d\varepsilon_{kl}\right) \quad [1.60]$$

Taking into account the Maxwell relation [1.57], we get:

$$dU = T\,ds + \frac{1}{2}c^{(s)}_{ijkl}\,d(\varepsilon_{ij}\varepsilon_{kl}) \quad [1.61]$$

After integration, the internal energy per unit volume of an elastic solid is expressed by:

$$U(s,\varepsilon_{ij}) = U_0(s) + \frac{1}{2}c^{(s)}_{ijkl}\varepsilon_{ij}\varepsilon_{kl} \quad [1.62]$$

where $U_0(s)$ is the internal energy of the non-deformed solid: $U_0(s) = U(s,\varepsilon_{ij}=0)$. Without specifying the thermodynamic conditions, the second term represents the elastic potential energy e_p (per unit volume), which is a quadratic form of the strains:

$$e_p = \frac{1}{2}c^{(s)}_{ijkl}\varepsilon_{ij}\varepsilon_{kl} \quad [1.63]$$

Depending on whether the deformation process is adiabatic or isothermal, this quantity represents the increase in either internal energy or free energy per unit volume of the elastic solid.

1.2. Isotropic solid

In the case of an isotropic solid, linear and elastic, the constitutive law involves only two characteristic constants for the propagation medium. The local equation of dynamics splits into two independent equations of propagation, whose solutions are as follows: for one, a succession of compressions and dilatations that propagate at the velocity V_L, and for the other, a shear wave with no change in volume that propagates at a velocity V_T smaller than V_L. Thus, two plane waves can propagate in an isotropic solid, one with a longitudinal polarization and the other with transverse polarizations. This decomposition is also found for waves with cylindrical symmetry.

1.2.1. *Constitutive equations*

In an *isotropic solid*, the stiffness tensor must be invariant for any change in axes (assumed to be orthonormal), such as a rotation or a symmetry with respect to a point or a plane. Only a scalar or the unit tensor δ_{ij} is unaffected by these orthogonal transformations. Consequently, each component c_{ijkl} can be expressed in terms of the components of the unit tensor. Due to the symmetry $\delta_{ij} = \delta_{ji}$, there are only three distinct combinations containing the four indices $ijkl$:

$$\delta_{ij}\delta_{kl}, \qquad \delta_{ik}\delta_{jl} \quad \text{and} \quad \delta_{il}\delta_{jk} \qquad [1.64]$$

thus, the stiffness tensor takes the form:

$$c_{ijkl} = \lambda\delta_{ij}\delta_{kl} + \mu_1\delta_{ik}\delta_{jl} + \mu_2\delta_{il}\delta_{jk} \qquad [1.65]$$

where λ, μ_1 and μ_2 are constants. In addition, the condition $c_{ijkl} = c_{jikl}$ requires that $\mu_1 = \mu_2 = \mu$. Other relations of symmetry being satisfied by this:

$$c_{ijkl} = \lambda\delta_{ij}\delta_{kl} + \mu\left(\delta_{ik}\delta_{jl} + \delta_{il}\delta_{jk}\right) \qquad [1.66]$$

the properties of an *isotropic* solid are defined by only *two independent constants*: the Lamé constants λ and μ. By substituting this expression for stiffnesses into the generalized Hooke's law [1.50], we obtain:

– for the normal stresses (σ_{11}, σ_{22} and σ_{33}):

$$\sigma_{ii} = c_{iikl}\varepsilon_{kl} = (\lambda\delta_{kl} + 2\mu\delta_{ik}\delta_{il})\varepsilon_{kl} \qquad [1.67]$$

that is:

$$\sigma_{ii} = \lambda(\varepsilon_{11} + \varepsilon_{22} + \varepsilon_{33}) + 2\mu\varepsilon_{ii} \qquad [1.68]$$

– for the tangential stresses (σ_{ij} with $i \neq j$):

$$\sigma_{ij} = \mu(\delta_{ik}\delta_{jl} + \delta_{il}\delta_{jk})\varepsilon_{kl} = 2\mu\varepsilon_{ij}, \qquad i \neq j \qquad [1.69]$$

Equations [1.68] and [1.69] can be grouped into a single equation:

$$\sigma_{ij}(\underline{x},t) = \lambda\delta_{ij}\varepsilon_{kk}(\underline{x},t) + 2\mu\varepsilon_{ij}(\underline{x},t)$$

or:

$$\underline{\underline{\sigma}}(\underline{x},t) = \lambda \, \text{Tr}\left[\underline{\underline{\varepsilon}}(\underline{x},t)\right]\underline{\underline{1}} + 2\mu\underline{\underline{\varepsilon}}(\underline{x},t) \qquad [1.70]$$

with $\underline{\underline{1}}$ being the identity tensor and $\text{Tr}[\underline{\underline{\varepsilon}}(\underline{x},t)]$ the local dilatation $\Theta(\underline{x},t)$.

Other couples of constants and parameters can be used. It is common to introduce the Young's modulus and Poisson's ratio. The Young's modulus E measures the resistance of a bar to a traction along its axis. The Poisson's ratio ν compares the transverse strain with the longitudinal strain, that is, along the direction of the applied force. These parameters are expressed as functions of the Lamé constants by:

$$E = \mu \frac{3\lambda + 2\mu}{\lambda + \mu} \quad \text{and} \quad \nu = \frac{\lambda}{2(\lambda + \mu)} \quad [1.71]$$

The bulk modulus K and the shear modulus G are also used. The bulk modulus expresses the material resistance to a change in volume under a hydrostatic compression. The shear modulus represents the material resistance to a change in shape. These parameters are given by:

$$K = \lambda + \frac{2}{3}\mu = \frac{E}{3(1 - 2\nu)} \quad \text{and} \quad G = \mu = \frac{E}{2(1 + \nu)} \quad [1.72]$$

In the case of a perfect fluid, since $G = 0$, only one constant is needed: the bulk modulus $K = \rho c^2$ for a liquid, with c being the speed of sound, and the inverse, the compressibility coefficient $\chi = 1/K$, for a gas.

For reasons of mechanical stability, the parameters E, K and G are necessarily positive: a traction $\sigma = F/S$ along the axis of a bar causes an elongation $\Delta \ell / \ell = \sigma/E > 0$, a compression $p = -\sigma$ causes a decrease in volume $\Delta V/V = -p/K < 0$, a shear stress $\sigma_{12} > 0$ creates a deformation $\varepsilon_{12} = \sigma_{12}/2\mu > 0$.

Given relations [1.71], the Poisson's ratio ν lies between -1 and 0.5. For a long time, the negative values allowed by the elasticity theory were excluded. Then in the 1980s, foams or honeycomb structures were developed whose Poisson's ratio approaches the limit of thermodynamic stability, $\nu = -1$. These materials, called auxetics, are very compressible: their behavior is the opposite of the quasi-incompressible behavior of rubber, whose Poisson's ratio is close to 0.5.

Table 1.1 gives the elastic parameters and the mass density of different isotropic materials. The Young's modulus and Lamé coefficients range from a few GPa, for certain polymers, to several hundred GPa, for metals such as tungsten or beryllium.

1.2.2. *Equations of propagation, wave decoupling*

In a homogeneous solid, the mechanical displacement $\underline{u}$ is the solution to the equation of motion [1.22]. Neglecting the convection term in the particular derivative (equation [1.19]), this is written as:

$$\rho \frac{\partial^2 \underline{u}}{\partial t^2} = \text{div } \underline{\underline{\sigma}} + \rho \underline{F} \quad \text{or} \quad \rho \frac{\partial^2 u_i}{\partial t^2} = \frac{\partial \sigma_{ij}}{\partial x_j} + \rho F_i \quad [1.73]$$

where ρ is the mass density of the solid and where the source term $\rho \underline{F}$ is a volume force. In the absence of a source ($\underline{F} = \underline{0}$) and considering the constitutive law [1.70] of an isotropic solid, the local equation of dynamics [1.73] takes the form of the Lamé–Navier equation:

$$\rho \frac{\partial^2 \underline{u}}{\partial t^2} = (\lambda + \mu)\,\underline{\mathrm{grad}}\,(\mathrm{div}\,\underline{u}) + \mu \Delta \underline{u} \qquad [1.74]$$

In this differential system, the three components of displacement are coupled. It is advantageous to transform it in order to obtain two independent equations.

Material	λ (GPa)	μ (GPa)	E (GPa)	ν	ρ (kg.m^{-3})
Pure metals					
Aluminium	57.4	26.1	70.2	0.344	2 700
Silver	83.5	30.3	82.8	0.367	10 500
Beryllium	15.6	145.6	305.2	0.048	1 846
Chrome	83.4	115.4	279.2	0.210	7 194
Copper	89.2	48.2	127.7	0.325	8 930
Nickel	132.3	84.4	220.3	0.305	8 907
Gold	146.9	27.8	78.9	0.420	19 280
Lead	41.8	5.56	16.0	0.441	11 340
Tantalum	143.1	72.8	193.9	0.331	16 670
Tungsten	222.3	169.0	434.0	0.284	19 250
Zinc	31.1	30.9	77.2	0.251	7 135
Metal alloys					
Steel	115.1	82.1	212/2	0.292	7 900
Stainless steel	96.8	75.4	193.3	0.281	7 800
Duralumin (AU4G)	58.1	27.8	74.4	0.338	2 800
Brass	87.5	37.5	101.2	0.350	8 500
Titanium (Ti6Al4V)	94	41	110/5	0.348	4 500
Zircaloy	104.3	52.1	139.0	0.333	9 360
Other materials					
Concrete*	11.1	16.7	40	0.2	2 400
Rubber	1.0	0.002	0.005	0.499	950
Epoxy° (araldite)	4.48	1.42	3.91	0.380	1 170
Granite	37.0	31.5	80.0	0.270	2 650
Bone* (tibia)	15.7	7.37	19.8	0.340	1 900
PMMA° (plexiglass)	4.46	2.10	5.63	0.340	1 188
Polyamide 6-6° (nylon)	5.07	1.38	3.84	0.393	1 140
PTFE° (teflon)	2.04	1.05	2.79	0.330	2 140
Fused silica	16.5	31.4	73.7	0.172	2 200
Sheet glass	24.7	29.7	72.8	0.227	2 500

Table 1.1. *Values of Lamé constants, Young's modulus, Poisson's ratio and mass density for various isotropic materials. Media that may be considered as heterogeneous at ultrasonic frequencies are marked by the superscript* * *and viscoelastic media are denoted by the superscript* °

1.2.2.1. *Helmholtz decomposition*

Knowing that the Laplacian of a vector is expressed as follows:

$$\Delta \underline{A} = \operatorname{grad}(\operatorname{div} \underline{A}) - \operatorname{rot}(\operatorname{rot} \underline{A}) \qquad [1.75]$$

the equation of motion [1.74] takes the form:

$$\rho \frac{\partial^2 \underline{u}}{\partial t^2} = (\lambda + 2\mu)\operatorname{grad}(\operatorname{div} \underline{u}) - \mu \operatorname{rot}(\operatorname{rot} \underline{u}) \qquad [1.76]$$

The Helmholtz decomposition is commonly used to obtain decoupled equations:

$$\underline{u} = \underline{u}_L + \underline{u}_T \quad \text{with} \quad \begin{cases} \underline{u}_L = \operatorname{grad} \phi \\ \underline{u}_T = \operatorname{rot} \underline{\psi} \end{cases} \qquad [1.77]$$

where ϕ is a scalar potential and $\underline{\psi}$ a vector potential. Given the relations:

$$\operatorname{rot} \operatorname{grad} \equiv \underline{0} \quad \text{and} \quad \operatorname{div} \operatorname{rot} \equiv 0 \qquad [1.78]$$

the field $\underline{u}_L$ is irrotational ($\operatorname{rot} \underline{u}_L = \underline{0}$), while the field $\underline{u}_T$ has zero divergence ($\operatorname{div} \underline{u}_T = 0$). Given these properties, equation [1.76] is written as:

$$\rho \frac{\partial^2 \underline{u}_L}{\partial t^2} + \rho \frac{\partial^2 \underline{u}_T}{\partial t^2} = (\lambda + 2\mu)\operatorname{grad}(\operatorname{div} \underline{u}_L) - \mu \operatorname{rot}(\operatorname{rot} \underline{u}_T) \qquad [1.79]$$

Applying the rotational, on the one hand, and the divergence, on the other hand, to this equation leads to two independent equations of propagation:

$$\frac{\partial^2 \underline{u}_L}{\partial t^2} - V_L^2 \operatorname{grad}(\operatorname{div} \underline{u}_L) = \underline{0} \quad \text{and} \quad \frac{\partial^2 \underline{u}_T}{\partial t^2} + V_T^2 \operatorname{rot}(\operatorname{rot} \underline{u}_T) = \underline{0} \qquad [1.80]$$

where the velocities V_L and V_T are defined by the relations:

$$V_L = \sqrt{\frac{\lambda + 2\mu}{\rho}} \quad \text{and} \quad V_T = \sqrt{\frac{\mu}{\rho}} \qquad [1.81]$$

The properties of the displacement fields $\underline{u}_L$ and $\underline{u}_T$, derived from [1.77]:

$$\operatorname{grad}(\operatorname{div} \underline{u}_L) = \Delta \underline{u}_L \quad \text{and} \quad \operatorname{rot}(\operatorname{rot} \underline{u}_T) = -\Delta \underline{u}_T \qquad [1.82]$$

lead to the d'Alembert equations for $\underline{u}_L$ and $\underline{u}_T$:

$$\frac{\partial^2 \underline{u}_L}{\partial t^2} - V_L^2 \Delta \underline{u}_L = \underline{0} \quad \text{and} \quad \frac{\partial^2 \underline{u}_T}{\partial t^2} - V_T^2 \Delta \underline{u}_T = \underline{0} \qquad [1.83]$$

Using the definition for the fields $\underline{u}_L$ and $\underline{u}_T$ as functions of the potentials ϕ and $\underline{\psi}$, these equations may be replaced by:

$$\frac{\partial^2 \phi}{\partial t^2} - V_L^2 \Delta \phi = 0 \quad \text{and} \quad \frac{\partial^2 \underline{\psi}}{\partial t^2} - V_T^2 \Delta \underline{\psi} = \underline{0} \qquad [1.84]$$

During these developments, the vector equation [1.74] with three unknowns (the three components of the displacement field $\underline{u}$) has been replaced by a scalar equation with a single unknown (the scalar potential ϕ) and a vector equation with three unknowns (the three components of the vector potential $\underline{\psi}$). In order to simplify the solution to certain problems, it is possible to impose an additional relation between these four potentials, such as the gauge condition:

$$\operatorname{div} \underline{\psi} = 0 \qquad [1.85]$$

Equations [1.83] express the independent propagation of:

– a longitudinal wave with the velocity V_L through successive compressions and dilatations, accompanied by a local variation of the volume V occupied by a given number of particles:

$$\frac{\Delta V}{V} = \operatorname{div} \underline{u} = \operatorname{div} \underline{u}_L = \Delta \phi \neq 0 \qquad [1.86]$$

– a transverse wave with the velocity V_T through the slip of parallel planes without any change of the volume, because $\operatorname{div} \underline{u}_T = 0$.

In the harmonic case, a double differentiation with respect to time is equivalent to multiplying by $-\omega^2$; the propagation equations [1.84] are reduced to two Helmholtz equations:

$$\begin{cases} \Delta \phi + k_L^2 \phi = 0 & \text{with} \quad k_L = \dfrac{\omega}{V_L} \\ \Delta \underline{\psi} + k_T^2 \underline{\psi} = 0 & \text{with} \quad k_T = \dfrac{\omega}{V_T} \end{cases} \qquad [1.87]$$

where k_L and k_T are the longitudinal and transverse wave numbers, respectively.

1.2.2.2. *Velocity of bulk waves*

For usual materials, since λ is positive, the inequality $V_T < V_L/\sqrt{2}$ is satisfied (the upper limit is reached for $\nu = 0$). In order to give orders of magnitude, longitudinal and transverse wave velocities of some common materials are given in Table 1.2. These velocities are derived from the values of elastic parameters in Table 1.1. A quick examination shows that these values are of several thousand m/s and that the transverse wave velocity is often of the order of $V_L/2$.

From the point of view of the propagation of elastic waves, an isotropic solid is, in fact, characterized by a single dimensionless parameter, the ratio of velocities $\kappa = V_T/V_L$ or its inverse V_L/V_T, or again the Poisson's ratio ν. Considering relation [1.71], it can be easily verified that:

$$\kappa = \frac{V_T}{V_L} = \sqrt{\frac{\mu}{\lambda + 2\mu}} = \sqrt{\frac{1 - 2\nu}{2(1 - \nu)}} \qquad [1.88]$$

and, conversely, that:

$$\nu = \frac{1 - 2\kappa^2}{2(1 - \kappa^2)} \qquad [1.89]$$

Two media having the same velocity ratio, that is, the same Poisson's ratio, are identical up to a homothety. Certain remarkable values are given in Table 1.3.

Material	V_T (m·s^{-1})	V_L (m·s^{-1})	$\kappa = V_T/V_L$	References
Pure metals				
Aluminum	3 111	6 374	0.488	(Briggs 1995)
Silver	1 698	3 704	0.458	(Briggs 1995)
Beryllium	8 880	12 890	0.689	(Briggs 1995)
Chrome	4 005	6 608	0.606	(Briggs 1995)
Copper	2 323	4 558	0.510	(Clorennec et al. 2007)
Nickel	3 078	5 814	0.529	(Briggs 1995)
Gold	1 200	3 240	0.370	(Briggs 1995)
Lead	700	2 160	0.324	(Briggs 1995)
Tantalum	2 090	4 162	0.502	(Clorennec et al. 2007)
Tungsten	2 963	5 395	0.549	(Clorennec et al. 2007)
Zinc	2 080	3 607	0.577	(Clorennec et al. 2007)
Metal alloys				
Steel	3 224	5 946	0.542	(Clorennec et al. 2007)
Stainless steel	3 110	5 635	0.552	(Clorennec et al. 2007)
Duralumin (AU4G)	3 150	6 370	0.495	(Clorennec et al. 2007)
Brass	2 100	4 372	0.480	(Briggs 1995)
Titanium (Ti6Al4V)	3 018	6 254	0.483	(Higuet et al. 2011)
Zircaloy	2 360	4 720	0.500	(Briggs 1995)
Other materials				
Concrete*	2 638	4 306	0.613	Typical values
Rubber°	42	1 050	0.04	(Kinsler et al. 2009)
Epoxy° (araldite)	1 100	2 500	0.44	Typical values
Granite	3 448	6 143	0.561	Typical values
Bone* (tibia)	1 970	4 000	0.492	(Briggs 1995)
PMMA° (plexiglass)	1 330	2 700	0.493	Typical values
Polyamide 6-6° (nylon)	1 100	2 620	0.420	Typical values
PTFE° (teflon)	700	1 390	0.504	Typical values
Fused silica	3 779	6 005	0.629	(Clorennec et al. 2007)
Sheet glass	3 440	5 800	0.594	(Raetz et al. 2015)

Table 1.2. *Bulk wave velocities for various isotropic materials. The media that may be considered as heterogeneous at ultrasonic frequencies are marked by the superscript* * *and viscoelastic media are marked by the superscript* °

In *soft media*, like biological tissues, the bulk modulus K is very large compared to the shear modulus μ. According to relations [1.71] and [1.72], the parameters for a quasi-incompressible soft solid are:

$$K \approx \lambda \gg \mu, \qquad E \approx 3\mu \quad \text{and} \quad \nu \approx 0.5 \qquad [1.90]$$

The velocity of compressional waves, close to that of water ($V_L = 1\,480$ m/s), is very high compared to the velocity of shear waves V_T, which is between 1 and 50 m/s. With $\rho = 1000$ kg/m^3 and $V_T = 5$ m/s, typical values are:

$$K \approx \rho V_L^2 = 2.2\,\text{GPa}, \qquad \mu = \rho V_T^2 = 25\,\text{kPa} \quad \text{and} \quad E \approx 75\,\text{kPa} \qquad [1.91]$$

Young's modulus E, which characterizes the stiffness of the medium, is directly derived from the shear wave velocity: $E \approx 3\rho V_T^2$. This is the principle underlying dynamic elastography equipment, using shear waves to obtain a quantitative image of the stiffness of biological tissues in the form of a mapping of the Young's modulus. The very high attenuation of shear waves limits the operating frequencies to a few hundred Hz.

Medium – behavior	ν	λ	K	κ	V_L
Rigid solid	0	$\ll \mu$	$2\mu/3$	$1/\sqrt{2}$	$\sqrt{2}V_T$
Fragile solid	0.1	$\mu/4$	$\approx \mu$	$2/3$	$3V_T/2$
Glasses	0.25	μ	$5\mu/3$	$1/\sqrt{3}$	$V_T\sqrt{3}$
Common metals	$1/3$	2μ	$8\mu/3$	$1/2$	$2V_T$
Incompressible solid	0.5	$\gg \mu$	λ	0	$\gg V_T$
Perfectly compressible	-1	$-2\mu/3$	0	$\sqrt{3}/2$	$2V_T/\sqrt{3}$

Table 1.3. *Remarkable values for the Poisson's ratio ν, the bulk modulus K and the bulk wave velocity ratio $\kappa = V_T/V_L$*

1.2.3. *Traveling, plane, sinusoidal waves*

A local disruption in the equilibrium conditions in a medium generally results in a disturbance which propagates: this is a *traveling wave*. The description of a wave involves the following parameters: velocity, wavelength and wave vector, whose definition does not depend on the nature of the vibration. If all the particles situated in parallel planes vibrate in phase, a *plane wave* propagates in a direction defined by the unit vector perpendicular to the wave planes. At the end of this section, we also review the benefits of the *complex representation* of sinusoidal functions.

1.2.3.1. *Traveling wave*

Let us observe the progression of a phenomenon by recording, at a fixed point, the variation over time of a characteristic quantity. At the point of abscissa x_0, the phenomenon is manifested by the level u of this quantity at time t_0. If the disturbance does not change during its propagation, for example, if it is not attenuated, and moves at a constant velocity V, in the direction of increasing x, then the level u is the same at the point x and instant t, such that:

$$x = x_0 + V(t - t_0) \quad [1.92]$$

that is:

$$x - Vt = x_0 - V t_0 \quad \text{or} \quad t - \frac{x}{V} = t_0 - \frac{x_0}{V} \quad [1.93]$$

Any function $u(x,t)$, which describes a phenomenon that propagates in the direction of increasing x only depends on the expression $x - Vt$ or $t - x/V$, because it takes the same value for any x and t, provided that these quantities are constant:

$$u(x,t) = f(x - Vt) = F(t - x/V) \quad [1.94]$$

If the propagation is in the direction of decreasing x:

$$u(x,t) = g(x + Vt) = G(t + x/V) \quad [1.95]$$

The disturbance resulting from the propagation of the two waves in the same medium and in opposite directions is:

$$u(x,t) = f(x - Vt) + g(x + Vt) = F(t - x/V) + G(t + x/V) \quad [1.96]$$

Among all possible departures from an equilibrium, sinusoidal vibrations around a mean value are very significant since all other disturbances result from the superposition of such oscillations. Thus, the elastic wave produced by a sinusoidal displacement of matter with period T and amplitude a at $x = 0$:

$$u = a\cos(\omega t) \quad \text{with} \quad \omega = 2\pi f = \frac{2\pi}{T} \quad [1.97]$$

is expressed, at a distance x, by[3]:

$$u = a\cos\left[\omega\left(t - \frac{x}{V}\right)\right] \quad \text{and} \quad u = a\cos\left[2\pi\left(\frac{t}{T} - \frac{x}{\lambda}\right)\right] \quad [1.98]$$

where the wavelength $\lambda = VT = V/f$ is the distance traveled by the vibration during the period T. As the phenomenon is unchanged after the time T, λ represents, at a given instant, the distance that separates two identical states of the solid – for instance, two consecutive maxima of the quantity u. It is also useful to write:

$$u = a\cos(\omega t - kx) \quad \text{where} \quad k = \frac{\omega}{V} = \frac{2\pi}{\lambda} \quad [1.99]$$

---

[3] u denotes the mechanical displacement, but it may also represent the acoustic pressure in a fluid.

is the wave number. The term $-kx$ measures, at a given time, the phase shift of the phenomenon at the point x with respect to the origin.

Let us call φ the overall phase of the wave: $\varphi = \omega t - kx$. The wave number k is the phase delay per unit length at a given time, while the angular frequency ω is the phase advance per unit time at a given point:

$$k = -\left.\frac{\partial \varphi}{\partial x}\right|_t \quad \text{and} \quad \omega = \left.\frac{\partial \varphi}{\partial t}\right|_x \qquad [1.100]$$

There thus exists a space-time correspondence:

period T		wavelength λ
angular frequency $\omega = 2\pi/T$	$\Longleftrightarrow$	wave number $k = 2\pi/\lambda$

The propagation velocity $V = \omega/k$ is called the phase velocity: it is the velocity at which an observer must move in order to see, at each instant, the vibration in the same phase state: $\omega(t - x/V) = \text{Const}$. The wave will then appear stationary to this observer.

1.2.3.2. *Plane wave*

When the particles of matter situated in parallel planes vibrate in phase, the motion at all points of these *wavefronts* is known if the distance x of the plane considered from the origin is given. The expression of the vibrational state of any point M, whose coordinates are given by the vector $\underline{x} = \underline{OM}$ joining this point to the origin, involves the unit vector $\underline{n}$ perpendicular to the wave planes (Figure 1.3). The abscissa x is the projection on $\underline{n}$ of the vector $\underline{x}$: $x = \underline{n}.\underline{x}$. Let us replace x with this value in expression [1.94], we get:

$$u(\underline{x},t) = F\left(t - \frac{\underline{n}\cdot\underline{x}}{V}\right) = F\left(t - \frac{n_1 x_1 + n_2 x_2 + n_3 x_3}{V}\right) \qquad [1.101]$$

by developing the scalar product in the system of orthonormal axes using the coordinates x_1, x_2, x_3 of the point $\underline{x}$ (components of the vector $\underline{x}$) and the direction cosines n_1, n_2, n_3 for the propagation direction (components of the vector $\underline{n}$).

In the case of a sinusoidal plane wave, the quantity u is expressed by:

$$u(\underline{x},t) = a\cos\left[\omega\left(t - \frac{\underline{n}.\underline{x}}{V}\right)\right] = a\cos(\omega t - \underline{k}.\underline{x}) \qquad [1.102]$$

by introducing the wave vector:

$$\underline{k} = \frac{\omega}{V}\underline{n} = k\underline{n} \qquad [1.103]$$

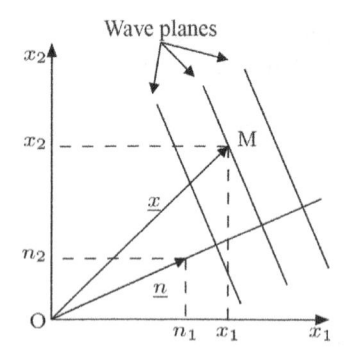

Figure 1.3. *At any point M marked by the vector $\underline{x}$, the vibration created by the wave propagating with the velocity V in the direction of the unit vector $\underline{n}$ is $u(\underline{x},t) = F(t - \underline{n}.\underline{x}/V)$*

The wave amplitude a was assumed to be constant. However, if the propagation occurs with attenuation, the amplitude depends on the observation point M, that is, on the distance x:

$$u(\underline{x},t) = a(x) \cos(\omega t - \underline{k}.\underline{x}) \qquad [1.104]$$

1.2.3.3. *Complex representation*

Despite being real, the quantities encountered in physics are often represented by complex numbers. This usage, which is very common for sinusoidal phenomena, is based on the fact that any linear relation with real coefficients between complex numbers can be applied to their real and imaginary parts separately. Equations in physics are very often linear, with real coefficients. The advantage of the complex representation:

$$u_c = a e^{i(\omega t - \varphi)} \qquad [1.105]$$

in the case of a sinusoidal quantity $u = a\cos(\omega t - \varphi) = \text{Re}[u_c]$ is due to the properties of the exponential function:

– in a sum of vibrations with the same frequency, the terms depending on time are written as factors: $\sum u_c = e^{i\omega t} \sum a e^{-i\varphi}$;

– differentiating with respect to time is equivalent to multiplying by $i\omega$: $\dfrac{du_c}{dt} = i\omega u_c$;

– the square of the amplitude a of the vibration is: $a^2 = u_c u_c^\star$, $u_c^\star$ being the complex conjugate function of u_c;

– the energy or power transported by a wave, equal to the time-average of the product of two sinusoidal quantities with period $T = 2\pi/\omega$:

$$u = a\cos(\omega t - \varphi) = \mathrm{Re}[u_c] \quad \text{and} \quad v = b\cos(\omega t - \psi) = \mathrm{Re}[v_c] \quad [1.106]$$

is expressed as:

$$P = \langle u(t)v(t)\rangle = \frac{1}{T}\int_{t_0}^{t_0+T} u(t)v(t)\,\mathrm{d}t = \frac{1}{2}\mathrm{Re}\left[u_c v_c^*\right] \quad [1.107]$$

Each of the two terms in the expansion of $\cos(\omega t - \underline{k}.\underline{x})$ can be taken as the complex representation of a traveling wave. We will often choose the exponential $e^{i(\underline{k}\cdot\underline{x}-\omega t)}$, which is more convenient for expressing the derivative with respect to the position $\underline{x}$.

1.2.4. *Polarization*

In the case of a time-harmonic plane wave propagating in the direction $\underline{n}$ with angular frequency ω, the associated mechanical displacement is defined by:

$$\underline{u}(\underline{x},t) = a\underline{p}\,e^{i(\underline{k}\cdot\underline{x}-\omega t)} \quad [1.108]$$

where a is the complex amplitude of displacement of the wave, $\underline{p}$ is the polarization vector, and $\underline{k} = k\underline{n}$ is the wave vector. The latter depends on the nature of the wave: longitudinal or transverse. The polarization vector $\underline{p}$, normalized to unity, indicates the direction of displacement of the particles during the passage of the wave. It thus depends on the nature of the wave and its propagation direction $\underline{n}$.

If we consider a longitudinal plane wave with the wave number $k_L = \omega/V_L$, the displacement is:

$$\underline{u}_L(\underline{x},t) = a_L\underline{p}_L\,e^{i(k_L\underline{n}\cdot\underline{x}-\omega t)} \quad [1.109]$$

Since the displacement field of a longitudinal wave is irrotational, we get:

$$\underline{\mathrm{rot}}\,\underline{u}_L = \underline{0} \quad \text{hence} \quad \underline{n}\wedge\underline{p}_L = \underline{0} \quad [1.110]$$

In the same way, the displacement of a transverse plane wave with wave number $k_T = \omega/V_T$ is given as:

$$\underline{u}_T(\underline{x},t) = a_T\underline{p}_T\,e^{i(k_T\underline{n}\cdot\underline{x}-\omega t)} \quad [1.111]$$

Since the displacement field of a transverse wave has zero divergence, we get:

$$\mathrm{div}\,\underline{u}_T = 0 \quad \text{hence} \quad \underline{n}\cdot\underline{p}_T = 0 \quad [1.112]$$

The polarization of the longitudinal wave is collinear with the propagation direction, since $\underline{n} \wedge \underline{u}_L = \underline{0}$, and the polarization of the transverse wave is perpendicular to the propagation direction, since $\underline{n} \cdot \underline{u}_T = 0$. Thus, in a given direction, for instance $\underline{n} = \underline{e}_1$, there can exist a longitudinal wave polarized in the propagation direction ($\underline{p}_L = \underline{e}_1$) and two transverse waves polarized perpendicular to this one ($\underline{p}_{T_1} = \underline{e}_2$ and $\underline{p}_{T_2} = \underline{e}_3$), with the same velocity V_T (Figure 1.4).

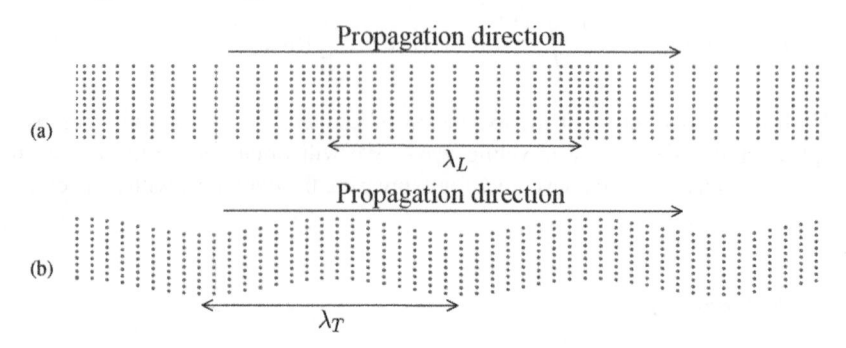

Figure 1.4. *a) Longitudinal wave: polarization and wave vector are parallel. At a given instant, the wave appears as a succession of compressions and dilatations. b) Transverse wave: polarization and wave vector are perpendicular. The planes perpendicular to the wave vector slip with respect to one another, while conserving the distance between them*

1.2.5. *Acoustic intensity*

By definition, the acoustic intensity is the average value of the component of the Poynting vector [1.43] along the propagation direction $\underline{n}$: $I = \langle \underline{J} \cdot \underline{n} \rangle$. In the harmonic case and using complex notation, we have:

$$I = -\frac{1}{2} \operatorname{Re}\left[(\underline{\underline{\sigma}} \cdot \underline{\dot{u}}^*) \cdot \underline{n}\right] = -\frac{1}{2} \operatorname{Re}\left[\sigma_{ij} \dot{u}_j^* n_i\right] \qquad [1.113]$$

Going back to the previous example of a longitudinal wave propagating in the x_1 direction, the displacement is written as:

$$\underline{u}(x_1, t) = a_L(1, 0, 0)^t e^{i(k_L x_1 - \omega t)} \qquad [1.114]$$

and the expression for the acoustic intensity is:

$$I_L = -\frac{1}{2} \operatorname{Re}\left[i\omega \sigma_{11} u_1^*\right] \qquad [1.115]$$

To compute this average, the stress σ_{11} has to be expressed:

$$\sigma_{11} = ik_L(\lambda + 2\mu) a_L e^{i(k_L x_1 - \omega t)} \qquad [1.116]$$

Since $\lambda + 2\mu = \rho V_L^2$, the acoustic intensity associated with a longitudinal wave is equal to:

$$I_L = \frac{1}{2}\omega^2 Z_L |a_L|^2 \qquad [1.117]$$

where $Z_L = \rho V_L$ is the acoustic impedance for this wave. In Chapter 2, it will be seen that this characteristic impedance plays an essential role in reflection and transmission phenomena at the interface between two media.

Similarly, for a transverse wave propagating in the x_1 direction and polarized along x_3, the displacement takes the form:

$$\underline{u}(x,t) = a_T (0,0,1)^t e^{i(k_T x_1 - \omega t)} \qquad [1.118]$$

and the time average of the component J_1 is given by:

$$I_T = -\frac{1}{2}\operatorname{Re}\left[i\omega \sigma_{13} u_3^\star \right] \qquad [1.119]$$

with:

$$\sigma_{13} = ik_T \mu a_T e^{i(k_T x_1 - \omega t)} \qquad [1.120]$$

Since $\mu = \rho V_T^2$, the expression for the acoustic intensity associated with a transverse wave is:

$$I_T = \frac{1}{2}\omega^2 Z_T |a_T|^2 \qquad [1.121]$$

where $Z_T = \rho V_T$ is the acoustic impedance of the transverse wave.

1.2.6. *Cylindrical and spherical waves*

The plane wave model considered until now is an approximation valid when all the points on a transducer, with large lateral dimensions compared to the wavelength of the emitted elastic waves, vibrate in phase according to the piston model described in Chapter 1 of Volume 2. In practice, other types of sources are encountered. In an isotropic solid, a point-like source, such as those used for oil exploration, generates spherical symmetric waves. A source line, created using a focused laser beam, for example, emits cylindrical symmetric waves. In order to analyze the propagation of these waves, it is natural to express the displacements and mechanical stresses, as well as the propagation equations, in the appropriate system of curvilinear coordinates. To have the necessary tools and mathematical formulae, refer to Appendix 1.

1.2.6.1. *Cylindrical waves*

Let R_c be the reference frame with cylindrical coordinates (r, θ, z) in Appendix 1. The Cartesian coordinates are defined in terms of the cylindrical coordinates by the relations [A1.1]. The field is contained in the xy plane and is therefore independent of the z coordinate. As in section 1.2.2.1, the solution to the propagation equation [1.76] is found by using the Helmholtz decomposition [1.77]. The number of potential functions can be reduced from four to three using the gauge invariance [1.85]:

$$\frac{\partial \psi_x}{\partial x} + \frac{\partial \psi_y}{\partial y} + \frac{\partial \psi_z}{\partial z} = 0 \quad [1.122]$$

When $\partial/\partial z = 0$, this equation is satisfied by introducing a function $\Phi_S(x, y)$ such that:

$$\psi_x = \frac{1}{k_T}\frac{\partial \Phi_S}{\partial y} \quad \text{and} \quad \psi_y = -\frac{1}{k_T}\frac{\partial \Phi_S}{\partial x} \quad [1.123]$$

Dividing by the wave number k_T ensures that the potential Φ_S is homogeneous to a squared length. ψ_x and ψ_y are the first two components of the rotational of the vector $\Phi_S \underline{e}_z / k_T$. Since the third component ψ_z is equal to another scalar function $\Phi_T(x, y)$, the vector potential $\underline{\psi}$ takes the form:

$$\underline{\psi} = \frac{1}{k_T} \underline{\mathrm{rot}}\, (\Phi_S \underline{e}_z) + \Phi_T \underline{e}_z \quad [1.124]$$

In the harmonic case, the potential Φ_S satisfies the Helmholtz equation: $\Delta \Phi_S = -k_T^2 \Phi_S$, so that:

$$\underline{\mathrm{rot}}\,[\underline{\mathrm{rot}}\,(\Phi_S \underline{e}_z)] = \underline{\mathrm{grad}}\left(\frac{\partial \Phi_S}{\partial z}\right) - \Delta(\Phi_S)\underline{e}_z = k_T^2 \Phi_S \underline{e}_z \quad [1.125]$$

The expressions for the components of the mechanical displacement $\underline{u} = \underline{\mathrm{grad}}\,\Phi_L + \underline{\mathrm{rot}}\,\underline{\psi}$ in cylindrical coordinates (Appendix 1), with $\partial/\partial z = 0$, are:

$$u_r(r) = \frac{\partial \Phi_L}{\partial r} + \frac{1}{r}\frac{\partial \Phi_T}{\partial \theta}, \quad u_\theta(r) = \frac{1}{r}\frac{\partial \Phi_L}{\partial \theta} - \frac{\partial \Phi_T}{\partial r} \quad \text{and} \quad u_z(r) = k_T \Phi_S \quad [1.126]$$

where Φ_L, Φ_T and Φ_S are scalar potentials that satisfy the equations

$$\Delta \Phi_L + k_L^2 \Phi_L = 0 \quad \text{and} \quad \Delta \Phi_M + k_T^2 \Phi_M = 0 \quad \text{with} \quad M = T, S \quad [1.127]$$

Given expression [A1.7] for the Laplacian using cylindrical coordinates, the Helmholtz equation with axial symmetry along z ($\partial/\partial\theta = 0$) is written as:

$$\frac{\partial^2 \Phi}{\partial r^2} + \frac{1}{r}\frac{\partial \Phi}{\partial r} + k^2 \Phi = 0 \quad [1.128]$$

Its solutions are of the type:

$$\Phi(r) = AH_0^{(1)}(kr) + BH_0^{(2)}(kr) \qquad [1.129]$$

where A and B are constants and where $H_0^{(1)}$ and $H_0^{(2)}$ are zero-order Hankel functions of the first and second kind. There is no analytical expression for Hankel functions. Nonetheless, when their argument is large, their asymptotic behavior is given by:

$$\lim_{x \to \infty} H_0^{(1)}(x) = \frac{1-i}{\sqrt{\pi x}} e^{ix} \quad \text{and} \quad \lim_{x \to \infty} H_0^{(2)}(x) = \frac{1+i}{\sqrt{\pi x}} e^{-ix} \qquad [1.130]$$

Thus, given the chosen time convention ($e^{-i\omega t}$), the function $H_0^{(1)}$ is associated with a divergent wave while the function $H_0^{(2)}$ is associated with a convergent wave.

Assuming that the three potentials Φ_L, Φ_T and Φ_S are independent of the variables θ and z, each component of the displacement can be expressed in terms of one of the three potentials:

$$\begin{cases} u_r(r, \theta) = \dfrac{\partial \Phi_L}{\partial r} \\ u_\theta(r, \theta) = -\dfrac{\partial \Phi_T}{\partial r} \\ u_z(r, \theta) = k_T \Phi_S \end{cases} \qquad [1.131]$$

Since these components are independent, there exist three types of cylindrical waves with axial symmetry. The first is a longitudinal wave modeled by the potential Φ_L. Its purely radial displacement expresses a breathing motion (Figure 1.5(a)). The second, modeled by the potential Φ_T, is a purely orthoradial transverse wave corresponding to a rotational motion (Figure 1.5(b)). The third, modeled by the potential Φ_S, is a transverse wave with a rectilinear polarization parallel to the z axis, unlike the first two waves, which are polarized in the propagation plane.

The expression for the time average for the radial component of the Poynting vector, that is, the acoustic intensity, is:

$$I = \langle \underline{J} . \underline{e}_r \rangle = -\frac{1}{2} \operatorname{Re}\left[\underline{\underline{\sigma}}\, \underline{\dot{u}}^*\right] \underline{e}_r = -\frac{\omega}{2} \operatorname{Re}\left[i\sigma_{rr} u_r^* + i\sigma_{r\theta} u_\theta^* + i\sigma_{rz} u_z^*\right]$$

$$[1.132]$$

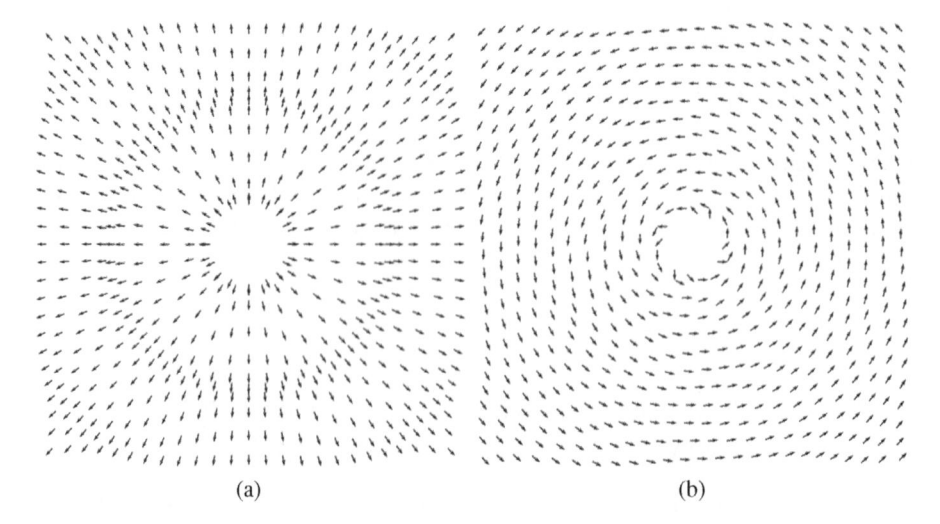

Figure 1.5. *Cylindrical waves: a) longitudinal and b) transverse*

Since $H_0'(x) = -H_1(x)$, the components of the displacement are expressed for each divergent wave by:

$$\begin{cases} u_r(r) = a_L \sqrt{\dfrac{\pi}{2}} H_1^{(1)}(k_L r) \\ u_\theta(r) = a_T \sqrt{\dfrac{\pi}{2}} H_1^{(1)}(k_T r) \\ u_z(r) = a_S \sqrt{\dfrac{\pi}{2}} H_0^{(1)}(k_T r) \end{cases} \qquad [1.133]$$

where the amplitudes a_L, a_T and a_S are homogeneous to a length. These displacements correspond to divergent waves of unit amplitude, exhibiting a geometric decrease in $1/\sqrt{r}$. Considering expressions [A1.9], the stresses associated with these three waves, that appear in the calculation of I, are expressed by:

$$\begin{cases} \sigma_{rr} = \sqrt{\dfrac{\pi}{2}} k_L a_L \left[(\lambda + 2\mu) H_0^{(1)}(k_L r) - \dfrac{2\mu}{k_L r} H_1^{(1)}(k_L r) \right] \\ \sigma_{r\theta} = \sqrt{\dfrac{\pi}{2}} \mu k_T a_T \left[H_0^{(1)}(k_T r) - \dfrac{2}{k_T r} H_1^{(1)}(k_T r) \right] \\ \sigma_{rz} = -\sqrt{\dfrac{\pi}{2}} \mu k_T a_S H_1^{(1)}(k_T r) \end{cases} \qquad [1.134]$$

Using the Wronskien formula (Abramowitz and Stegun 1965):

$$\left[H_0^{(1)}(x) \right]^* H_1^{(1)}(x) - \left[H_1^{(1)}(x) \right]^* H_0^{(1)}(x) = -\dfrac{2i}{\pi x} \qquad [1.135]$$

shows that the acoustic intensity of each of the three types of waves is given by:

$$I_L = \frac{Z_L \omega^2}{2k_L r}|a_L|^2 \quad \text{and} \quad I_M = \frac{Z_T \omega^2}{2k_T r}|a_M|^2 \quad \text{for} \quad M = T, S \quad [1.136]$$

similar to expressions [1.117] and [1.121] for a plane wave up to a geometric decay term of $1/r$.

1.2.6.2. Spherical waves

Let R_s be the reference frame for spherical coordinates (r, θ, φ) in Appendix 1. The Cartesian coordinates are defined in terms of the spherical coordinates by the relations [A1.10]. The solution to the propagation equation [1.76] is found by using the Helmholtz decomposition [1.77]. The case of spherical longitudinal and transverse waves is studied separately, with the common hypothesis of spherical symmetry.

1.2.6.2.1. Longitudinal wave

Given the expression [A1.16] of the Laplacian and for a wave with spherical symmetry, equation [1.84] for the potential $\Phi = \Phi_L$ is written as:

$$\frac{\partial^2 \Phi_L}{\partial r^2} + \frac{2}{r}\frac{\partial \Phi_L}{\partial r} + k_L^2 \phi = 0 \quad [1.137]$$

This equation takes the form of the Helmholtz equation:

$$\frac{\partial^2 (r\Phi_L)}{\partial r^2} + k_L^2 (r\Phi_L) = 0 \quad [1.138]$$

whose general solution is the sum of a divergent wave and a convergent wave:

$$\Phi_L(r) = \frac{a_L}{r} e^{ik_L r} + \frac{b_L}{r} e^{-ik_L r} \quad [1.139]$$

Considering the expression [A1.11] for the gradient and the spherical symmetry, the displacement is given by:

$$\underline{u}_L = \underline{\text{grad}}\ \Phi_L = \left[(ik_L r - 1)a_L \frac{e^{ik_L r}}{r^2} - (ik_L r + 1)b_L \frac{e^{-ik_L r}}{r^2}\right] \underline{e}_r \quad [1.140]$$

As for cylindrical waves, the purely radial polarization of longitudinal waves with spherical symmetry expresses a dilatational motion.

1.2.6.2.2. Transverse wave

Given that the displacement of a transverse wave has zero divergence, the simplest form of the vector potential is $\underline{\psi} = \Phi_T \underline{e}$ with Φ_T being a scalar potential, leading to the displacement:

$$\underline{u}_T = \underline{\text{rot}}\,(\Phi_T \underline{e}) = \underline{\text{grad}}\ \Phi_T \wedge \underline{e} + \Phi_T\ \underline{\text{rot}}\ \underline{e} \quad [1.141]$$

In a spherical symmetry problem where all directions are equivalent, the unit vector $\underline{e}$ can be chosen constant in the Cartesian coordinate system (x_1, x_2, x_3). Consequently, the term $\underline{\text{rot}}\,\underline{e}$ is zero. Given relation [A1.11], the spherical symmetry requires that the displacement is expressed by:

$$\underline{u}_T = \frac{\partial \Phi_T}{\partial r}\underline{e}_r \wedge \underline{e} \qquad [1.142]$$

Since the basis of the spherical coordinates is mobile, the vector $\underline{e}$ is not constant in this basis – it depends on the angles θ and φ. The displacement associated with the transverse wave does indeed have a polarization (in the $(\underline{e}_\theta, \underline{e}_\varphi)$ plane) that is orthogonal to the propagation direction $\underline{e}_r$. However, this displacement does not have spherical symmetry: there is no monopole (or isotropic) transverse wave.

1.3. Anisotropic solid

The anisotropy of a crystal arises from its perfectly ordered structure. Crystals can be divided into seven crystalline systems, which are further subdivided into 32-point symmetry classes (section A2.2.3). A crystal belonging to one system must necessarily possess certain elements of symmetry. For example, a crystal belonging to the trigonal system has a third-order (triad) axis and a crystal of the cubic system has four triad axes and three binary axes. Since the physical properties of a crystal are identical for two equivalent positions in a symmetry operation, the number of constants required to describe these properties is reduced.

In industrial materials, such as metals, composites, piezoelectric ceramics or metamaterials, the anisotropy is a result of the manufacturing process, for example, the rolling of steel sheets, the laying of the sheets of wires for composites, the preferred orientation of crystallites for zircaloy used in the nuclear industry, or for AM1 type alloys used in aeronautics. Before analyzing the propagation of elastic waves in these materials, it is necessary to assign them a class of symmetry: for instance, orthotropic or transverse isotropic, which optimally models their mechanical properties.

1.3.1. *Symmetry and elasticity tensor*

Given the Maxwell relation [1.57] and the symmetry of the stress and strain tensors, the components c_{ijkl} satisfy the following relations:

$$c_{ijkl} = c_{jikl} = c_{ijlk} = c_{klij} \qquad [1.143]$$

Because of these symmetries, the number of independent components of the stiffness tensor reduces from $3^4 = 81$ to only 21. The Voigt notation makes it possible to convert the index pairs (i, j) into a single index I as follows:

(i,j)	$(1,1)$	$(2,2)$	$(3,3)$	$(2,3)$ or $(3,2)$	$(1,3)$ or $(3,1)$	$(1,2)$ or $(2,1)$
I	1	2	3	4	5	6

If the convention $\sigma_I = \sigma_{ij}$ is adopted for the stresses, we must write:

$$\varepsilon_1 = \varepsilon_{11}, \quad \varepsilon_2 = \varepsilon_{22}, \quad \varepsilon_3 = \varepsilon_{33}, \quad \varepsilon_4 = 2\varepsilon_{23},$$
$$\varepsilon_5 = 2\varepsilon_{13} \quad \text{and} \quad \varepsilon_6 = 2\varepsilon_{12}$$
[1.144]

so that the stress–strain relationship can be written as:

$$\sigma_I = C_{IJ}\varepsilon_J \quad \text{with} \quad I, J = 1, 2, ...6 \qquad [1.145]$$

For example, in the expansion [1.50] of the stress σ_{11}, terms such as $c_{1123}\varepsilon_{23}$ and $c_{1132}\varepsilon_{32}$ are identical, which leads to $2c_{1123}\varepsilon_{23}$. To obtain the same contribution in the stress $\sigma_1 = \sigma_{11}$, deduced from [1.145], that is $\sigma_1 = C_{14}\varepsilon_4$ with $C_{14} = c_{1123}$, we must write $\varepsilon_4 = 2\varepsilon_{23}$. Thus, Hooke's law in matrix form is given by:

$$\begin{pmatrix} \sigma_{11} \\ \sigma_{22} \\ \sigma_{33} \\ \sigma_{23} \\ \sigma_{13} \\ \sigma_{12} \end{pmatrix} = \begin{pmatrix} C_{11} & C_{12} & C_{13} & C_{14} & C_{15} & C_{16} \\ C_{12} & C_{22} & C_{23} & C_{24} & C_{25} & C_{26} \\ C_{13} & C_{23} & C_{33} & C_{34} & C_{35} & C_{36} \\ C_{14} & C_{24} & C_{34} & C_{44} & C_{45} & C_{46} \\ C_{15} & C_{25} & C_{35} & C_{45} & C_{55} & C_{56} \\ C_{16} & C_{26} & C_{36} & C_{46} & C_{56} & C_{66} \end{pmatrix} \begin{pmatrix} \varepsilon_{11} \\ \varepsilon_{22} \\ \varepsilon_{33} \\ 2\varepsilon_{23} \\ 2\varepsilon_{13} \\ 2\varepsilon_{12} \end{pmatrix} \qquad [1.146]$$

Materials may present certain symmetries that reduce the number of independent constants for a given class (Royer and Dieulesaint 1996). If the transformation corresponds to a point symmetry (symmetry center, symmetry plane, direct or inverse axis of rotation), the physical properties of the crystal are indistinguishable in the two systems of axes.

The stresses and strains are second-rank tensors, that is, by definition (section A2.3.2), in a change of basis, their components are transformed as follows:

$$\sigma'_{ij} = \alpha_i^k \alpha_j^l \sigma_{kl} \quad \text{and} \quad \varepsilon'_{ij} = \alpha_i^k \alpha_j^l \varepsilon_{kl} \qquad [1.147]$$

where α_i^k are the components of the change of basis matrix $\boldsymbol{\alpha}$. If, due to its symmetry elements, the mechanical properties of the material are identical in the reference frames $\mathbb{R}$ and $\mathbb{R}'$, Hooke's law is expressed with the same stiffness tensor c_{ijkl}:

$$\sigma_{ij} = c_{ijkl}\varepsilon_{kl} \quad \text{and} \quad \sigma'_{ij} = c_{ijkl}\varepsilon'_{kl} \qquad [1.148]$$

The development of equations [1.147] and [1.148] leads to relations between the stiffnesses c_{ijkl}, which reduce the number of independent constants.

A *triclinic* material, which has no symmetry, is characterized by the 21 elastic constants of the matrix [1.146].

A *monoclinic* material has a symmetry plane, for example, the $(\underline{e}_1, \underline{e}_2)$ plane. Then, the matrix of transition from reference frame $\mathbb{R} = (\underline{e}_1, \underline{e}_2, \underline{e}_3)$ to reference frame $\mathbb{R}' = (\underline{e}_1, \underline{e}_2, -\underline{e}_3)$ is:

$$\alpha = \begin{pmatrix} 1 & 0 & 0 \\ 0 & 1 & 0 \\ 0 & 0 & -1 \end{pmatrix} \qquad [1.149]$$

Since the matrix α is diagonal, relations [1.147] lead to:

$$\begin{cases} \sigma'_{ij} = \sigma_{ij} & \text{if} \quad i,j = 1,2 \\ \sigma'_{23} = -\sigma_{23}, \quad \sigma'_{13} = -\sigma_{13} \quad \text{and} \quad \sigma'_{33} = \sigma_{33} \end{cases} \qquad [1.150]$$

and to identical relations for the strains. Using Voigt notation, we get:

$$\begin{pmatrix} \sigma_{11} \\ \sigma_{22} \\ \sigma_{33} \\ -\sigma_{23} \\ -\sigma_{13} \\ \sigma_{12} \end{pmatrix} = \begin{pmatrix} C_{11} & C_{12} & C_{13} & C_{14} & C_{15} & C_{16} \\ C_{12} & C_{22} & C_{23} & C_{24} & C_{25} & C_{26} \\ C_{13} & C_{23} & C_{33} & C_{34} & C_{35} & C_{36} \\ C_{14} & C_{24} & C_{34} & C_{44} & C_{45} & C_{46} \\ C_{15} & C_{25} & C_{35} & C_{45} & C_{55} & C_{56} \\ C_{16} & C_{26} & C_{36} & C_{46} & C_{56} & C_{66} \end{pmatrix} \begin{pmatrix} \varepsilon_{11} \\ \varepsilon_{22} \\ \varepsilon_{33} \\ -2\varepsilon_{23} \\ -2\varepsilon_{13} \\ 2\varepsilon_{12} \end{pmatrix} \qquad [1.151]$$

and a comparison with equation [1.146] leads to the following six relations:

$$\begin{cases} C_{14}\varepsilon_{23} + C_{15}\varepsilon_{13} = 0 \\ C_{24}\varepsilon_{23} + C_{25}\varepsilon_{13} = 0 \\ C_{34}\varepsilon_{23} + C_{35}\varepsilon_{13} = 0 \\ C_{14}\varepsilon_{11} + C_{24}\varepsilon_{22} + C_{34}\varepsilon_{33} + 2C_{46}\varepsilon_{12} = 0 \\ C_{15}\varepsilon_{11} + C_{25}\varepsilon_{22} + C_{35}\varepsilon_{33} + 2C_{56}\varepsilon_{12} = 0 \\ C_{46}\varepsilon_{23} + C_{56}\varepsilon_{13} = 0 \end{cases} \qquad [1.152]$$

Since these equations are true for any value of the strains ε_{ij}, the components $C_{14}, C_{15}, C_{24}, C_{25}, C_{34}, C_{35}, C_{46}$ and C_{56} are necessarily zero. The stiffness tensor involves only 13 independent constants:

$$\underline{\underline{C}} = \begin{pmatrix} C_{11} & C_{12} & C_{13} & 0 & 0 & C_{16} \\ C_{12} & C_{22} & C_{23} & 0 & 0 & C_{26} \\ C_{13} & C_{23} & C_{33} & 0 & 0 & C_{36} \\ 0 & 0 & 0 & C_{44} & C_{45} & 0 \\ 0 & 0 & 0 & C_{45} & C_{55} & 0 \\ C_{16} & C_{26} & C_{36} & 0 & 0 & C_{66} \end{pmatrix} \quad \begin{array}{l} \text{monoclinic} \\ (\text{M} \perp x_3) \end{array} \qquad [1.153]$$

A material with three symmetry planes is called *orthotropic* (or orthorhombic). A similar calculation applied to the two other symmetry planes shows that the components C_{16}, C_{26}, C_{36} and C_{45} are also zero. An orthotropic material is characterized by nine independent elastic constants:

$$\underline{\underline{C}} = \begin{pmatrix} C_{11} & C_{12} & C_{13} & 0 & 0 & 0 \\ C_{12} & C_{22} & C_{23} & 0 & 0 & 0 \\ C_{13} & C_{23} & C_{33} & 0 & 0 & 0 \\ 0 & 0 & 0 & C_{44} & 0 & 0 \\ 0 & 0 & 0 & 0 & C_{55} & 0 \\ 0 & 0 & 0 & 0 & 0 & C_{66} \end{pmatrix} \quad \text{orthotropic} \qquad [1.154]$$

An orthotropic material with axial symmetry is called *transverse isotropic*. In this case, the number of independent elastic constants is five. If the isotropic axis is the $\underline{e}_1$ axis, the expression for the basis change matrix is:

$$\alpha = \begin{pmatrix} 1 & 0 & 0 \\ 0 & \cos\varphi & -\sin\varphi \\ 0 & \sin\varphi & \cos\varphi \end{pmatrix} \quad \forall \varphi \in [0, 2\pi] \qquad [1.155]$$

In that case, we get:

$$C_{12} = C_{13}, \quad C_{22} = C_{33}, \quad C_{55} = C_{66} \quad \text{and} \quad C_{44} = \frac{C_{22} - C_{23}}{2} \qquad [1.156]$$

implying the following form for the stiffness tensor:

$$\underline{\underline{C}} = \begin{pmatrix} C_{11} & C_{12} & C_{12} & 0 & 0 & 0 \\ C_{12} & C_{22} & C_{23} & 0 & 0 & 0 \\ C_{12} & C_{23} & C_{22} & 0 & 0 & 0 \\ 0 & 0 & 0 & \dfrac{C_{22} - C_{23}}{2} & 0 & 0 \\ 0 & 0 & 0 & 0 & C_{66} & 0 \\ 0 & 0 & 0 & 0 & 0 & C_{66} \end{pmatrix} \quad \begin{array}{l} \text{transverse isotropic} \\ (\text{axis} // Ox_1) \end{array}$$

[1.157]

A material with *cubic* symmetry has a stiffness tensor which is invariant through a circular permutation of the basis vectors in which it is defined. This means that if the stiffness tensor is defined in the reference frame $\mathbb{R}$ with basis $(\underline{e}_1, \underline{e}_2, \underline{e}_3)$, then this tensor is the same in the reference frames $\mathbb{R}'$ with basis $(\underline{e}_2, \underline{e}_3, \underline{e}_1)$ and $\mathbb{R}''$ with basis $(\underline{e}_3, \underline{e}_1, \underline{e}_2)$. The matrix for the transition from reference frame $\mathbb{R}$ to $\mathbb{R}'$ and from $\mathbb{R}'$ to $\mathbb{R}''$ is:

$$\alpha = \begin{pmatrix} 0 & 1 & 0 \\ 0 & 0 & 1 \\ 1 & 0 & 0 \end{pmatrix} \qquad [1.158]$$

The elastic constants, therefore, satisfy the equalities:

$$C_{11} = C_{22} = C_{33}, \quad C_{44} = C_{55} = C_{66} \quad \text{and} \quad C_{12} = C_{13} = C_{23} \qquad [1.159]$$

imposing the following form for the stiffness tensor:

$$\underline{\underline{C}} = \begin{pmatrix} C_{11} & C_{12} & C_{12} & 0 & 0 & 0 \\ C_{12} & C_{11} & C_{12} & 0 & 0 & 0 \\ C_{12} & C_{12} & C_{11} & 0 & 0 & 0 \\ 0 & 0 & 0 & C_{44} & 0 & 0 \\ 0 & 0 & 0 & 0 & C_{44} & 0 \\ 0 & 0 & 0 & 0 & 0 & C_{44} \end{pmatrix} \quad \text{cubic} \qquad [1.160]$$

Finally, in the case of an *isotropic solid*, these components are given by relation [1.66] in terms of the Lamé constants λ and μ. By giving the pairs (ij) and (kl), values from 1 to 6, we get:

$$\begin{cases} C_{11} = C_{22} = C_{33} = \lambda + 2\mu \\ C_{12} = C_{13} = C_{23} = \lambda \\ C_{44} = C_{55} = C_{66} = \mu = \dfrac{C_{11} - C_{12}}{2} \end{cases} \qquad [1.161]$$

The 12 other moduli c_{ijkl} are zero as they have an odd number of distinct indices $ijkl$ (see Appendix A2.5). By expressing all the components in terms of C_{11} and C_{12}, the table is similar to [1.160] with C_{44} equal to $(C_{11} - C_{12})/2$. Table 1.4 indicates the relations between the different pairs of characteristic elastic parameters of an isotropic solid: the Lamé constants λ and μ, Young's modulus E and Poisson's ratio ν, the bulk modulus K and shear modulus $G = \mu$, and the general constants C_{11} and C_{12}.

	E, ν	E, μ	λ, μ	C_{11}, C_{12}
λ	$\dfrac{E\nu}{(1+\nu)(1-2\nu)}$	$\dfrac{\mu(E-2\mu)}{3\mu - E}$	λ	C_{12}
μ	$\dfrac{E}{2(1+\nu)}$	μ	μ	$\dfrac{C_{11} - C_{12}}{2}$
E	E	E	$\dfrac{\mu(3\lambda + 2\mu)}{\lambda + \mu}$	$C_{11} - \dfrac{2C_{12}^2}{C_{11} + C_{22}}$
K	$\dfrac{E}{3(1-2\nu)}$	$\dfrac{\mu E}{3(3\mu - E)}$	$\lambda + \dfrac{2}{3}\mu$	$\dfrac{C_{11} + 2C_{12}}{3}$
ν	ν	$\dfrac{E - 2\mu}{2\mu}$	$\dfrac{\lambda}{2(\lambda + \mu)}$	$\dfrac{C_{12}}{C_{11} + C_{12}}$

Table 1.4. *Relations between the various elastic constants and the Poisson's ratio ν for an isotropic solid*

It must be noted that the stiffness tensor is given in the basis of a unique reference frame $\mathbb{R}$. Studying an anisotropic material in the basis of another reference frame $\mathbb{R}'$ requires to express the stiffness tensor in this new basis. For this, it is sufficient to use the transition matrix α between the bases of both reference frames. Nonetheless, a material with a specific type of symmetry in the reference frame $\mathbb{R}$ needs not necessarily have the same symmetry in the reference frame $\mathbb{R}'$. The components of the elasticity tensor in the basis of the reference frame $\mathbb{R}'$ are expressed as a function of those in the reference frame $\mathbb{R}$ by the relation [A2.40]:

$$c'_{ijkl} = \alpha_i^p \alpha_j^q \alpha_k^r \alpha_l^s c_{pqrs} \qquad [1.162]$$

Furthermore, there exist other classes of symmetry, such as the tetragonal classes (with an axis A_4) and trigonal classes (with an axis A_3). These types of symmetry are particularly important for piezoelectric materials. The reduction in the number of independent elastic constants is carried out in Appendix A2.5 for all classes of symmetry. The results are given in Table 1.5 for the seven crystalline system and the isotropic solid.

Table 1.6 gives the stiffness constants and mass density of anisotropic solids categorized by crystalline systems.

Some composites can be modeled by a particular class of symmetry, like orthotropic or transverse isotropic. Let us take the case of a composite material composed of a laminated assembly of carbon fibers embedded in a heat hardened epoxy resin. In a unidirectional composite (Figure 1.6(a)), the layers (or plies) are stacked in a fixed direction (for example x_1) and the elastic constants are those of a solid with orthotropic symmetry. In a cross-ply composite (Figure 1.6(b)), this direction varies, for example, by 45° from one ply to the next (0/45°/90°/-45° layout). The composite thus includes an eightfold axis of symmetry, perpendicular to the plies, equivalent from the point of view of the elastic constants to a sixfold axis, and it behaves like a homogeneous solid as long as the wavelength is large compared to the thickness of a ply. This type of composite, of which there are several varieties, is used as a propagation medium for elastic waves during the ultrasonic control of their homogeneity. This non-destructive evaluation, thanks to which the mutual adherence between the layers can be verified in particular, requires a knowledge of the wave propagation in the composite. The benefit of these materials, whose use extends in domains such as aeronautics and automobile industries, is the gain in mass ($\rho = 1550$ kg·m^{-3}) obtained for the same stiffness.

It is possible to invert Hooke's law in order to express strains as a function of stresses:

$$\varepsilon_{ij} = s_{ijkl} \sigma_{kl} \qquad [1.163]$$

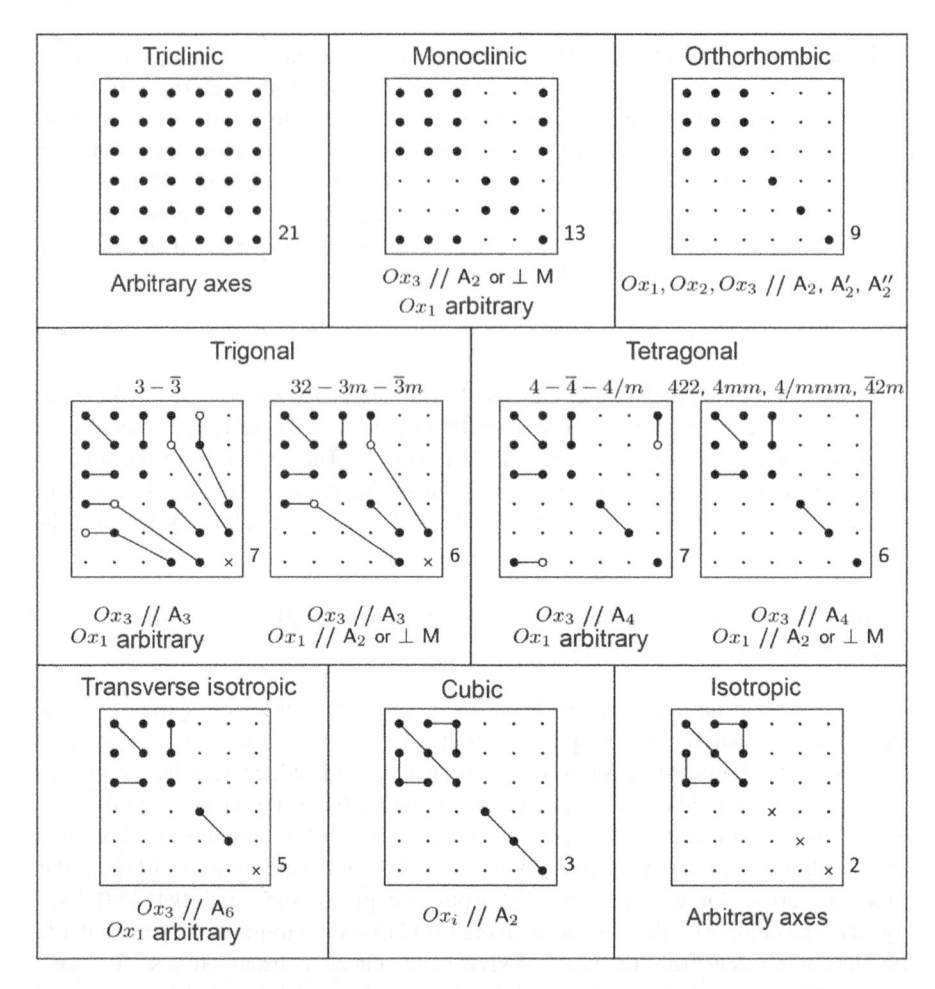

Table 1.5. *Components of the stiffness tensor C_{IJ} for crystals, according to their symmetry. Components:* ● *non-zero,* · *zero,* ●—● *equal,* ●—○ *opposite,* × *equal to* $(C_{11} - C_{12})/2$. *The symmetry with respect to the main diagonal is not mentioned. The number of independent components is given at the bottom right of each tensor*

The *flexibility constants* or *compliances* s_{ijkl} form a fourth-rank tensor with the same properties of symmetry as the tensor c_{ijkl}:

$$s_{ijkl} = s_{jikl} \quad \text{and} \quad s_{ijkl} = s_{ijlk} \qquad [1.164]$$

Similarly, the solution to system [1.145] in terms of the strains ε_I leads to:

$$\varepsilon_I = S_{IJ}\sigma_J \qquad [1.165]$$

where the matrix S_{IJ} is the inverse of the matrix C_{IJ}:

$$S_{IJ} = (C_{IJ})^{-1} \quad \text{or} \quad S_{IJ}C_{JK} = \delta_{IK} \qquad [1.166]$$

δ_{IK} being the six-dimensional Kronecker symbol. To relate the constants S_{IJ} with s_{ijkl}, we must compare the developments of equations [1.163] and [1.165], knowing that $\varepsilon_I = 2\varepsilon_{ij}$, if $I > 3$, and that the term $s_{ijkl}\sigma_{kl}$ with $J = (k,l) > 3$ are repeated in the sum [1.163]. In general, the compliance constants can be written as:

$$S_{IJ} = 2^p s_{ijkl} \qquad [1.167]$$

where p is the number of indices greater than 3 in the pair (I, J). In an orthonormal reference frame, the reduction of the compliance tensor s_{ijkl} is identical to that of the tensor c_{ijkl}. Nonetheless, the relations between the components S_{IJ} may be different. For example, in the trigonal or hexagonal system, and in isotropic media, the relation $C_{66} = (C_{11} - C_{12})/2$, that is $c_{1212} = (c_{1111} - c_{1122})/2$ implies the same relation between corresponding compliances s_{ijkl}. However, since $S_{66} = 4s_{1212}$, $S_{11} = s_{1111}$ and $S_{12} = s_{1122}$, it comes: $S_{66} = 2(S_{11} - S_{12})$.

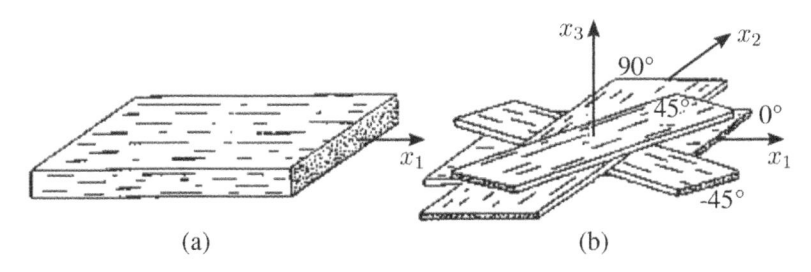

Figure 1.6. *Composite made up of carbon fibers embedded in a thermoset epoxy resin. a) Single ply (pre-impregnated layer) and unidirectional composite, b) cross-ply composite: fiber orientation (0/45°/90°/–45° plies) for a four layer stacking (a spatial period)*

The constitutive law of the anisotropic solid, assumed to be linear, being known, it is now possible to establish the propagation equation and to determine the velocity and polarization of the elastic plane waves.

	Material	ρ	C_{11}	C_{22}	C_{33}	C_{44}	C_{55}	C_{66}	C_{12}	C_{13}	C_{23}	C_{14}	C_{16}	References
Cubic	AsGa	5 307	118.8			59.4			53.8			0	0	(Bateman et al. 1959)
	Bi$_{12}$GeO$_{20}$	9230	128			25.5			30.5			0	0	(Slobodnik and Sethares 1972)
	Copper	8 900	170			75.8			122.5			0	0	(Chang and Himmel 1966)
	Gold	19 280	192.5			42.4			163			0	0	(Chang and Himmel 1966)
	Silicon	2329	165.6			79.5			63.9			0	0	(Hall 1967)
Trans. iso.	Carbon-epoxy* (b)	1550	43.4		13.5	5.4		16.8	9.8	6.2		0	0	
	PZT-4 Ceramic	7 500	139		115	25.6		30.5	78	74		0	0	(Jaffe and Berlincourt 1965)
	Zinc Oxide	5 676	209.7		210.9	42.5		44.3	121.1	105.1		0	0	(Bateman 1962)
	Titanium	4 506	162.4		180.7	46.7		35.2	92	69		0	0	(Fisher and Renken 1964)
Tetra.	Lead molybdate	6950	109.2		91.7	26.7		33.7	68.3	52.8		0	13.6	(Coquin et al. 1971)
	Paratellurite	6 000	56		106	26.5		66	51	22		0	0	(Ohmachi and Uchida 1970)
	Rutile	4 250	273		484	125		194	176	149		0	0	(Verma 1960)
Trigonal	Alumina	3 986	497		498	147		167	163	111		-23.5	0	
	Lithium niobate	4 700	203		245	60		75	53	75		9	0	(Warner et al. 1967)
	Lithium tantalate	7 450	233		275	94		93	47	80		-11	0	(Warner et al. 1967)
	Quartz	2 648	86.7		107.2	57.9		39.9	7	11.9		-17.9	0	(Bechmann 1958)
Ortho.	Austenitic steel	7840	250	250	250	117	91.5	70	112	180	138	0	0	
	Carbon-Epoxy* (a)	1 560	143.8	13.3	13.3	3.6	5.7	5.7	6.2	6.2	6.5	0	0	(Lanceleur 1993)
	Polyester-glass *	1 560	13.6	20.4	19.7	6.6	6.5	9.4	4.4	3.8	3.7	0	0	

Table 1.6. *Mass density (in kg.m^{-3}) and stiffness constants (in GPa) for some anisotropic solids. The materials marked by an asterisk are composites. For transverse isotropic materials $C_{66} = (C_{11} - C_{12})/2$. The stiffness constants are given in the reference frame XYZ defined in Appendix 2*

1.3.2. *Propagation equation, phase velocity, polarization*

In a homogeneous solid, the mechanical displacement $\underline{u}$ is the solution of the equation of motion [1.73], that is, for the component u_i:

$$\rho \frac{\partial^2 u_i}{\partial t^2} = \frac{\partial \sigma_{ij}}{\partial x_j} \qquad [1.168]$$

Under the hypothesis of small strains, the stress tensor is related to the strain tensor by the generalized Hooke's law:

$$\sigma_{ij} = c_{ijkl}\varepsilon_{kl} = c_{ijkl}\frac{\partial u_k}{\partial x_l} \qquad [1.169]$$

then, equation [1.168] can be written as:

$$\rho \frac{\partial^2 u_i}{\partial t^2} = c_{ijkl}\frac{\partial^2 u_k}{\partial x_j \partial x_l} \qquad [1.170]$$

Let us find a solution in the form of a time-harmonic plane wave propagating in the direction $\underline{n}$:

$$u_i(\underline{x},t) = p_i e^{i(\underline{k}\cdot\underline{x}-\omega t)} \qquad [1.171]$$

where p_i are the components of the normalized polarization vector, ω is the angular frequency and $\underline{k} = (\omega/V)\underline{n}$ the wave vector, V being the phase velocity of the wave. Introducing this solution into equation [1.170] leads to:

$$\rho\omega^2 p_i = c_{ijkl}k_j k_l p_k \quad \text{or} \quad c_{ijkl}n_j n_l p_k = \rho V^2 p_i \qquad [1.172]$$

and then to the *Christoffel equation*:

$$\Gamma_{ik}p_k - \rho V^2 p_i = 0 \quad \text{or} \quad (\underline{\underline{\Gamma}} - \rho V^2 \underline{\underline{1}})\underline{p} = \underline{0} \qquad [1.173]$$

where the components of the Christoffel tensor are given by:

$$\Gamma_{ik} = c_{ijkl}n_j n_l \qquad [1.174]$$

The eigenvalues ρV_M^2 (with M = 1,2,3) are the roots to the characteristic equation:

$$\det\left(\Gamma_{ij} - \rho V^2 \delta_{ij}\right) = 0 \quad \text{or} \quad \det\left(\underline{\underline{\Gamma}} - \rho V^2 \underline{\underline{1}}\right) = 0 \qquad [1.175]$$

and the eigenvectors $\underline{p}_M$ are then obtained by solving the equation:

$$\left(\underline{\underline{\Gamma}} - \rho V_M^2 \underline{\underline{1}}\right)\underline{p}_M = \underline{0} \qquad [1.176]$$

Thus, for a given propagation direction $\underline{n}$, there exist three pairs of eigenvalues and eigenvectors, each associated with a phase velocity and a polarization. Moreover, since the Christoffel tensor is symmetric, its eigenvalues are real and its eigenvectors are orthogonal. The wave whose polarization $\underline{p}_1$ is closest to the vector $\underline{n}$ is called quasi-longitudinal. Its velocity is usually larger than that of the other two waves, called quasi-transverse, with polarizations $\underline{p}_2$ and $\underline{p}_3$. The three polarizations are orthogonal (Figure 1.7).

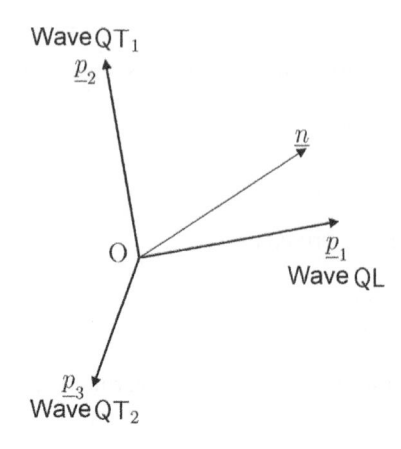

Figure 1.7. *In general, three plane waves propagate in any direction $\underline{n}$ of an anisotropic solid*

The six components of the symmetric Christoffel tensor $\Gamma_{il} = c_{ijkl} n_j n_k$ are expanded as follows:

$$\Gamma_{il} = c_{i11l} n_1^2 + c_{i22l} n_2^2 + c_{i33l} n_3^2 + (c_{i12l} + c_{i21l}) n_1 n_2 \\ + (c_{i13l} + c_{i31l}) n_1 n_3 + (c_{i23l} + c_{i32l}) n_2 n_3 \quad [1.177]$$

or in the Voigt notation:

$$\begin{cases} \Gamma_{11} = C_{11} n_1^2 + C_{66} n_2^2 + C_{55} n_3^2 + 2 C_{16} n_1 n_2 + 2 C_{15} n_1 n_3 + 2 C_{56} n_2 n_3 \\ \Gamma_{22} = C_{66} n_1^2 + C_{22} n_2^2 + C_{44} n_3^2 + 2 C_{26} n_1 n_2 + 2 C_{46} n_1 n_3 + 2 C_{24} n_2 n_3 \\ \Gamma_{33} = C_{55} n_1^2 + C_{44} n_2^2 + C_{33} n_3^2 + 2 C_{45} n_1 n_2 + 2 C_{35} n_1 n_3 + 2 C_{34} n_2 n_3 \\ \Gamma_{12} = C_{16} n_1^2 + C_{26} n_2^2 + C_{45} n_3^2 + (C_{12} + C_{66}) n_1 n_2 + (C_{14} + C_{56}) n_1 n_3 \\ \quad\quad + (C_{46} + C_{25}) n_2 n_3 \\ \Gamma_{13} = C_{15} n_1^2 + C_{46} n_2^2 + C_{35} n_3^2 + (C_{14} + C_{56}) n_1 n_2 + (C_{13} + C_{55}) n_1 n_3 \\ \quad\quad + (C_{36} + C_{45}) n_2 n_3 \\ \Gamma_{23} = C_{56} n_1^2 + C_{24} n_2^2 + C_{34} n_3^2 + (C_{46} + C_{25}) n_1 n_2 + (C_{36} + C_{45}) n_1 n_3 \\ \quad\quad + (C_{23} + C_{44}) n_2 n_3 \end{cases}$$

[1.178]

The propagation direction $\underline{n}$ is defined either by the three components n_1, n_2, n_3, or by the two polar angles θ and φ, as shown in Figure 1.8:

$$n_1 = \sin\theta \cos\varphi, \qquad n_2 = \sin\theta \sin\varphi \qquad \text{and} \qquad n_3 = \cos\theta \qquad [1.179]$$

If the orthonormal reference frame $Ox_1x_2x_3$ respects the convention of crystallographic axes X, Y, Z, inherent to the symmetry of the material (see Figure A2.8 of Appendix 2), the parameters C_{IJ} in the developments [1.178] are directly the elastic constants given in Table 1.6. If this is not the case, one or more rotations of the constants must be carried out to shift from the system of the crystallographic axes to the reference frame $Ox_1x_2x_3$.

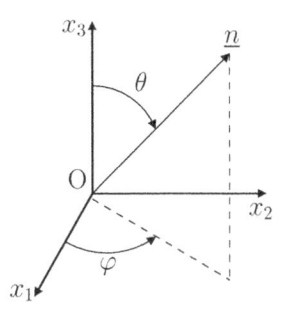

Figure 1.8. *The propagation direction $\underline{n}$ is defined by the angles θ and φ*

In practice, the secular equation can only be solved analytically for specific directions or planes, for which at least two of the three non-diagonal components of the Christoffel tensor are zero. The determinant is then factorized into a product of polynomials of first degree or second degree in V^2.

1.3.3. *Propagation in an orthotropic material*

As an illustration, let us study the propagation of plane waves in an orthotropic material in one of its symmetry planes (Figure 1.9). The material is characterized by the nine independent components of the stiffness tensor [1.154]. The propagation direction $\underline{n}$ is defined by the angle $\varphi \in [0, 2\pi]$ with respect to the binary axis X, carried by x_1:

$$\underline{n} = (\cos\varphi, \sin\varphi, 0)^t \qquad [1.180]$$

The non-zero components of the Christoffel tensor are thus given by:

$$\begin{cases} \Gamma_{11} = C_{11}n_1^2 + C_{66}n_2^2 = C_{11}\cos^2\varphi + C_{66}\sin^2\varphi \\ \Gamma_{12} = (C_{12} + C_{66})n_1 n_2 = (C_{12} + C_{66})\sin\varphi \cos\varphi \\ \Gamma_{22} = C_{66}n_1^2 + C_{22}n_2^2 = C_{66}\cos^2\varphi + C_{22}\sin^2\varphi \\ \Gamma_{33} = C_{55}n_1^2 + C_{44}n_2^2 = C_{55}\cos^2\varphi + C_{44}\sin^2\varphi \end{cases} \qquad [1.181]$$

Since the Christoffel tensor takes the form:

$$\Gamma_{il} = \begin{pmatrix} \Gamma_{11} & \Gamma_{12} & 0 \\ \Gamma_{12} & \Gamma_{22} & 0 \\ 0 & 0 & \Gamma_{33} \end{pmatrix} \qquad [1.182]$$

the solution of the characteristic equation [1.175] leads to the expressions for the phase velocities:

$$\begin{cases} V_1 = \sqrt{\dfrac{1}{2\rho}\left[\Gamma_{11} + \Gamma_{22} + \sqrt{(\Gamma_{11} - \Gamma_{22})^2 + 4\Gamma_{12}^2}\right]} \\ V_2 = \sqrt{\dfrac{1}{2\rho}\left[\Gamma_{11} + \Gamma_{22} - \sqrt{(\Gamma_{11} - \Gamma_{22})^2 + 4\Gamma_{12}^2}\right]} \\ V_3 = \sqrt{\dfrac{\Gamma_{33}}{\rho}} \end{cases} \qquad [1.183]$$

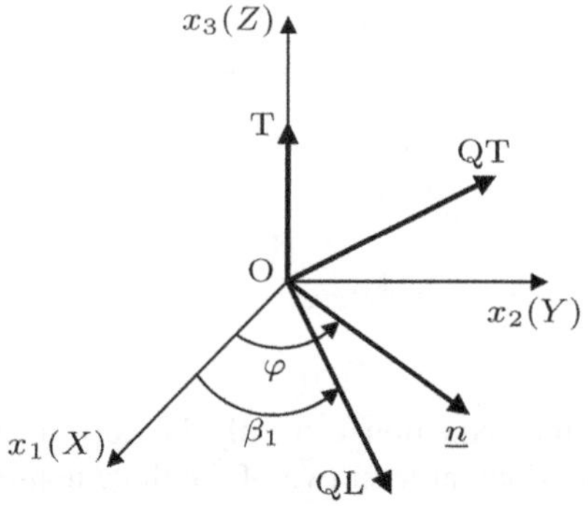

Figure 1.9. *Propagation in the symmetry plane XY of an orthotropic solid. The quasi-longitudinal wave QL and quasi-transverse wave QT are polarized in the XY plane, and the transverse wave T is polarized along the perpendicular direction Z*

The first velocity is associated with a quasi-longitudinal wave, and the second with a quasi-transverse wave. These waves, polarized in the $x_1 x_2$ plane, are only purely longitudinal or transverse in the directions 0, $\pi/2$, π and $3\pi/2$ for which the component Γ_{12} is canceled. Let β_1 and $\beta_2 = \beta_1 + \pi/2$, be the angles of the polarization vectors $\underline{p}_1$ and $\underline{p}_2$ with the Ox_1 axis. For the quasi-longitudinal wave, the Christoffel equation:

$$\left(\Gamma_{11} - \rho V_1^2\right) p_1 + \Gamma_{12} p_2 = 0 \qquad [1.184]$$

which is written as:

$$\left[\frac{\Gamma_{11} - \Gamma_{22}}{2} - \sqrt{\left(\frac{\Gamma_{11} - \Gamma_{22}}{2}\right)^2 + \Gamma_{12}^2}\right] p_1 + \Gamma_{12} p_2 = 0 \qquad [1.185]$$

leads to:

$$\tan \beta_1 = \frac{p_2}{p_1} = a + \sqrt{a^2 + 1} \quad \text{with} \quad a = \frac{\Gamma_{22} - \Gamma_{11}}{2\Gamma_{12}} \quad [1.186]$$

It is worthwhile to introduce the double angle which, considering the indeterminacy of π on the tangent, provides the two orthogonal polarizations β_1 and β_2:

$$\tan(2\beta_i) = -\frac{1}{a} = \frac{2\Gamma_{12}}{\Gamma_{11} - \Gamma_{22}} \quad \text{with} \quad i = 1, 2 \quad [1.187]$$

The third velocity is associated with a purely transverse wave whose polarization $\underline{p}_3 = (0, 0, 1)^t$ is perpendicular to the vector $\underline{n}$ for any angle φ.

1.3.4. *Group velocity and energy velocity*

In an isotropic solid, the phase velocity $V = \omega/k$ is constant, the propagation of longitudinal and transverse waves is non-dispersive and the direction of propagation of energy, that is, the acoustic ray, is parallel to the wave vector $\underline{k}$. Different phenomena may cause the dispersive effect. For example, we will see (section 1.5.3) that in a dissipative medium, the attenuation and velocity of elastic waves depend on frequency. In a waveguide, the existence of one or more characteristic dimensions results in a nonlinear dispersion relation $\omega(k)$. The spectral components of a complex wave no longer propagate at the same phase velocity. A wave packet, whose frequency spectrum is centered around an angular frequency ω_0, corresponding to the wave number k_0, thus propagates with the velocity:

$$V^g = \left.\frac{d\omega}{dk}\right|_{k_0} \quad [1.188]$$

called the group velocity, which is different from V (section 4.1.2).

In an anisotropic solid, the phase velocity depends on the propagation direction $\underline{n}$ and the acoustic ray is deviated – it is parallel to the group velocity vector, defined by:

$$\underline{V}^g = \frac{d\omega}{d\underline{k}} \quad \text{with components} \quad V_i^g = \frac{\partial \omega}{\partial k_i} \quad [1.189]$$

Given that $\underline{k} = k\underline{n}$ and $\omega = kV(\underline{n})$, this velocity is the derivative of the phase velocity with respect to the propagation direction:

$$\underline{V}^g = \frac{dV}{d\underline{n}} \quad \text{that is} \quad V_i^g = \frac{\partial V}{\partial n_i} \quad [1.190]$$

Considering relation [1.172], the square of the phase velocity is expressed by:

$$V^2 = \frac{1}{\rho} c_{ijkl} n_j n_l p_k p_i = \frac{1}{\rho} c_{jklp} p_j p_l n_k n_p \quad [1.191]$$

and its derivative with respect to n_i is given by:

$$2V\frac{\partial V}{\partial n_i} = \frac{1}{\rho}c_{jklp}p_jp_l(n_k\delta_{ip} + n_p\delta_{ik}) = \frac{2}{\rho}c_{ijkl}p_jp_ln_k \qquad [1.192]$$

Thus, the final expression for the components of the group velocity is:

$$V_i^g = \frac{1}{\rho V}c_{ijkl}p_jp_kn_l \qquad [1.193]$$

It is instructive to compare this velocity to the *energy velocity* which, by definition, is equal to the Poynting vector divided by the total energy density per unit volume. In the case of a plane wave with unit amplitude, defined by the real part of [1.171]: $u_i = p_i\cos(\underline{k}.\underline{x} - \omega t)$, the kinetic energy density [1.40] and potential energy density [1.63]:

$$e_k(\underline{x},t) = \frac{1}{2}\rho(\dot{u}_i)^2 \quad \text{and} \quad e_p(\underline{x},t) = \frac{1}{2}c_{ijkl}\frac{\partial u_i}{\partial x_j}\frac{\partial u_k}{\partial x_l} \qquad [1.194]$$

are expressed by:

$$\begin{cases} e_k(\underline{x},t) = \frac{1}{2}\rho\omega^2 p_i^2 \sin^2(\underline{k}.\underline{x} - \omega t) \\ e_p(\underline{x},t) = \frac{1}{2}c_{ijkl}k_jk_lp_ip_k \sin^2(\underline{k}.\underline{x} - \omega t) \end{cases} \qquad [1.195]$$

Multiplying equation [1.172] by p_i:

$$c_{ijkl}k_jk_lp_ip_k = \rho\omega^2 p_i^2 \qquad [1.196]$$

shows that the kinetic and potential energy densities of a traveling plane wave are equal at all points in space and at each time. Since $p_i^2 = 1$, the total mechanical energy density is given by:

$$e_m(\underline{x},t) = e_k + e_p = \rho\omega^2 \sin^2(\underline{k}.\underline{x} - \omega t) \qquad [1.197]$$

In the case of a plane wave, the Poynting vector, defined by relation [1.43]:

$$J_i = -\sigma_{ij}\dot{u}_j = -c_{ijkl}\frac{\partial u_k}{\partial x_l}\frac{\partial u_j}{\partial t} \qquad [1.198]$$

is expressed as:

$$J_i = \omega c_{ijkl}k_lp_kp_j \sin^2(\underline{k}.\underline{x} - \omega t) \qquad [1.199]$$

The *energy transport velocity* associated with the plane wave is defined as the ratio of the Poynting vector to the mechanical energy density. Its components are given by the relation:

$$V_i^e = \frac{J_i}{e_m} = \frac{1}{\rho V}c_{ijkl}p_kp_jn_l \qquad [1.200]$$

A comparison with formula [1.193] shows that, in the case of a plane wave propagating in an anisotropic medium, the energy transport velocity $\underline{V}^e$ is equal to the group velocity $\underline{V}^g$. This property is not valid in the case of a dissipative material. Taking into account expression [1.172], the scalar product of the energy velocity by the propagation direction:

$$\underline{V}^e . \underline{n} = V_i^e n_i = \frac{c_{ijkl} n_i n_l p_k p_j}{\rho V} \qquad [1.201]$$

takes the form:

$$\underline{V}^e . \underline{n} = V \qquad [1.202]$$

The projection of the energy velocity vector on the propagation direction is thus equal to the phase velocity (Figure 1.10). This also implies that the energy velocity is always larger than or equal to the phase velocity of the considered wave.

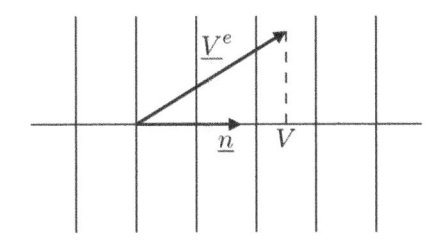

Figure 1.10. *The projection of the energy velocity vector $\underline{V}^e$ on the propagation direction $\underline{n}$ is equal to the phase velocity V*

By definition, the *acoustic intensity* is the average value of the Poynting vector component along the propagation direction:

$$I = \langle \underline{J} . \underline{n} \rangle = \langle J_i n_i \rangle \qquad [1.203]$$

According to equation [1.200] $J_i = V_i^e e_m$ and considering relation [1.202]:

$$I = V_i^e n_i \langle e_m \rangle = V \langle e_m \rangle \qquad [1.204]$$

For a sinusoidal plane wave of amplitude a, the average value of the total mechanical energy [1.197] is equal to $\frac{1}{2} \rho \omega^2 a^2$, so that the expression for the acoustic intensity in an anisotropic solid:

$$I = \frac{1}{2} \rho \omega^2 V a^2 = \frac{1}{2} Z \omega^2 a^2 \qquad [1.205]$$

is similar to the expression for a longitudinal wave [1.117] or a transverse wave [1.121] propagating in an isotropic solid. However, there are still two essential differences: the direction of the Poynting vector, that is, the direction of the acoustic ray, is generally not parallel to the wave vector, and the acoustic impedance Z depends on the propagation direction. Let us also note that the acoustic intensity is always positive, while the average value of the energy flux $\langle \underline{J} . \underline{l} \rangle$ is positive or negative depending on whether the energy exits or enters through the surface element of unit normal $\underline{l}$.

1.3.5. *Slowness surface and wave surface*

In many problems, especially those involving planar interfaces (reflection-transmission, guided wave, etc), it is judicious to introduce the slowness $s = V^{-1}$ (the inverse of the phase velocity) as well as the slowness vector:

$$\underline{s} = \frac{\underline{n}}{V} = \frac{\underline{k}}{\omega} \qquad [1.206]$$

carried by the propagation direction $\underline{n}$. Since $k_i = s_i \omega$, a time-harmonic plane wave with complex amplitude a and unit polarization $\underline{p}$ is thus written as:

$$\underline{u} = a\underline{p}\, e^{i(k_j x_j - \omega t)} = a\underline{p}\, e^{i\omega(s_j x_j - t)} \qquad [1.207]$$

Two characteristic surfaces are useful to illustrate the propagation of elastic waves in an anisotropic solid.

The *slowness surface* represents, in polar coordinates, the slowness along the propagation direction (see Figure 1.11(a)). It is the locus of the extremity of the phase slowness vector or wave vector up to a homothety, plotted from a fixed origin. This makes it analogous to the index surface in optics. It is composed of three sheets: the sheet of the quasi-longitudinal wave is inside the sheets of the two quasi-transverse waves. The slowness surface plays an important role in reflection and transmission problems (Chapter 2). Moreover, its normal indicates the direction of the energy transport. To demonstrate this property, let us form the scalar product of the energy velocity $\underline{V}^e$ with the vector $d\underline{s} = \frac{\partial \underline{s}}{\partial n_j} dn_j$, tangent to the slowness surface:

$$\underline{V}^e . d\underline{s} = V_i^e\, dn_j \frac{\partial}{\partial n_j}\left(\frac{n_i}{V}\right) = V_i^e \left(\delta_{ij} - \frac{n_i}{V}\frac{\partial V}{\partial n_j}\right) \frac{dn_j}{V} \qquad [1.208]$$

or, since the phase velocity is the projection of the energy velocity vector on the propagation direction $V_i^e n_i = V$:

$$\underline{V}^e . d\underline{s} = \left(V_j^e - \frac{\partial V}{\partial n_j}\right) \frac{dn_j}{V} = 0 \qquad [1.209]$$

Since this orthogonality relation applies to all vectors $d\underline{s}$ of the plane tangent to the slowness surface, the energy velocity is, at any point, normal to the slowness surface.

When the propagation direction varies, the *wave surface* is the locus of the extremity of the energy velocity vector drawn from the origin (see Figure 1.11(b)). The ray vector, joining the origin to a point on the wave surface, represents the distance traveled by the elastic energy in the unit time. Consequently, the wave surface is the set of points reached, at unit time, by the vibration emitted, at time zero, by a point source located at the origin. This is also an equi-phase surface, since all the points begin to oscillate at the same time. On the other hand, since $\underline{V}^e \cdot d\underline{s} = 0$, the relation [1.202], which is written as $\underline{V}^e \cdot \underline{s} = 1$, shows by differentiation that $\underline{s} \cdot d\underline{V}^e = 0$. The propagation direction $\underline{n}$ of a plane wave is therefore perpendicular to all the vectors $d\underline{V}^e$ in the plane tangent to the wave surface at the end of the vector $\underline{V}^e$. The energy velocity takes the form $\underline{V}^e = V^e \underline{n}^e$, where $\underline{n}^e$ is the unit vector in the direction of propagation of the energy and $\underline{V}^e = \underline{V}^g$.

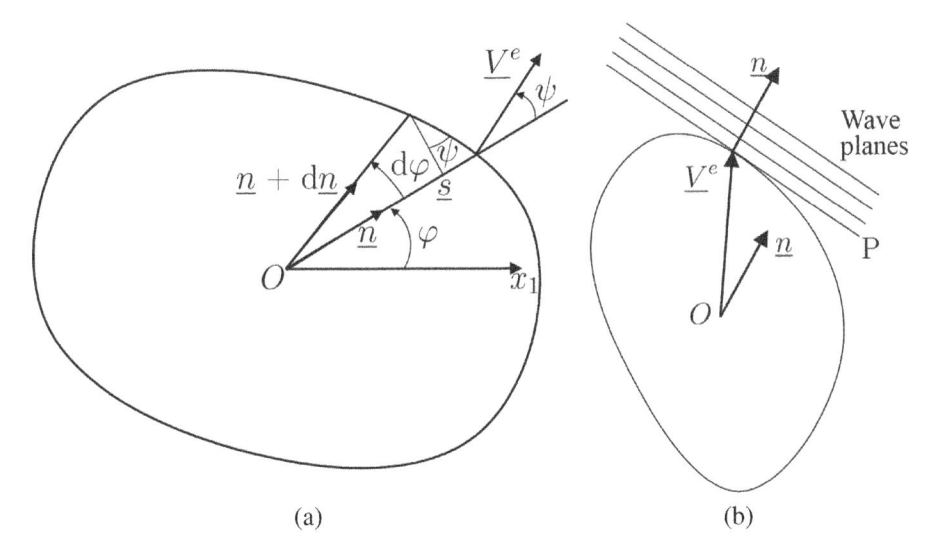

Figure 1.11. *Characteristic surfaces: a) Slowness surface: the energy velocity $\underline{V}^e$ is normal to the slowness surface at all points. b) Wave surface: the wave plane P is tangent to the wave surface at the extremity of the energy velocity vector $\underline{V}^e$*

The wave surface is not as regular as it appears in Figure 1.11(b). The graphical construction in Figure 1.12 shows that the curvature changes in the slowness surface lead to foldings in the wave surface. Moving from A to B along the slowness curve, the corresponding point on the wave surface moves from A' (maximum energy velocity without deviation) to D' (maximum negative deviation of the energy vector), passing through the point B' (minimum energy velocity without deviation). When moving from point A to point C along the slowness curve, the energy deviation is positive, so the corresponding portion on the wave surface is symmetric. The energy velocity at C is the same as at B, and since $\varphi_e = 0$, the points B' and C' coincide. Finally, the wave surface takes the form of a "crescent horn".

In general, only the intersections of the slowness surface with a plane are drawn. If this plane is a symmetry plane for elastic properties (mirror or a plane perpendicular to an axis of even order), it contains the normal to the slowness surface, therefore the energy vector. In this case, there exist simple relations between the energy velocity $V^e(\varphi_e)$, the phase velocity $V(\varphi)$ and the angle of deviation of energy $\psi = \varphi_e - \varphi$. The angle ψ in the small triangle in Figure 1.11(a) is defined by its tangent:

$$\tan \psi = -\frac{ds}{s\,d\varphi} = \frac{1}{V}\frac{dV}{d\varphi} \quad \text{because} \quad s(\varphi) = \frac{1}{V(\varphi)} \qquad [1.210]$$

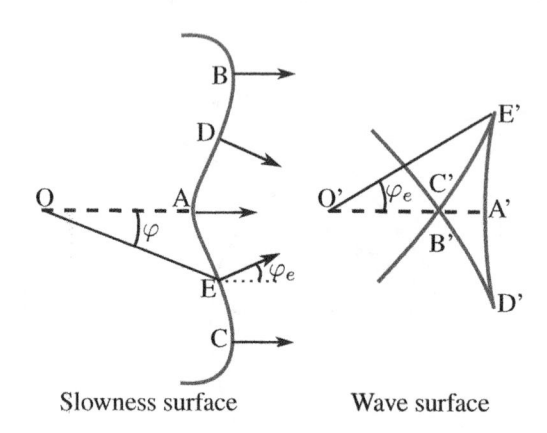

Figure 1.12. *Section of the wave surface. A change in curvature of the slowness surface gives rise, for a given direction, to several values of the energy velocity. To the inflection points D and E on the slowness surface correspond the peaks D' and E' of the two "horns" on the wave surface. For a color version of this figure, see www.iste.co.uk/royer/waves1.zip*

The minus sign indicates that φ_e is larger than φ if $s(\varphi)$ is a decreasing function of φ. According to formula [1.202]:

$$V^e = \frac{V}{\cos \psi} = V\sqrt{1 + \tan^2 \psi} \quad \text{or} \quad V^e = \sqrt{V^2(\varphi) + \left(\frac{dV}{d\varphi}\right)^2}$$

[1.211]

the energy velocity is deduced from variations of the phase velocity as a function of the polar angle φ. Moreover, upon differentiating the relation $V^e \cos \psi = V$, we obtain:

$$dV^e \cos \psi - V^e \sin \psi \, d\varphi_e + V^e \sin \psi \, d\varphi = dV \qquad [1.212]$$

Considering relation [1.210], which can be written as $dV = V^e \sin\psi\, d\varphi$:

$$\tan\psi = \frac{1}{V^e}\frac{dV^e}{d\varphi_e} \quad \text{and} \quad V = \frac{V^e(\varphi_e)}{\sqrt{1+\left(\dfrac{1}{V^e}\dfrac{dV^e}{d\varphi_e}\right)^2}} \qquad [1.213]$$

the phase velocity is linked to the variations of the energy velocity as a function of the angle φ_e.

The phase slowness and wave surfaces in the XY plane for a copper crystal with cubic symmetry are represented in Figures 1.13(a) and 1.13(b), respectively. The transverse wave polarized along the Z axis presents a slowness surface and a wave surface that are perfectly circular, because the phase velocity does not depend on the propagation direction. Conversely, the characteristic curves of the QL and QT waves polarized in the XY plane present more or less complex forms due to the anisotropy of the material.

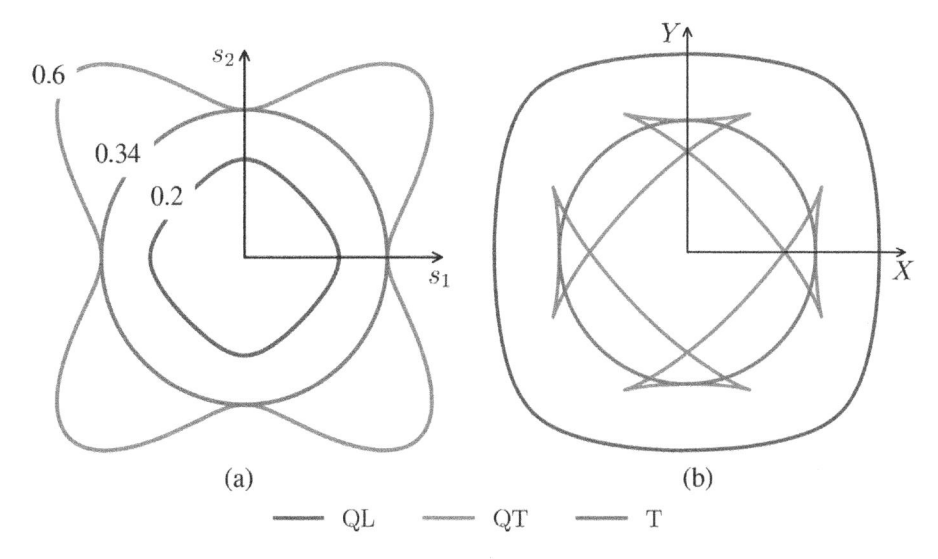

Figure 1.13. *a) Slowness surface (s/km) and b) wave surface for a copper crystal with cubic symmetry in the plane of crystallographic axes XY. The stiffness constants used are those in Table 1.6. For a color version of this figure, see www.iste.co.uk/ royer/waves1.zip*

The characteristic curves may be very different in another plane. Figure 1.14 shows, for the same copper crystal, a cross-section of the slowness surface and of the wave surface by a plane $X'Y'$ perpendicular to the [111] axis. In a cubic crystal, this axis (Z'), carried by the diagonal of the cube in Figure A2.7, is a third-order axis of symmetry. The stiffness tensor C'_{IJ} in the reference frame $X'Y'Z'$ is similar to that [A2.69] of a crystal with trigonal symmetry. The constants c'_{ijkl} in this reference frame are derived from the constants c_{ijkl} in the system of crystallographic axes XYZ by the relation [1.162] where α is the change of basis matrix. In addition to the symmetry through rotation of $\pi/3$, due to the triad axis and the invariance ($+x$, $-x$) of the propagation, it appears that the quasi-longitudinal wave is almost isotropic. This property is due to the fact that the constant $C'_{14} = -25.6$ GPa is small compared to the difference $C'_{11} - C'_{44} = 178.3$ GPa.

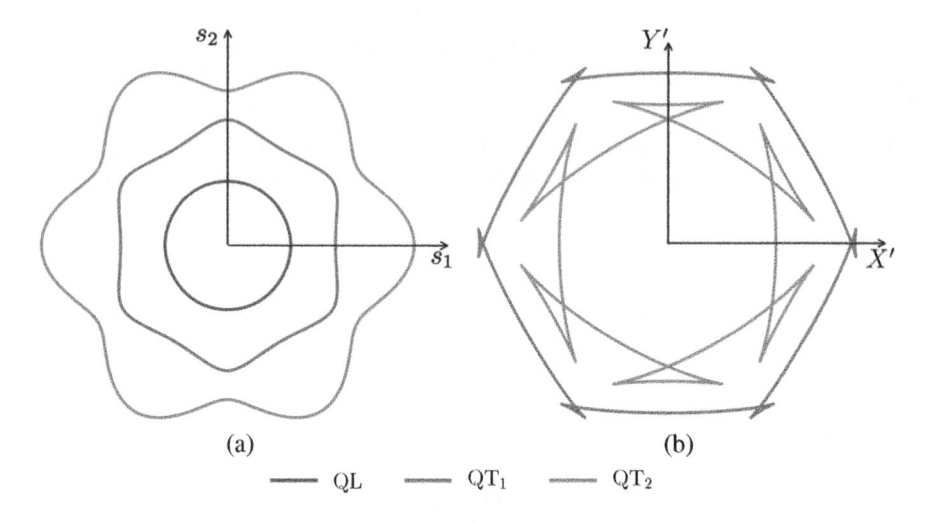

Figure 1.14. *Cross-section of (a) the slowness surface and (b) the wave surface by a plane perpendicular to the [111] axis of a copper crystal. The energy vector is not contained in the $X'Y'$ plane since this plane is not a mirror and only the energy velocity for the two quasi-transverse waves has been represented. For a color version of this figure, see www.iste.co.uk/royer/waves1.zip*

1.3.5.1. *Anisotropy factor*

In a medium with cubic symmetry, $C_{66} = C_{55} = C_{44}$ and $C_{22} = C_{11}$, the velocity of the quasi-transverse wave is extremal in the [100] and [110] directions corresponding to $\varphi = 0$ and $\varphi = \pi/4$; according to expressions [1.181] and [1.183]:

– for $\varphi = 0$, $\Gamma_{12} = 0$, hence $V_2[100] = \sqrt{\dfrac{C_{44}}{\rho}}$;

– for $\varphi = \dfrac{\pi}{4}$, $\Gamma_{11} = \Gamma_{22} = \dfrac{C_{11} + C_{66}}{2}$, hence $V_2[110] = \sqrt{\dfrac{C_{11} - C_{12}}{2\rho}}$.

The ratio of these two velocities:

$$\frac{V_2[100]}{V_2[110]} = \sqrt{\frac{2C_{44}}{C_{11} - C_{12}}} \qquad [1.214]$$

characterizes the anisotropy of a cubic crystal. The curves of copper (Figure 1.13(a)) correspond to the most frequent case where the *anisotropy factor*, defined by:

$$A = \frac{2C_{44}}{C_{11} - C_{12}} \qquad [1.215]$$

is larger than one ($A = 3.2$). When A is smaller than unity, the distortion of the slowness curves is reversed. This is the case for the Bismuth Germanium Oxide ($Bi_{12}GeO_{20}$), a piezoelectric material (Figure 1.16). Of course, $A = 1$ for an isotropic material. Let us note that the slowness surface of the longitudinal wave is concave. This property remains true for any anisotropic solid, whatever its symmetry.

1.3.5.2. *3D slowness surfaces*

Slowness surfaces have complex shapes in the case of anisotropic materials and are spherical for isotropic media. The slowness surfaces for quasi-longitudinal and quasi-transverse waves in a copper crystal are given in Figure 1.15. However, this three-dimensional representation is not very common insofar as it is not quantitative. The intersection with a particular plane is often preferred (see Figure 1.13(a)).

(a) Quasi-longitudinal wave   (b) Quasi-transverse wave 2

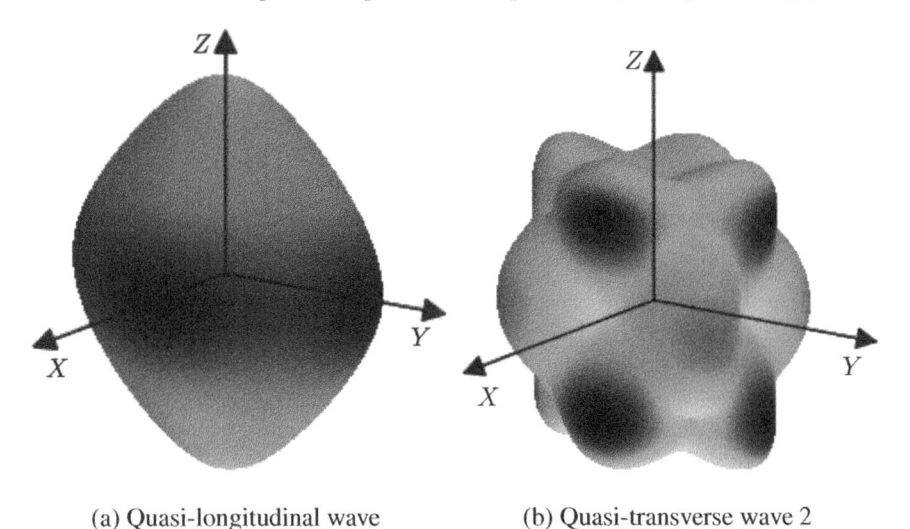

Figure 1.15. *3D slowness surface of quasi-longitudinal and quasi-transverse waves calculated for a copper crystal. For a color version of this figure, see www.iste.co.uk/royer/waves1.zip*

1.4. Piezoelectric solid

A solid is piezoelectric if it becomes electrically polarized under the action of a mechanical force (direct effect) and if it is deformed when an electric field E is applied to it (inverse effect).

The direct piezoelectric effect, discovered by Pierre and Jacques Curie in 1880, expresses the appearance of charges, that is, of an electric induction D, under the effect of a strain[4] S, that is, in a one-dimensional model:

$$D = \varepsilon E + eS \qquad [1.216]$$

The inverse piezoelectric effect, deduced the following year by Gabriel Lippmann, indicates that a piezoelectric solid placed in an electric field is deformed. The mechanical stress is expressed by a relation that generalizes Hooke's law:

$$\sigma = cS - eE \qquad [1.217]$$

REMARK.–

– The coefficients of proportionality of the two effects are equal and opposite.

– At zero strain, an electric field creates a stress $\sigma = -eE$.

– At zero stress, an electric field creates a strain $S = eE/c$.

– In fact, the constant e is one of the components of a third-rank tensor, since the induction, like the electric field, is a first-rank tensor (vector) and the strain, like stress, is a second-rank tensor.

Once the tensor equations for piezoelectricity established, we will study the propagation of elastic waves in a piezoelectric solid.

1.4.1. *Constitutive equations*

The tensor equations for piezoelectricity, generalizing the one-dimensional models described above, are derived from thermodynamic considerations analogous to those developed in sections 1.1.3 and 1.1.4. To the work produced by mechanical forces $dw_m = \rho F_i \, du_i$ (formula [1.37]), we must add the work produced by electric forces: $dw_e = \Phi \, d\rho_e$, where Φ is the electric potential and ρ_e is the charge density per unit volume. The power p_s, supplied per unit volume by mechanical and electrical sources, is given by:

$$p_s(\underline{x}, t) = \rho F_i \dot{u}_i + \Phi \dot{\rho}_e \qquad [1.218]$$

---

[4] To avoid any confusion with the permittivity ε of the medium, the strain tensor is denoted by $\underline{S}$ in this section.

Considering the propagation equation [1.73] and the Gauss equation $\rho_e = \operatorname{div} \underline{D}$, we obtain:

$$p_s = \rho \ddot{u}_i \dot{u}_i - \frac{\partial \sigma_{ij}}{\partial x_j} \dot{u}_i + \Phi \frac{\partial \dot{D}_j}{\partial x_j} \qquad [1.219]$$

or again:

$$p_s = \rho \ddot{u}_i \dot{u}_i + \sigma_{ij} \frac{\partial \dot{u}_i}{\partial x_j} - \dot{D}_j \frac{\partial \Phi}{\partial x_j} + \frac{\partial}{\partial x_j}\left(-\sigma_{ij}\dot{u}_i + \Phi \dot{D}_j\right) \qquad [1.220]$$

As in section 1.1.3, it can be noted that the first term is the time derivative of the kinetic energy density e_k and that the last term is the divergence of the Poynting vector. The expression for the Poynting vector:

$$J_j = -\sigma_{ij}\dot{u}_i + \Phi \dot{D}_j \qquad \text{or} \qquad \underline{J} = -\underline{\underline{\sigma}}\, \underline{\dot{u}} + \Phi \underline{\dot{D}} \qquad [1.221]$$

generalizes the definition [1.43] to the case of a piezoelectric material.

By introducing the strains S_{ij} and the electric field $E_j = -\partial \Phi / \partial x_j$, equation [1.220]:

$$p_s = \frac{\mathrm{d}e_k}{\mathrm{d}t} + \sigma_{ij}\frac{\mathrm{d}S_{ij}}{\mathrm{d}t} + E_j \frac{\mathrm{d}D_j}{\mathrm{d}t} + \frac{\partial J_j}{\partial x_j} \qquad [1.222]$$

reveals the elastic and electric potential energy densities, defined by the variation:

$$\mathrm{d}e_p(\underline{x},t) = \sigma_{ij}\, \mathrm{d}S_{ij} + E_j\, \mathrm{d}D_j \qquad [1.223]$$

Equations [1.44] and [1.46], which express the law of conservation of energy, are still valid in a piezoelectric solid, provided that the electric terms are added to the potential energy and Poynting vector. Similarly, expression [1.54] for the differential of internal energy per unit volume:

$$\mathrm{d}U = T\, \mathrm{d}s + \sigma_{ij}\, \mathrm{d}S_{ij} + E_i\, \mathrm{d}D_i \qquad [1.224]$$

shows that U is a function of the entropy s, the strains S_{ij} and the electric induction D_i: $U = U(s, S_{ij}, D_i)$. To generalize equations [1.216] and [1.217], it is useful to bring out the variables s, S_{ij} and E_i by introducing the (free) electric enthalpy:

$$H = U - E_i D_i \qquad [1.225]$$

whose variation:

$$\mathrm{d}H = T\, \mathrm{d}s + \sigma_{ij}\, \mathrm{d}S_{ij} - D_i\, \mathrm{d}E_i \qquad [1.226]$$

is an exact differential. It then follows that the stresses and the electric induction are partial derivatives of H:

$$\sigma_{ij} = \left.\frac{\partial H}{\partial S_{ij}}\right|_{s, E_i} \qquad \text{and} \qquad D_i = -\left.\frac{\partial H}{\partial E_i}\right|_{s, S_{ij}} \qquad [1.227]$$

To obtain linear relationships between the four variables $\underline{\underline{\sigma}}$, $\underline{\underline{S}}$, $\underline{D}$ and $\underline{E}$, we must start from the most general quadratic form for the enthalpy:

$$H = \frac{1}{2}c^E_{ijkl}S_{ij}S_{kl} - \frac{1}{2}\varepsilon^S_{ij}E_iE_j - e_{kij}E_kS_{ij} \qquad [1.228]$$

In this expression, c^E_{ijkl} are the components of the stiffness tensor, defined at constant entropy and electric field, and ε^S_{ij} are the permittivity at constant entropy and strain. The parameters e_{kij} express the coupling between the electric field and the strains.

The derivatives [1.227] of enthalpy lead to the tensor equations for piezoelectricity:

$$\begin{cases} \sigma_{ij} = c^E_{ijkl}S_{kl} - e_{kij}E_k \\ D_i = \varepsilon^S_{ij}E_j + e_{ijk}S_{jk} \end{cases} \qquad [1.229]$$

The second equation expresses the direct piezoelectric effect and the first equation expresses the inverse effect. The coefficients of proportionality for both effects are opposite and the inverse piezoelectric effect is a thermodynamic consequence of the direct effect.

REMARK.–

– It is easy to verify that due to the symmetries of tensors c_{ijkl} and ε_{ij}, the factors $1/2$ disappear during the derivation of enthalpy [1.228].

– The first two terms of the enthalpy do not change sign. They represent, respectively, the mechanical and electrical energies. The third term has no defined sign. It expresses the coupling between the mechanical and electrical phenomena: there is no piezoelectric energy.

– The exponent E indicates that the elastic modulus links the stresses and strains when the electric field is maintained constant. Indeed, since the material is piezoelectric, the electrical conditions modify the values of the mechanical constants – they should, therefore, be specified. Similarly, the coefficient ε_{ij} which is involved in equation [1.229] is the permittivity at constant strain ε^S_{ij}. It is different from the usual permittivity ε^σ_{ij} of the material, free of any stress ($\underline{\underline{\sigma}} = \underline{\underline{0}}$).

Since the strain tensor is symmetric, the tensor e_{ijk} is also symmetric with respect to the two indices j and k. This pair only takes six values, in accordance with Voigt's notation (section 1.3.1), referenced by the index J:

$$e_{iJ} = e_{ijk} = e_{ikj} \qquad [1.230]$$

The piezoelectric constants are the components of a tensor with three rows ($i = 1, 2, 3$) and six columns ($J = 1, 2, 3, 4, 5, 6$). The number of piezoelectric constants is therefore reduced from 27 (any third-rank tensor) to 18:

$$\underline{\underline{e}} = \begin{pmatrix} e_{11} & e_{12} & e_{13} & e_{14} & e_{15} & e_{16} \\ e_{21} & e_{22} & e_{23} & e_{24} & e_{25} & e_{26} \\ e_{31} & e_{32} & e_{33} & e_{34} & e_{35} & e_{36} \end{pmatrix} \quad [1.231]$$

Relations [1.229] constitute a first system of constitutive equations, which are written in matrix notation as:

$$\begin{cases} \sigma_I = c^E_{IJ} S_J - e_{iI} E_i \\ D_i = \varepsilon^S_{ik} E_k + e_{iJ} S_J \end{cases} \quad \text{with} \quad i, k = 1, 2, 3 \quad \text{and} \quad I, J = 1, 2 ... 6 \quad [1.232]$$

These expressions, which provide the mechanical stress and electric induction as functions of the electric field and strains, are useful to study the propagation of elastic waves (section 1.4.3). Depending on the independent variables chosen, the piezoelectricity equations can take three other forms. For example, to express the strains and the electric induction as functions of the stresses and the electric field, the variables must be changed by introducing a new thermodynamic function:

$$G = H - \sigma_{jk} S_{jk} = U - \sigma_{jk} S_{jk} - E_i D_i \quad [1.233]$$

Considering relation [1.226], the exact differential:

$$\mathrm{d}G = T \, \mathrm{d}s - S_{jk} \, \mathrm{d}\sigma_{jk} - D_i \, \mathrm{d}E_i \quad [1.234]$$

shows that G is a function of the entropy s and of the variables σ_{jk} and E_i: $G = G(s, \sigma_{jk}, E_i)$, whose partial derivatives are:

$$S_{jk} = -\left.\frac{\partial G}{\partial \sigma_{jk}}\right|_{s, E_i} \quad \text{and} \quad D_i = -\left.\frac{\partial G}{\partial E_i}\right|_{s, \sigma_{jk}} \quad [1.235]$$

The most general quadratic form for G:

$$G = -\frac{1}{2} s^E_{jklm} \sigma_{jk} \sigma_{lm} - \frac{1}{2} \varepsilon^\sigma_{ij} E_i E_j - d_{ijk} E_i \sigma_{jk} \quad [1.236]$$

leads after derivation to the following linear relations:

$$\begin{cases} S_{jk} = s^E_{jklm} \sigma_{lm} + d_{ijk} E_i \\ D_i = \varepsilon^\sigma_{ij} E_j + d_{ijk} \sigma_{jk} \end{cases} \quad [1.237]$$

To transpose these equations in matrix notation:

$$\begin{cases} S_J = s^E_{JK} \sigma_K + d_{iJ} E_i \\ D_i = \varepsilon^\sigma_{ij} E_j + d_{iJ} \sigma_J \end{cases} \quad [1.238]$$

and satisfy the definitions for σ_J and S_J (section 1.3.1), we must write:

$$d_{iJ} = d_{ijk} \quad \text{if} \quad J = 1, 2, 3 \quad \text{and} \quad d_{iJ} = 2d_{ijk} \quad \text{if} \quad J = 4, 5, 6 \tag{1.239}$$

The moduli d_{iJ} are identical and have the same sign for direct and inverse effects. They express the proportionality relation between the strains and the components of the electric field, or again, between the mechanical displacement and a potential difference applied between two electrodes. They are of the order of 10^{-10} to 10^{-12} m/V.

The piezoelectric constants e_{iJ}, d_{iJ} and mechanical constants c_{JK}^E and s_{JK}^E are not independent. Introducing equation [1.238] for the strain S_J in the first equation [1.232], we obtain:

$$\sigma_I = c_{IJ}^E s_{JK}^E \sigma_K + \left(c_{IJ}^E d_{iJ} - e_{iI}\right) E_i \tag{1.240}$$

that is since $\sigma_I = \delta_{IK}\sigma_K$:

$$c_{IJ}^E s_{JK}^E = \delta_{IK} \quad \text{and} \quad e_{iI} = c_{IJ}^E d_{iJ} \tag{1.241}$$

The first relation shows that s_{JK}^E is the inverse of c_{JK}^E (as in a non-piezoelectric material). The second relation allows the calculation of the constants e_{iI} from the moduli d_{iJ} or, conversely, using the formula:

$$d_{iJ} = s_{IJ}^E e_{iI} \tag{1.242}$$

In order to link the two permittivities ε_{ij}^σ and ε_{ij}^S, let us transfer the expression of the stress ($\sigma_K = C_{KJ}S_J - e_{jK}E_j$) taken from [1.232] into the second equation [1.238]:

$$D_i = \varepsilon_{ij}^\sigma E_j + d_{iK}\left(C_{JK}^E S_J - e_{jK}E_j\right) \tag{1.243}$$

On identifying this equation with the second equation in [1.232], we obtain:

$$\varepsilon_{ij}^\sigma - \varepsilon_{ij}^S = d_{iK}e_{jK} = d_{iK}C_{JK}^E d_{jJ} \tag{1.244}$$

The permittivity of the rigidly bound solid is difficult to measure; it is calculated using the previous relation, from the piezoelectric moduli and the usual permittivity of the solid free of any stress. The difference between ε_{ij}^σ and ε_{ij}^S is important for strongly piezoelectric materials, such as lithium niobate (LiNbO$_3$):

$$\varepsilon_{11}^\sigma = 74.3 \times 10^{-11} \text{ F.m}^{-1} \quad \text{and} \quad \varepsilon_{11}^S = 38.9 \times 10^{-11} \text{ F.m}^{-1} \tag{1.245}$$

The systems [1.232] and [1.238] are the most widely used. The two other pairs of state equations, in which the independent electric variable is D_i, are obtained by considering the internal energy $U(s, S_{jk}, D_i)$ and mechanical free enthalpy $F(s, \sigma_{jk}, D_i) = U - \sigma_{jk} S_{jk}$. Thus, based on the variables D and S:

$$\begin{cases} \sigma_{jk} = c^D_{jklm} S_{lm} - h_{ijk} D_i \\ E_i = \beta^S_{ij} D_j - h_{ijk} S_{jk} \end{cases} \quad [1.246]$$

c^D_{jklm} is the stiffness tensor at constant electric induction and β^S_{ij} is the inverse of the dielectric tensor ε^S_{ij}. The piezoelectric coefficient h_{ijk} defined by:

$$h_{ijk} = -\left.\frac{\partial E_i}{\partial S_{jk}}\right|_{s,D} = -\left.\frac{\partial \sigma_{jk}}{\partial D_i}\right|_{s,S} = -\frac{\partial^2 U}{\partial D_i \partial S_{jk}} \quad [1.247]$$

is written in matrix notation as $h_{iJ} = h_{ijk}$. The piezoelectric modulus g_{ijk} of the last system of equations:

$$\begin{cases} S_{jk} = s^D_{jklm} \sigma_{lm} + g_{ijk} D_i \\ E_i = \beta^\sigma_{ij} D_j - g_{ijk} \sigma_{jk} \end{cases} \quad [1.248]$$

is the opposite of the second derivative of the enthalpy F with respect to the independent variables D_i and σ_{jk}:

$$g_{ijk} = -\frac{\partial^2 F}{\partial D_i \partial \sigma_{jk}} \quad [1.249]$$

In matrix notation: $g_{iJ} = g_{ijk}$ if $J \leq 3$ and $g_{iJ} = 2g_{ijk}$ if $J > 3$.

1.4.2. *Reduction in the number of independent piezoelectric constants*

We will use the method exposed in section A2.4 in Appendix 2. The condition of invariance of the components of the piezoelectric tensor in a symmetry operation corresponding to the change of basis matrix α is, according to relation [A2.36]:

$$e_{ijk} = \alpha^l_i \alpha^m_j \alpha^n_k e_{lmn} \quad [1.250]$$

Let us examine the effect of point symmetry of crystals, restricting ourselves to those for which the matrix α is diagonal.

– *Center of symmetry*: since $\alpha^k_i = -\delta^k_i$, the relation [1.250] becomes:

$$e_{ijk} = (-1)^3 e_{ijk} \quad \longrightarrow \quad e_{ijk} = 0, \quad \forall i, j, k \quad [1.251]$$

The piezoelectric tensor is 0 in the 11 centrosymmetric classes:

$$\bar{1},\ 2/m,\ mmm,\ \bar{3},\ \bar{3}m,\ 4/m,\ 4/mmm,\ 6/m,\ 6/mmm,\ m3,\ m3m \quad [1.252]$$

– *Plane of symmetry*: the change of axis matrix is again diagonal. Assuming (Ox_3) perpendicular to the mirror plane, it is given by [A2.38]. The condition [1.250], which can be written as:

$$e_{ijk} = \alpha_i^i \alpha_j^j \alpha_k^k e_{ijk} \quad \longrightarrow \quad e_{ijk} = (-1)^n e_{ijk} \quad [1.253]$$

where n is the number of occurrences of the index 3 in the permutation (ijk), implies that the constants with one or three indices equal to 3 are zero. This reduces to 10 the number of moduli e_{iJ} for the class m:

$$e_{iJ} = \begin{pmatrix} e_{11} & e_{12} & e_{13} & 0 & 0 & e_{16} \\ e_{21} & e_{22} & e_{23} & 0 & 0 & e_{26} \\ 0 & 0 & 0 & e_{34} & e_{35} & 0 \end{pmatrix} \quad \text{class } m \quad (M \perp x_3) \quad [1.254]$$

– *Direct binary (or dyad) axis*: the rotation matrix of angle π around (Ox_3) is given by [A2.37]. Equation [1.253] results in components whose value is zero if they have an odd number of indices equal to 1 and 2, that is, an even number of indices equal to 3, so that the table of class 2 is complementary to that of class m:

$$e_{iJ} = \begin{pmatrix} 0 & 0 & 0 & e_{14} & e_{15} & 0 \\ 0 & 0 & 0 & e_{24} & e_{25} & 0 \\ e_{31} & e_{32} & e_{33} & 0 & 0 & e_{36} \end{pmatrix} \quad \text{class } 2 \quad (A_2 // x_3) \quad [1.255]$$

– *Multiple binary axes*: the non-zero piezoelectric coefficients for crystals belonging to the class 222 have an odd number for each index; this is the case only for e_{123}, e_{213}, e_{312}:

$$e_{iJ} = \begin{pmatrix} 0 & 0 & 0 & e_{14} & 0 & 0 \\ 0 & 0 & 0 & 0 & e_{25} & 0 \\ 0 & 0 & 0 & 0 & 0 & e_{36} \end{pmatrix} \quad \text{class } 222 \quad [1.256]$$

For the class $2mm$, the mirror planes perpendicular to (Ox_1) and (Ox_2) cancel the coefficients $e_{14} = e_{123}$, $e_{25} = e_{213}$ and $e_{36} = e_{312}$ in Table [1.255] having an odd number of indices 1 and 2:

$$e_{iJ} = \begin{pmatrix} 0 & 0 & 0 & 0 & e_{15} & 0 \\ 0 & 0 & 0 & e_{24} & 0 & 0 \\ e_{31} & e_{32} & e_{33} & 0 & 0 & 0 \end{pmatrix} \quad \text{class } 2mm \quad [1.257]$$

To establish the form of the piezoelectric tensor for crystals with three-, four- or sixfold axes of symmetry, it is possible to apply the method used in section A2.5 for the stiffness tensor. Table 1.7 summarizes the results for the reduction in the number of components of the tensors expressing the elastic, electric and piezoelectric properties of crystals belonging to the 32-point symmetry classes. The reference axes are those given in Figure A2.8 and the elements of this table are arranged according to the following model:

$$\begin{array}{|cccccc|ccc|}
\hline
C_{11} & C_{12} & C_{13} & C_{14} & C_{15} & C_{16} & e_{11} & e_{21} & e_{31} \\
C_{12} & C_{22} & C_{23} & C_{24} & C_{25} & C_{26} & e_{12} & e_{22} & e_{32} \\
C_{13} & C_{23} & C_{33} & C_{34} & C_{35} & C_{36} & e_{13} & e_{23} & e_{33} \\
C_{14} & C_{24} & C_{34} & C_{44} & C_{45} & C_{46} & e_{14} & e_{24} & e_{34} \\
C_{15} & C_{25} & C_{35} & C_{45} & C_{55} & C_{56} & e_{15} & e_{25} & e_{35} \\
C_{16} & C_{26} & C_{36} & C_{46} & C_{56} & C_{66} & e_{16} & e_{26} & e_{36} \\
\hline
e_{11} & e_{12} & e_{13} & e_{14} & e_{15} & e_{16} & \varepsilon_{11} & \varepsilon_{12} & \varepsilon_{13} \\
e_{21} & e_{22} & e_{23} & e_{24} & e_{25} & e_{26} & \varepsilon_{12} & \varepsilon_{22} & \varepsilon_{23} \\
e_{31} & e_{32} & e_{33} & e_{34} & e_{35} & e_{36} & \varepsilon_{13} & \varepsilon_{23} & \varepsilon_{33} \\
\hline
\end{array}$$
[1.258]

in which the electrical and mechanical conditions imposed on the material are not specified. They must, naturally, match the constitutive equations. The constants presented are C_{IJ}^E, ε_{ij}^S and e_{iJ}. For another set of constants this would be, for example, C_{JK}^D, β_{ij}^S and h_{iJ}.

The results of the reduction in the number of constants are applicable to any third-rank tensor symmetric in j and k, and therefore to the components e_{ijk}, d_{ijk}, h_{ijk} and g_{ijk}. This is also true for constants with two indices $e_{iJ} = e_{ijk}$ and $h_{iJ} = h_{ijk}$. However, in order to find the relations between the moduli d_{iJ} or g_{iJ}, formula [1.239] must be taken into account – for example, for crystals of class 32: $e_{26} = -e_{11}$, but $d_{26} = -2d_{11}$.

In the following section, we will describe the procedure to follow in order to predict waves propagating in a piezoelectric crystal.

1.4.3. *Plane waves in a piezoelectric crystal*

In a piezoelectric solid, the interdependence between electrical and mechanical quantities implies a coupling between elastic waves and electromagnetic waves. Electrical terms are introduced in the equation of dynamics and mechanical terms are introduced in Maxwell's equations. In principle, the problem of propagation is treated by simultaneously solving these coupled equations. The solutions obtained are mixed waves, that is, elastic waves with velocity V, accompanied by an electromagnetic field, and electromagnetic waves with velocity $c \cong 10^5 V$, accompanied by a mechanical strain. The magnetic field of the first type of wave, created by the electric field traveling with a velocity V, very small compared to c, is almost zero. The magnetic energy is negligible compared to the elastic energy. For the second type of wave, it is the elastic energy which is very small compared to electromagnetic energy. Thus, even in a strongly piezoelectric material, the interaction between the three elastic waves and the two electromagnetic waves is weak, because their velocities are very different. Consequently, their propagation can be treated independently.

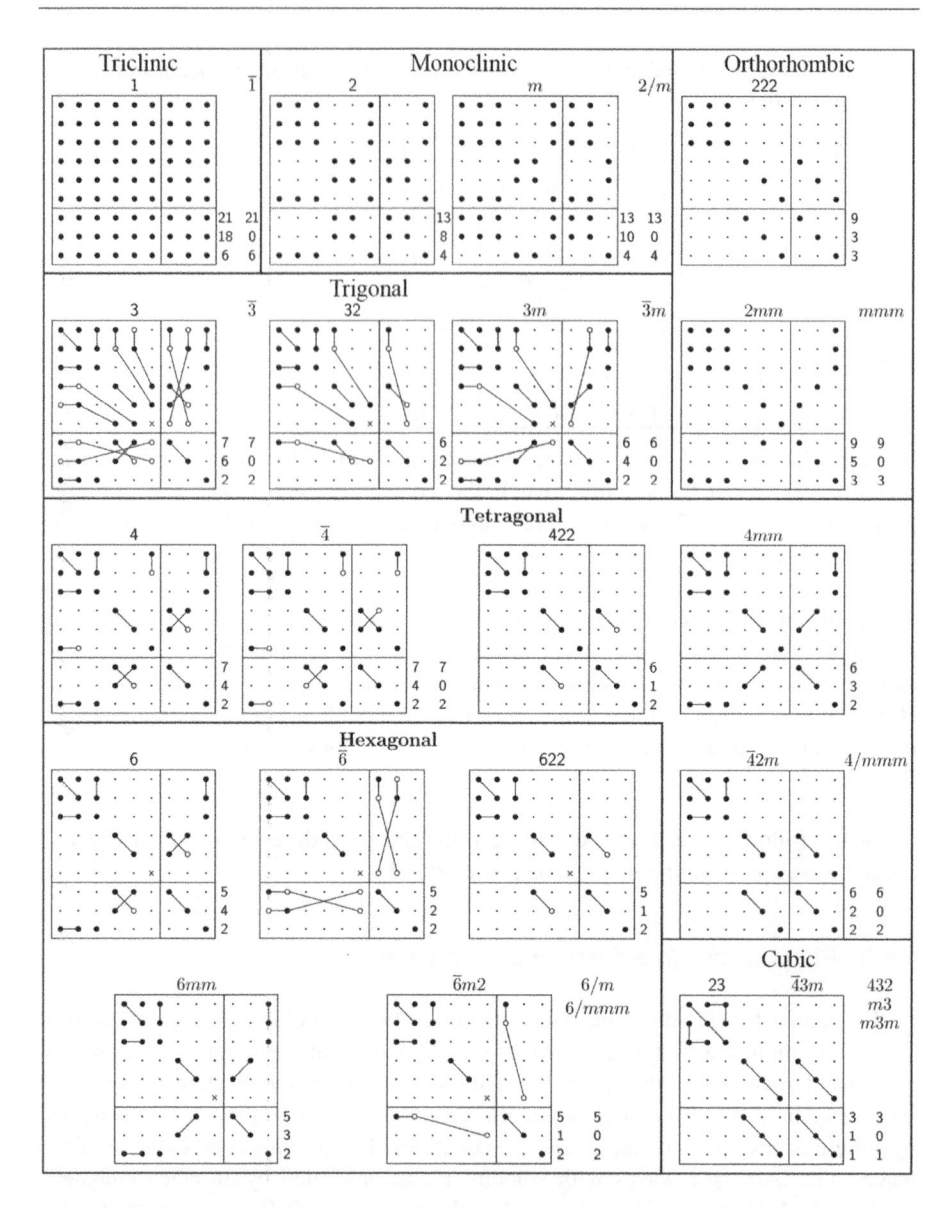

Table 1.7. Components of the elastic, piezoelectric and dielectric tensors: • and ○ non-zero, · zero, •–• equal, •–○ opposite, × equal to $(C_{11} - C_{22})/2$. The symmetry with respect to the main diagonal is not mentioned. The numbers of independent elastic, piezoelectric and dielectric constants are indicated at the right of each case

The propagation of elastic waves in a piezoelectric solid is studied using the hypothesis that the electric field appears static *vis-a-vis* the electromagnetic phenomena. According to the Maxwell–Faraday equation:

$$\underline{\text{rot }} \underline{E} = -\frac{\partial B}{\partial t} \approx 0 \quad [1.259]$$

the electric field derives from a scalar potential Φ, as in electrostatics:

$$\underline{E} = -\underline{\text{grad }} \Phi \quad \text{hence} \quad E_j = -\frac{\partial \Phi}{\partial x_j} \quad [1.260]$$

In the framework of this *quasi-static approximation*, the second electrical quantity involved is the electric induction (or displacement) $\underline{D}$. It is the solution to the Maxwell–Gauss equation:

$$\text{div } \underline{D} = \rho_e \quad \text{hence} \quad \frac{\partial D_i}{\partial x_i} = \rho_e \quad [1.261]$$

where ρ_e is the volume density of free electric charges. In the case of an insulating material, ρ_e is zero, indicating that the electric displacement has zero divergence. In this case, and considering equation [1.229], we obtain:

$$D_i = e_{ijk}\frac{\partial u_j}{\partial x_k} - \varepsilon_{ij}^S\frac{\partial \Phi}{\partial x_j} \quad \text{or} \quad e_{ijk}\frac{\partial^2 u_j}{\partial x_i \partial x_k} - \varepsilon_{ij}^S\frac{\partial^2 \Phi}{\partial x_i \partial x_j} = 0 \quad [1.262]$$

Introducing the expression for the stress:

$$\sigma_{ij} = c_{ijkl}^E \frac{\partial u_l}{\partial x_k} + e_{kij}\frac{\partial \Phi}{\partial x_k} \quad [1.263]$$

in [1.73], the equation for motion is written as:

$$\rho\frac{\partial^2 u_i}{\partial t^2} = c_{ijkl}^E \frac{\partial^2 u_l}{\partial x_j \partial x_k} + e_{kij}\frac{\partial^2 \Phi}{\partial x_j \partial x_k} \quad [1.264]$$

The solutions are found in the form of plane waves propagating in the direction $\underline{n}$:

$$\underline{u}(\underline{x},t) = \underline{p}\, e^{ik(\underline{n}\cdot\underline{x}-Vt)} \quad \text{and} \quad \Phi(\underline{x},t) = \Phi_0 e^{ik(\underline{n}\cdot\underline{x}-Vt)} \quad [1.265]$$

where $\underline{p}$ is the normalized polarization vector, Φ_0 the amplitude of the electric potential, $k = \omega/V$ the wave number and V the phase velocity. Substituting these solutions into equations [1.262] and [1.264] leads to the relations:

$$\begin{cases} \rho V^2 p_i = c_{ijkl}^E p_l n_j n_k + e_{kij}\Phi_0 n_j n_k \\ e_{ijk}p_j n_i n_k - \varepsilon_{ij}^S \Phi_0 n_i n_j = 0 \end{cases} \quad [1.266]$$

The second equation yields the amplitude of the electric potential:

$$\Phi_0 = \frac{\gamma_j p_j}{\varepsilon^S} \quad \text{with} \quad \gamma_j = e_{ijk} n_i n_k \quad \text{and} \quad \varepsilon^S = \varepsilon_{ij}^S n_i n_j \quad [1.267]$$

Finally, the propagation equation takes the form of the Christoffel equation:

$$\tilde{\Gamma}_{il} p_l - \rho V^2 p_i = 0 \quad \text{or} \quad \left(\underline{\tilde{\Gamma}} - \rho V^2 \underline{\underline{1}}\right) \underline{p} = \underline{0} \quad [1.268]$$

where the components of the modified Christoffel tensor are given by:

$$\tilde{\Gamma}_{il} = c_{ijkl}^E n_j n_k + \frac{\gamma_i \gamma_l}{\varepsilon^S} \quad [1.269]$$

The piezoelectric effect is involved through the intermediary of the components:

$$\gamma_l = e_{jkl} n_j n_k = e_{11l} n_1^2 + e_{22l} n_2^2 + e_{33l} n_3^2 + (e_{12l} + e_{21l}) n_1 n_2 \\ + (e_{13l} + e_{31l}) n_1 n_3 + (e_{23l} + e_{32l}) n_2 n_3 \quad [1.270]$$

that is, for each value of the index l:

$$\begin{cases} \gamma_1 = e_{11} n_1^2 + e_{26} n_2^2 + e_{35} n_3^2 + (e_{16} + e_{21}) n_1 n_2 \\ \qquad + (e_{15} + e_{31}) n_1 n_3 + (e_{25} + e_{36}) n_2 n_3 \\ \gamma_2 = e_{16} n_1^2 + e_{22} n_2^2 + e_{34} n_3^2 + (e_{12} + e_{26}) n_1 n_2 \\ \qquad + (e_{14} + e_{36}) n_1 n_3 + (e_{24} + e_{32}) n_2 n_3 \\ \gamma_3 = e_{15} n_1^2 + e_{24} n_2^2 + e_{33} n_3^2 + (e_{14} + e_{25}) n_1 n_2 \\ \qquad + (e_{13} + e_{35}) n_1 n_3 + (e_{23} + e_{34}) n_2 n_3 \end{cases} \quad [1.271]$$

The solution, analogous to that in section 1.3.2, still leads to the propagation of three waves, whose polarizations are orthogonal. Depending on the symmetry of the crystal and the propagation direction, these three waves (or just one or two of them) are piezoelectrically active. Their phase velocity is increased by the piezoelectric effect and the electric field $E_k = -\partial \Phi / \partial x_k$ associated with these waves is longitudinal, since each wave plane is equipotential.

In the following sections, we will first examine the propagation in the simple case of a crystal with cubic symmetry, in order to define the electromechanical coupling coefficient. After having shown the physical meaning of this coefficient, we will examine the case of a crystal belonging to the trigonal system employed to generate and detect elastic waves. The elastic and piezoelectric parameters used for numerical simulations are given in Tables 1.6 and 1.8, respectively.

Material	Piezoelectric constants (C.m^{-2})				Dielectric cts (10^{-11} F.m^{-1})	References
AsGa	$e_{14} = -0.16$				$\varepsilon_{11}^S = 9.73 = \varepsilon_{33}^S$	(Arlt and Quadflieg 1968)
Bi$_{12}$GeO$_{20}$	$e_{14} = 0.99$				$\varepsilon_{11}^S = 34.2 = \varepsilon_{33}^S$	(Slobodnik and Sethares 1972)
ZnO	$e_{15} = -0.59$	$e_{33} = 1.14$		$e_{31} = -0.61$	7.38	7.83	(Jaffe and Berlincourt 1965)
PZT-4	$e_{15} = 12.7$	$e_{33} = 15.1$		$e_{31} = -5.2$	650	560	(Jaffe and Berlincourt 1965)
LiNbO$_3$	$e_{15} = 3.7$	$e_{22} = 2.5$	$e_{31} = 0.2$	$e_{33} = 1.3$	38.9	25.7	(Warner et al. 1967)
LiTaO$_3$	$e_{15} = 2.6$	$e_{22} = 1.6$	$e_{31} \approx 0$	$e_{33} = 1.9$	36.3	38.1	(Warner et al. 1967)
Quartz	$e_{11} = 0.171$			$e_{14} = -0.0406$	3.92	4.1	(Bechmann 1958)

Table 1.8. *Piezoelectric constants (C/m^2) and dielectric constants (10^{-11} F/m) of different materials*

1.4.3.1. *Propagation in a material of cubic symmetry*

According to Table 1.7, the non-zero piezoelectric and dielectric constants are given as:

$$e_{14} = e_{25} = e_{36} \quad \text{and} \quad \varepsilon_{11}^S = \varepsilon_{22}^S = \varepsilon_{33}^S \qquad [1.272]$$

The developments [1.271] are reduced to:

$$\gamma_1 = 2e_{14}n_2n_3, \quad \gamma_2 = 2e_{14}n_1n_3, \quad \text{and} \quad \gamma_3 = 2e_{14}n_1n_2 \qquad [1.273]$$

In the (001) plane where $n_1 = \cos\varphi$, $n_2 = \sin\varphi$ and $n_3 = 0$, only the component $\gamma_3 = e_{14}\sin(2\varphi)$ is non-zero. The Christoffel tensor is of the form [1.182] and only the component Γ_{33} is modified by the piezoelectricity:

$$\tilde{\Gamma}_{33} = \Gamma_{33} + \frac{\gamma_3^2}{\varepsilon_{11}^S} = C_{44}^E + \frac{e_{14}^2}{\varepsilon_{11}^S}\sin^2(2\varphi) \qquad [1.274]$$

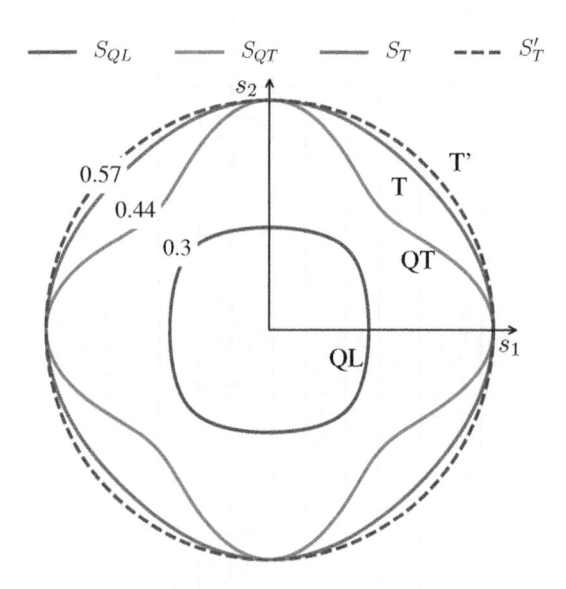

Figure 1.16. *Slowness curves (s/km) in the XY plane for bismuth germanium oxide. Only the transverse wave T is piezoelectric. The curve T' (dashed line) does not take into account the piezoelectricity. For a color version of this figure, see www.iste.co.uk/royer/waves1.zip*

The phase velocities V_1 and V_2 for quasi-longitudinal and quasi-transverse waves, polarized in the (001) plane, are unchanged, except for the fact that the elastic constants c_{ijkl} determined at constant entropy must be replaced by elastic constants

c^E_{ijkl} at constant entropy and electric field. Conversely, the phase velocity of the transverse wave polarized in the [001] direction is affected by the piezoelectricity:

$$V_3 = \sqrt{\frac{\tilde{\Gamma}_{33}}{\rho}} = \sqrt{\frac{C_{44}}{\rho}}\sqrt{1 + \frac{e_{14}^2}{\varepsilon_{11}^S C_{44}^E}\sin^2(2\varphi)} \quad [1.275]$$

The magnitude of this change is related to the value of the *electromechanical coupling coefficient* K, defined by the relation:

$$K^2 = \frac{e_{14}^2}{\varepsilon_{11}^S C_{44}^E + e_{14}^2} \quad \text{hence} \quad V_3 = \sqrt{\frac{C_{44}^E}{\rho}}\sqrt{1 + \frac{K^2}{1-K^2}\sin^2(2\varphi)} \quad [1.276]$$

ORDER OF MAGNITUDE.– In the case of bismuth germanium oxide (Bi$_{12}$GeO$_{20}$), whose slowness curves are shown in Figure 1.16, $K = 0.32$. The maximal relative variation of the velocity V_3 is $\Delta V_3/V_3 \approx K^2/2 \approx 5$ %.

1.4.3.2. *Electromechanical coupling coefficient*

According to its definition, the electromechanical coupling coefficient K expresses the influence of the piezoelectricity on the velocity V of the elastic waves:

$$K^2 = \frac{V^2 - (V')^2}{V^2} \quad [1.277]$$

where V' is the velocity calculated without taking into account the piezoelectricity with $\underline{\underline{C}} \equiv \underline{\underline{C}}^E$. Let us show that this coefficient measures the ability of a piezoelectric material to generate elastic waves.

By multiplying the Christoffel equation [1.268] by the component p_i of the polarization, assumed to be normalized ($p_i p_i = 1$), we obtain:

$$\rho V^2 = \Gamma_{il} p_l p_i + \frac{(\gamma_i p_i)^2}{\varepsilon^S} \quad \text{and} \quad \rho(V')^2 = \Gamma_{il} p'_l p'_i \quad [1.278]$$

Since the polarizations p_i and p'_i are close (identical if the mode is purely longitudinal or transverse), the square of the electromechanical coupling coefficient takes the general form:

$$K^2 = \frac{(\gamma_i p_i)^2}{\varepsilon^S \Gamma_{il} p_l p_i + (\gamma_i p_i)^2} = \frac{e^2}{\varepsilon^S c^E + e^2} \quad [1.279]$$

where $e = \gamma_i p_i$ is homogeneous to a piezoelectric constant and $c^E = \Gamma_{il} p_l p_i$ to a stiffness. This coefficient is always less than unity. In the harmonic case, it is expressed as a function of the average elastic and electric potential energy densities. For a plane wave of unit amplitude:

$$\underline{u}(\underline{x},t) = \underline{p}\cos(\underline{k}.\underline{x} - \omega t) \quad \text{and} \quad \Phi(\underline{x},t) = \frac{\gamma_i p_i}{\varepsilon^S}\cos(\underline{k}.\underline{x} - \omega t) \quad [1.280]$$

we obtain:

$$\begin{cases} e_p^{(ac)} = \dfrac{1}{2} c_{ijkl} \dfrac{\partial u_i}{\partial x_j} \dfrac{\partial u_l}{\partial x_k} = \dfrac{1}{2} \Gamma_{il} p_i p_l k^2 \sin^2(\underline{k}.\underline{x} - \omega t) \\ e_p^{(el)} = \dfrac{1}{2} \varepsilon_{jk}^S \dfrac{\partial \Phi}{\partial x_j} \dfrac{\partial \Phi}{\partial x_k} = \dfrac{1}{2} \dfrac{(\gamma_i p_i)^2}{\varepsilon^S} k^2 \sin^2(\underline{k}.\underline{x} - \omega t) \end{cases} \quad [1.281]$$

and because $e = \gamma_i p_i$:

$$\left\langle e_p^{(ac)} \right\rangle = \frac{1}{4} c^E k^2 \quad \text{and} \quad \left\langle e_p^{(el)} \right\rangle = \frac{1}{4} \frac{e^2}{\varepsilon^S} k^2 \quad [1.282]$$

According to relation [1.279]:

$$K^2 \approx \frac{\left\langle e_p^{(el)} \right\rangle}{\left\langle e_p^{(ac)} \right\rangle + \left\langle e_p^{(el)} \right\rangle} \quad [1.283]$$

the square of the electromechanical coupling coefficient is equal to the ratio of the electric potential energy to the total potential energy transported by the plane wave. Therefore, this factor gives us information on the piezoelectric material's ability to generate acoustic waves. Indeed, regardless of its polarization, the electric field associated with a plane wave is longitudinal. Through the inverse piezoelectric effect, plane acoustic waves can be generated by applying an electric field perpendicularly to the faces of the plate.

1.4.3.3. *Propagation in a crystal of trigonal system*

In order to illustrate the variations in the electromechanical coupling coefficient, let us examine the propagation in the YZ plane (that is, $x_2 x_3$) of lithium niobate (LiNbO$_3$), a ferroelectric crystal belonging to class $3m$ of the trigonal system. With $n_1 = 0$, $n_2 = \sin\theta$ and $n_3 = \cos\theta$, the developments [1.178] show that the components Γ_{12} and Γ_{13} are zero and that the other components are equal to:

$$\begin{cases} \Gamma_{11} = C_{66} n_2^2 + C_{44} n_3^2 + 2C_{14} n_2 n_3 = C_{66} \sin^2\theta + C_{44} \cos^2\theta + C_{14} \sin(2\theta) \\ \Gamma_{22} = C_{11} n_2^2 + C_{44} n_3^2 - 2C_{14} n_2 n_3 = C_{11} \sin^2\theta + C_{44} \cos^2\theta - C_{14} \sin(2\theta) \\ \Gamma_{33} = C_{44} n_2^2 + C_{33} n_3^2 = C_{44} \sin^2\theta + C_{33} \cos^2\theta \\ \Gamma_{23} = -C_{14} n_2^2 + (C_{13} + C_{44}) n_2 n_3 = -C_{14} \sin^2\theta + (C_{13} + C_{44}) \sin\theta \cos\theta \end{cases}$$

[1.284]

The components γ_i are given by equations [1.271]. With $n_1 = 0$ and given the tensor e_{iJ} in Table 1.7, γ_1 is zero, while:

$$\begin{cases} \gamma_2 = e_{22} n_2^2 + e_{34} n_3^2 + (e_{24} + e_{32}) n_2 n_3 = e_{22} \sin^2\theta + (e_{24} + e_{31}) \sin\theta \cos\theta \\ \gamma_3 = e_{24} n_2^2 + e_{33} n_3^2 + (e_{23} + e_{34}) n_2 n_3 = e_{24} \sin^2\theta + e_{33} \cos^2\theta \end{cases}$$

[1.285]

The non-zero components of the modified Christoffel tensor are:

$$\tilde{\Gamma}_{11} = \Gamma_{11}, \quad \tilde{\Gamma}_{22} = \Gamma_{22} + \frac{\gamma_2^2}{\varepsilon^S}, \quad \tilde{\Gamma}_{33} = \Gamma_{33} + \frac{\gamma_3^2}{\varepsilon^S} \quad \text{and} \quad \tilde{\Gamma}_{23} = \Gamma_{23} + \frac{\gamma_2 \gamma_3}{\varepsilon^S}$$

[1.286]

with:

$$\varepsilon^S = \varepsilon_{22}^S \sin^2 \theta + \varepsilon_{33}^S \cos^2 \theta \qquad [1.287]$$

The piezoelectricity does not modify the form of the Christoffel tensor, nor the velocity $V_3 = \sqrt{\Gamma_{11}/\rho}$ of the transverse wave polarized along x_1. On the contrary, velocities V_1 and V_2 of quasi-longitudinal and quasi-transverse waves, given by:

$$2\rho V_M^2 = \tilde{\Gamma}_{22} + \tilde{\Gamma}_{33} - (-1)^M \sqrt{\left(\tilde{\Gamma}_{22} - \tilde{\Gamma}_{33}\right)^2 + 4\tilde{\Gamma}_{23}^2} \quad \text{with} \quad M = 1, 2$$

[1.288]

depend on the piezoelectric constants.

The slowness curves are plotted in Figure 1.17. The dashed curves, calculated using the same elastic constants and ignoring the piezoelectricity, show the importance of this effect on the propagation of quasi-longitudinal and quasi-transverse waves. The magnitude of the gap between the two curves corresponding to one of these waves reveals the crystalline cuts able to excite this wave. Indeed, in its simplest form, the transducer (converter of electric energy into elastic energy and *vice versa*) is a plate or a disk whose faces are metallized (Volume 2, section 3.1). Any electric voltage applied between these electrodes produces a field that causes the plate to vibrate if the crystal cut gives rise to an electromechanical coupling. Thus, a field $\underline{E}_L$ directed along OL is associated with a longitudinal wave propagating in this direction ($\beta = 36°$). Since the piezoelectric effect is reciprocal, a field $\underline{E}_L$, created between two electrodes perpendicular to OL, generates a longitudinal wave. The normal to the faces of the plate makes an angle of 36° with the Y axis. If these two faces are free, the plate is a thickness resonator that vibrates with a high quality factor. If one of the faces is bonded to a solid, the loaded resonator, that is, the transducer, partially transmits its longitudinal vibrations into the solid. The two other curves $1/V_2$ indicate the excitation of a (quasi-) transverse wave by a plate chosen to have a cut in the direction $+163°$ from the Y-axis: the plate undergoes shear-thickness vibrations.

These observations are confirmed by the curves in Figure 1.18, which represent the electromechanical coupling coefficients K_L and K_T of the quasi-longitudinal and quasi-transverse waves as functions of the direction in the YZ plane. For the direction $+36°$ with respect to the Y-axis: $K_L = 0.49$ and $K_T = 0$, for the direction $Y+163°$:

$K_T = 0.62$ and $K_L = 0$. For these two angles, the coupling with the other wave is zero: the transducer is selective, it only emits a single type of wave (longitudinal or transverse).

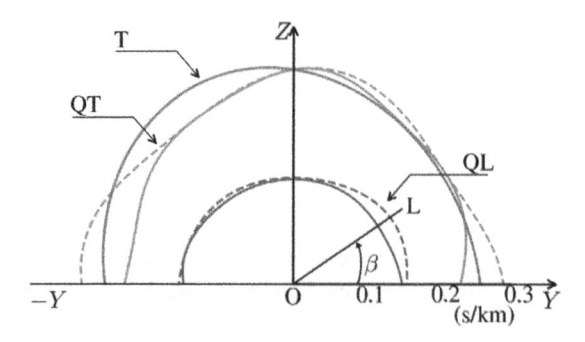

Figure 1.17. *Cross-section of the slowness surface of lithium niobate (class $3m$) by the YZ plane. The curves in dashed lines do not take into account the piezoelectricity. The transverse wave is piezoelectrically inactive. For a color version of this figure, see www.iste.co.uk/royer/waves1.zip*

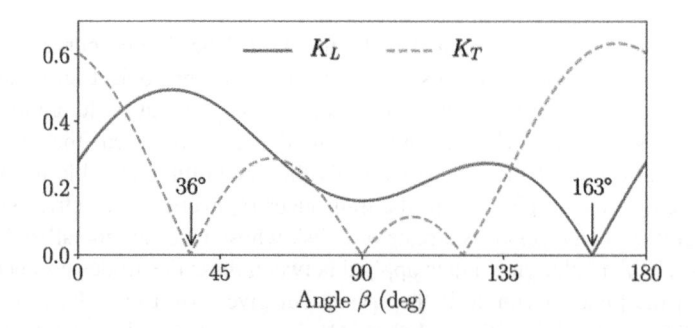

Figure 1.18. *Electromechanical coupling coefficients K_L and K_T as functions of the propagation direction in the YZ plane of lithium niobate. For a color version of this figure, see www.iste.co.uk/royer/waves1.zip*

1.5. Viscoelastic media

In the ultrasonic frequency range, many materials can no longer be considered as being purely elastic: losses through viscoelastic dissipation have to be taken into account. However, the rheology depends on the material. For example, metallic materials such as steel or aluminum are very weakly dissipative in the MHz range, while polymers or resins are particularly dissipative at these frequencies. There are many models that describe rheological behaviors, the degree of complexity increasing with the number of physical parameters to take into account. This section presents the general modeling of viscoelasticity, as well as the two simplest models used to describe the rheology of a viscoelastic material.

If the characteristic parameters of the purely elastic materials are real, taking into account viscoelastic dissipation leads to the introduction of complex values. The density is not affected by this dissipation and remains a real quantity, even if complex values for the mass density are sometimes used to model heterogeneous media, such as poroelastic materials (Norris 2017). A direct consequence of using complex elastic parameters is that the wave number is also complex and that its imaginary part expresses the attenuation of the wave.

1.5.1. *Constitutive equation of linear viscoelasticity*

The stress–strain relation [1.50] indicates that an elastic solid instantaneously responds to any solicitation: it stores potential energy (equation [1.42]) without dissipation. This "ideal" behavior is limited to solicitations that are slow with respect to the characteristic times that reflect the phenomena of internal relaxation. The dissipation of a part of the stored energy induces a time-delayed response. The stresses at time t depend on the strains at this instant and also on the instants t' up to this time. The constitutive equation for a linear viscoelastic solid is the result of these two contributions (Christensen 2003):

$$\sigma_{ij}(\underline{x},t) = c^0_{ijkl}\varepsilon_{kl}(\underline{x},t) + \int_{-\infty}^{t} c'_{ijkl}(t-t')\frac{\partial \varepsilon_{kl}(\underline{x},t')}{\partial t'}\,\mathrm{d}t' \qquad [1.289]$$

The constants c^0_{ijkl} express the purely elastic behavior of the material, while the functions $c'_{ijkl}(t)$ model its viscoelastic behavior, the latter being causal functions (Christensen 2003). In the harmonic case, the stresses and strains take the form:

$$\sigma_{ij}(\underline{x},t) = \sigma_{ij}(\underline{x})e^{-i\omega t} \quad \text{and} \quad \varepsilon_{kl}(\underline{x},t) = \varepsilon_{kl}(\underline{x})e^{-i\omega t} \qquad [1.290]$$

and equation [1.289] can be written as:

$$\sigma_{ij}(\underline{x})e^{-i\omega t} = \left[c^0_{ijkl}e^{-i\omega t} - i\omega \int_{-\infty}^{t} c'_{ijkl}(t-t')e^{-i\omega t'}\,\mathrm{d}t'\right]\varepsilon_{kl}(\underline{x}) \qquad [1.291]$$

By changing the variable $\zeta = t - t'$, the stress tensor takes a form similar to Hooke's law:

$$\sigma_{ij}(\underline{x}) = \tilde{c}_{ijkl}\varepsilon_{kl}(\underline{x}) \qquad [1.292]$$

by writing:

$$\tilde{c}_{ijkl} = c^0_{ijkl} - i\omega \int_0^{+\infty} c'_{ijkl}(\zeta)e^{i\omega\zeta}\,\mathrm{d}\zeta \qquad [1.293]$$

Given the definition of the Fourier transform $\hat{f}(\omega)$ of a function $f(t)$:

$$\hat{f}(\omega) = \int_{-\infty}^{+\infty} f(t) e^{i\omega t} \, dt \qquad [1.294]$$

the expression for the coefficient $\tilde{c}_{ijkl}$ is:

$$\tilde{c}_{ijkl} = c^0_{ijkl} - i\omega \hat{c}_{ijkl}(\omega) \qquad [1.295]$$

This coefficient is a *dynamic modulus of viscoelasticity*, composed of an elastic part (c^0_{ijkl}) independent of frequency, and a viscoelastic part ($\hat{c}_{ijkl}$), which is the Fourier transform of the relaxation function c'_{ijkl}. A similar reasoning may be used for an isotropic solid, by introducing the generalized Lamé constants $\tilde{\lambda}$ and $\tilde{\mu}$:

$$\underline{\sigma}(\underline{x}) = \tilde{\lambda} \, \mathrm{Tr} \left[\underline{\varepsilon}(\underline{x}) \right] \underline{\underline{1}} + 2 \tilde{\mu} \underline{\varepsilon}(\underline{x}) \qquad [1.296]$$

or other pairs of elastic constants, such as the Young's modulus or the bulk modulus and the shear modulus.

1.5.2. *Simple rheological models*

In this section, the dynamic moduli of elasticity $\tilde{c}_{ijkl}$ are represented by the modulus $\tilde{C}$, which may have different expressions according to the modeling of viscoelasticity considered. As a reminder, the dynamical modulus is such that:

$$\tilde{C}(\omega) = C_0 - i\omega \hat{C}(\omega) \qquad [1.297]$$

where C_0 is an elastic modulus that is independent of time (and, therefore, of frequency) and where $\hat{C}(\omega)$ is the time Fourier transform of the causal relaxation function $C'(t)$. It is important to note that the unit of C_0 is the pascal, while the unit for the modulus $\hat{C}$ is the pascal.second: C_0 is thus homogeneous to a stiffness and $\hat{C}$ to a viscosity.

The simplest rheological model used to represent viscoelastic losses in solid materials is the Kelvin–Voigt model, for which:

$$C'(t) = \eta \delta(t) \qquad \text{or} \qquad \hat{C}(\omega) = \eta \qquad [1.298]$$

where $\delta(t)$ is the Dirac function and η is a viscosity. Consequently, the expression for the dynamic modulus is:

$$\tilde{C}(\omega) = C_0 - i\omega \eta \qquad [1.299]$$

From the phenomenological point of view, it corresponds to a spring of stiffness C_0 and a dashpot of damping coefficient η associated in parallel. The real part of the dynamic modulus is the elastic modulus C_0 and the imaginary part is the viscous term $\omega\eta$, which thus linearly depends on frequency. This modeling, which is valid as long as this term is small compared to the elastic modulus, is used for metallic materials and alloys in the conventional ultrasonic frequency range (between 10 kHz and 10 MHz). For materials that are less rigid, such as plastics or some polymers, this kind of modeling cannot be used beyond the MHz.

The second model is the Maxwell model, which corresponds to a spring and a dashpot associated in series. Therefore, the expression for the instantaneous elastic modulus is:

$$C'(t) = C_\infty H(t) e^{-t/\tau} \qquad \text{hence} \qquad \hat{C}(\omega) = \frac{C_\infty \tau}{1 - i\omega\tau} \qquad [1.300]$$

where $H(t)$ is the Heaviside step function, C_∞ an elastic modulus independent of the frequency and τ a relaxation time. The dynamic modulus $\tilde{C}$ is then given by:

$$\tilde{C}(\omega) = C_1(\omega) - iC_2(\omega) \qquad \text{with} \qquad \begin{cases} C_1(\omega) = C_0 + C_\infty \dfrac{(\omega\tau)^2}{1 + (\omega\tau)^2} \\ C_2(\omega) = C_\infty \dfrac{\omega\tau}{1 + (\omega\tau)^2} \end{cases}$$

[1.301]

The real component C_1 of the dynamic modulus depends on the frequency and is limited by two asymptotic values: one low frequency value C_0 and one high frequency value $C_0 + C_\infty$ (Figure 1.19). The imaginary component C_2 tends to $C_\infty \omega\tau$ at low frequency and toward $C_\infty/(\omega\tau)$ at high frequency. The highest value $(C_\infty/2)$ is reached when the angular frequency is equal to the inverse of the relaxation time, that is, for $\omega\tau = 1$. Since the component C_1 is mainly linked to the phase velocity of the wave and the component C_2 to the attenuation, when the frequency is such that $\omega\tau \approx 1$, the wave can be dispersive and highly attenuated.

The Kelvin–Voigt and the Maxwell models are the simplest models that describe the viscoelasticity of solid materials. These models may, therefore, prove to be limited in many situations. Consequently, generalized Maxwell models with several relaxation times or fractional derivative models are used; however, this increases the number of parameters to be defined. Further, temperature may play a crucial role in the characterization of the rheology, since for certain materials (e.g. polymers or some resins) the relaxation times are strongly dependent on temperature.

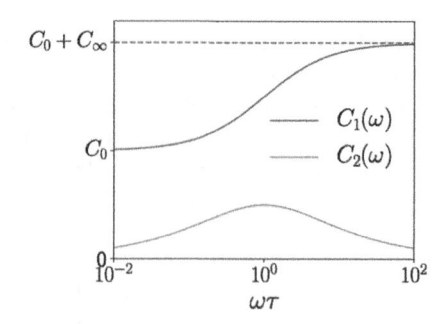

Figure 1.19. *Maxwell model. The real and imaginary components of an elastic modulus as functions of the parameter $\omega\tau$. At low frequency, the Maxwell model tends toward the Kelvin–Voigt model, by writing $\eta = C_\infty \tau$. For a color version of this figure, see www.iste.co.uk/royer/waves1.zip*

1.5.3. *Velocity and attenuation in a viscoelastic medium*

Since the general expression for Hooke's law is not modified by taking into account the viscoelasticity of the material, the calculations in section 1.2.2 remain unchanged. The wave numbers for longitudinal and transverse waves are therefore given by:

$$\tilde{k}_L = \omega\sqrt{\frac{\rho}{\tilde{\lambda} + 2\tilde{\mu}}} \quad \text{and} \quad \tilde{k}_T = \omega\sqrt{\frac{\rho}{\tilde{\mu}}} \qquad [1.302]$$

These complex wave numbers take the form $\tilde{k}_M = k_M + i\alpha_M$. A time-harmonic plane wave propagating in the direction x_1 is therefore associated with the term:

$$\underline{u}(x_1, t) = a\underline{p}\, e^{i(\tilde{k}_M x_1 - \omega t)} = a\underline{p}\, e^{-\alpha_M x_1} e^{i(k_M x_1 - \omega t)} \quad \text{with} \quad M = L, T \qquad [1.303]$$

The imaginary party α_M of the wave number is necessarily positive, while the real part k_M is related to the phase velocity by $\omega = k_M V_M$.

The velocity and attenuation of the transverse wave are expressed as functions of the real and imaginary parts of the shear modulus: $\tilde{\mu} = \mu_1 - i\mu_2$. The square of the transverse wave number takes the form:

$$\tilde{k}_T^2 = k_T^2 - \alpha_T^2 + 2ik_T\alpha_T = \frac{\rho\omega^2(\mu_1 + i\mu_2)}{\mu_1^2 + \mu_2^2} \qquad [1.304]$$

The equality of the real and imaginary parts of this equation leads to the relations:

$$k_T^2 - \alpha_T^2 = \frac{\rho\omega^2\mu_1}{\mu_1^2 + \mu_2^2} \quad \text{and} \quad 2k_T\alpha_T = \frac{\rho\omega^2\mu_2}{\mu_1^2 + \mu_2^2} \qquad [1.305]$$

It is important to highlight that the imaginary part of the shear modulus is necessarily negative. The solution of these equations leads to the following expressions for the phase velocity and for the attenuation coefficient as functions of the parameters μ_1 and μ_2:

$$\tilde{V}_T = \sqrt{\frac{2\left(\mu_1^2 + \mu_2^2\right)}{\rho\left(\mu_1 + \sqrt{\mu_1^2 + \mu_2^2}\right)}} \quad \text{and} \quad \alpha_T = \frac{\mu_2 k_T}{\mu_1 + \sqrt{\mu_1^2 + \mu_2^2}} \quad [1.306]$$

By writing $\varepsilon = \mu_2/\mu_1$, these expressions lead to:

$$\tilde{V}_T = \sqrt{\frac{\mu_1}{\rho}} \sqrt{\frac{2\left(1 + \varepsilon^2\right)}{1 + \sqrt{1 + \varepsilon^2}}} \quad \text{and} \quad \alpha_T = \frac{k_T \varepsilon}{1 + \sqrt{1 + \varepsilon^2}} \quad [1.307]$$

or, since $\varepsilon^2 \ll 1$:

$$\tilde{V}_T \approx V_T \sqrt{\frac{1 + \varepsilon^2}{1 + \varepsilon^2/4}} \approx V_T \left(1 + \frac{3}{8}\varepsilon^2\right) \quad \text{and} \quad \alpha_T \lambda_T \approx \pi \varepsilon \quad [1.308]$$

Although the calculation was carried out in the case of a transverse wave, the result is valid for any type of wave:

$$\tilde{V} \approx V \left[1 + \frac{3}{8}\left(\frac{\alpha\lambda}{\pi}\right)^2\right] \quad [1.309]$$

In the case of the *Kelvin–Voigt* model, the shear modulus is written as $\tilde{\mu} = \mu - i\omega\eta_s$, where the static modulus $\mu = \mu_1$ and the viscosity coefficient η_s are constant. The attenuation is then given by the approximate expression:

$$\alpha_T \approx \frac{\eta_s \omega^2}{2\rho V_T^3} \quad [1.310]$$

In a first approximation, the attenuation coefficient is proportional to ω^2, signifying that the attenuation has a quadratic dependence with the frequency. Since $\lambda_T \approx 2\pi V_T/\omega$, the attenuation per wavelength is proportional to the angular frequency:

$$\alpha_T \lambda_T \approx \frac{\pi \eta_s \omega}{\rho V_T^2} \approx \frac{\pi \eta_s \omega}{\mu} \quad [1.311]$$

This dimensionless quantity shows that a part of the quadratic dependence of the attenuation per unit length (α_T in m^{-1}) is due to the variation of wavelength with frequency.

> ORDER OF MAGNITUDE.– Despite a very large attenuation per wavelength $\alpha\lambda = 0.2$ corresponding to 17 dB for a path of 10λ, the relative variation of the velocity is very small: 1.5×10^{-3}.

For longitudinal waves, the choice of the elastic parameter is made between the first Lamé constant and the bulk modulus, which is preferred for liquids. By modeling the complex bulk modulus in the form $\tilde{K} = K - i\omega\eta_v$, with K the static modulus and η_v the bulk viscosity coefficient, and since:

$$\tilde{\lambda} + 2\tilde{\mu} = \tilde{K} + \frac{4}{3}\tilde{\mu} = K + \frac{4}{3}\mu - i\omega\left(\eta_v + \frac{4}{3}\eta_s\right) \quad [1.312]$$

the attenuation of longitudinal waves takes the form:

$$\alpha_L = \frac{\eta_v + \frac{4}{3}\eta_s}{2V_L\left(K + \frac{4}{3}\mu\right)}\omega^2 \quad [1.313]$$

Between two points separated by a distance d, the ratio of the amplitudes of the wave is $r = e^{\alpha d}$, and the ratio of the powers is equal to r^2. To enlarge the field of measurements, it is common to convert these ratios into a logarithmic scale. Several units are used to express the attenuation (a_{tt}):

– Neper (Np), by taking the natural logarithm of r:

$$a_{tt}(\text{Np}) = \ln r = \alpha d$$

the attenuation coefficient α is thus expressed in Np/m;

– decibel (dB) by taking the decimal logarithm of the power ratio r^2:

$$a_{tt}(\text{dB}) = 10\log\left(r^2\right) = 20\log(e)\alpha d$$

the coefficient α is thus evaluated in dB/m. The two units are linked by:

$$\alpha(\text{dB/m}) = 20\log(e)\alpha(\text{Np/m}) \approx 8.7\alpha(\text{Np/m})$$

In experiments, the elastic parameters are often unknown and measurements of phase velocity and attenuation are carried out. In materials with low attenuation, the phase velocity is assumed to be constant and the attenuation is often modeled by a power law (Szabo and Wu 2000):

$$\alpha_M = \alpha_{0M} + \alpha_{1M}\left(\frac{f}{f_0}\right)^{\beta_M} \quad \text{with} \quad M = L, T \quad [1.314]$$

where α_{0M} and α_{1M} are attenuation coefficients that are independent of frequency, f_0 is the center frequency of the measuring range and β_M an adjustable parameter, ranging from 0 to 2. For example, in the case of PMMA the coefficients for the longitudinal wave obtained by the immersion method (described in section 2.4.2.4) for a center frequency $f_0 = 5$ MHz are: $\alpha_{0L} = 0.59$ dB/mm, $\alpha_{1L} = 0.96$ dB/mm and $\beta_L = 1.09$. For polyurethane, these coefficients are equal to $\alpha_{0L} = 0.98$ dB/mm, $\alpha_{1L} = 2.16$ dB/mm and $\beta_L = 1.46$.

1.5.4. *Time–temperature superposition principle*

The time–temperature superposition principle states that a polymer material has a rheological behavior that depends on temperature, similar to the rheological behavior depending on time or inversely similar to that depending on frequency. This principle makes it possible to determine the rheological behavior of a material over a frequency interval much wider than that of the measurements. For the shear modulus μ, it is formally expressed by the relation between the values at two temperatures T_1 and T_2:

$$\mu(t,T_1) = b_T \mu(t/a_T, T_2) \quad \text{or} \quad \tilde{\mu}(f,T_1) = b_T \tilde{\mu}(a_T f, T_2) \quad [1.315]$$

This expression means that the complex modulus $\tilde{\mu}$ has the same behavior at a temperature T_1 over a frequency interval $[f_{min}, f_{max}]$ than at a temperature T_2 over a frequency interval $[a_T f_{min}, a_T f_{max}]$, up to an amplitude b_T. In experiments, the measurements are restricted to a frequency range, limited by the means for generating elastic waves. As an illustration, the curves for the imaginary component of the shear modulus are plotted in Figure 1.20(a) for five temperatures, increasing from T_1 to T_5. Among these, let us take T_3 as the reference temperature. There exists a coefficient $a_T(T, T_{ref} = T_3)$, such that the experimental curves for $T < T_3$ can be shifted toward the higher frequencies and those for $T > T_3$ toward the lower frequencies (Figure 1.20(b)). In the same way, the coefficient $b_T(T, T_{ref} = T_3)$ makes it possible to shift the curves vertically in order to obtain a master curve, which extends over a frequency range larger than that in experiments (Figure 1.20(c)).

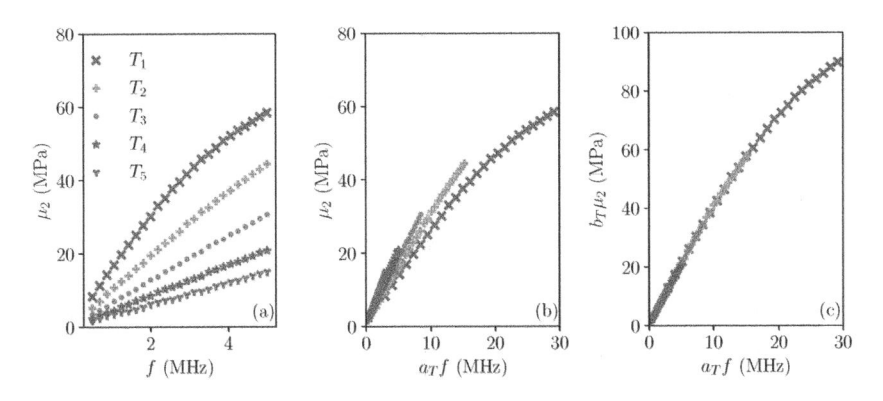

Figure 1.20. *Imaginary component of the shear modulus of a polymer as a function of frequency. For a color version of this figure, see www.iste.co.uk/royer/waves1.zip*

The coefficients a_T and b_T are empirically determined from measurements at different temperatures. The parameter b_T acts on the complex shear modulus, both on its real and imaginary parts. Since the ratio μ_2/μ_1 is independent of the parameter

b_T, the coefficient a_T is then obtained by analyzing this ratio. This one is compared to the expression given by Williams *et al.* (1995):

$$\log a_T = \frac{-K_1(T - T_{ref})}{K_2 + T - T_{ref}} \qquad [1.316]$$

where the characteristic parameters of the material, K_1 and K_2, are determined at the reference temperature. The coefficient b_T is obtained in a second step, but its measurement is more complicated.

1.5.5. *Newtonian fluid*

The objective of this section is to highlight a similarity of behavior between a viscous Newtonian fluid and an isotropic solid. A perfect fluid responds instantaneously to any excitation, whatever its frequency, without any dissipation of energy. Acoustic waves propagate in a perfect fluid without attenuation. In a *real fluid*, there are several mechanisms that cause progressive and irreversible dissipation of a part of the acoustic energy into heat; the *viscosity* associated with the relative movement of adjacent particles (internal friction) or with the friction of the fluid on the walls; the *thermal conductivity* between the compressed parts of the fluid, whose temperature rises, and the dilated parts, whose temperature decreases; the molecular relaxation that expresses a coupling and, therefore, an energy exchange between the acoustic wave and the internal motions (translation, vibration, rotation) of the molecules of a polyatomic fluid. Chemical or structural processes are also possible. These phenomena are all the more complex as they depend on the temperature and ambient pressure, on the presence of impurities and also on the acoustic pressure and frequency of the wave. Each of these effects is expressed by a relaxation time, which measures, at a given frequency, the attenuation per wavelength. Depending on the nature of the fluid or the frequency of the acoustic wave, one (or two) of the processes play a dominant role. The fluid is then characterized by one or two relaxation times.

The internal frictions in the fluid are manifested by the appearance of mechanical tractions, which depend on the orientation of the surface element on which they act. A stress tensor $\tilde{\sigma}_{ij}$ for the viscous fluid should be added to the term resulting from the acoustic pressure:

$$\sigma_{ij}(\underline{x},t) = -p_a(\underline{x},t)\delta_{ij} + \tilde{\sigma}_{ij}(\underline{x},t) \qquad [1.317]$$

These internal frictions only appear if the different regions of the fluid move relative to each other. Each component $\tilde{\sigma}_{ij}$, therefore, depends only on the gradient of the particle velocity $\partial v_i / \partial x_j$. In the case of an infinitesimal rotation of the fluid, this tensor is antisymmetric. As for an isotropic solid (section 1.2.1), the viscous stress tensor for a Newtonian fluid involves only two viscosity coefficients: η_s and η_v.

– The first coefficient expresses the tangential viscous stresses:

$$\tilde{\sigma}_{ij} = 2\eta_s \dot{\varepsilon}_{ij} \quad \text{with} \quad i \neq j \quad [1.318]$$

The coefficient η_s, which reflects the shear viscosity that appears during the relative sliding of two adjacent layers of the fluid, can be measured through the resistance of the fluid to shear motions.

– The second coefficient expresses the normal viscous stresses ($\tilde{\sigma}_{11}$, $\tilde{\sigma}_{22}$ and $\tilde{\sigma}_{33}$) as a function of the dilatation rate $\text{div}\,\underline{v} = \dot{\Theta}$:

$$\tilde{\sigma}_{ii} = \eta_v \,\text{div}\,\underline{v} = \eta_v \dot{\Theta} \quad [1.319]$$

The bulk viscosity or second viscosity η_v, related to the variations in the fluid volume, is much more difficult to measure.

Equations [1.318] and [1.319] are grouped into a single equation similar to [1.70]:

$$\tilde{\sigma}_{ij} = \delta_{ij}(\eta_v - \frac{2}{3}\eta_s)\,\text{div}\,\underline{v} + 2\eta_s \dot{\varepsilon}_{ij} \quad [1.320]$$

Thermodynamic considerations demonstrate that the two viscosity coefficients are positive.

In order to calculate the total stress σ_{ij}, it is necessary to express the pressure p_a. Assuming that the acoustic wave is a weak disturbance from the reference state (defined by the pressure p_0 and the mass density ρ_0), the difference $\rho - \rho_0$ is small compared to ρ_0. It is possible to linearize the state equation $p(\rho)$ and the conservation equations. Retaining only the first term of the Taylor expansion of $p(\rho)$, the acoustic pressure p_a is given by:

$$p_a = p - p_0 = c_0^2(\rho - \rho_0) \quad \text{where} \quad c_0^2 = \left.\frac{\partial p}{\partial \rho}\right|_{s,\rho_0} \quad [1.321]$$

is the square of the speed of sound in the fluid. The subscript s indicates that the transformations are assumed to be adiabatic and reversible, that is, isentropic. Given the linearized continuity equation [1.15], we obtain ($\rho \approx \rho_0$):

$$\frac{\partial p_a}{\partial t} + \rho_0 c_0^2 \,\text{div}\,\underline{v} = 0 \quad [1.322]$$

By integrating with respect to time and introducing the bulk modulus $K = \rho_0 c_0^2$, it appears that the acoustic pressure is proportional to the dilatation $\Theta = \text{div}\,\underline{u}$:

$$p_a = -K\,\text{div}\,\underline{u} \quad [1.323]$$

In the harmonic case ($\underline{v} = -i\omega \underline{u}$), the total stress takes the form:

$$\sigma_{ij} = \delta_{ij}\left[K - i\omega(\eta_v - \frac{2}{3}\eta_s)\right]\operatorname{div}\underline{u} - 2i\omega\eta_s\varepsilon_{ij} \qquad [1.324]$$

similar to [1.296], by introducing the dynamic modulus:

$$\tilde{\lambda} = K - i\omega\left(\eta_v - \frac{2}{3}\eta_s\right) \quad \text{and} \quad \tilde{\mu} = -i\omega\eta_s \qquad [1.325]$$

The constitutive equation of a viscous Newtonian fluid is thus written in a form similar to that of an isotropic solid. The modeling of the viscosity of a Newtonian fluid is, therefore, similar to the Kelvin–Voigt model for a solid (Nagy and Nayfeh 1996). Consequently, the velocities of the longitudinal and transverse waves are, respectively, expressed as:

$$\tilde{V}_L = c_0\sqrt{1 - i\omega\frac{\eta_v + \frac{4}{3}\eta_s}{K}} \quad \text{and} \quad \tilde{V}_T = \frac{1-i}{2}\sqrt{\frac{2\omega\eta_s}{\rho_0}} \qquad [1.326]$$

Using relation [1.313] and since $\mu = 0$, the expression for the attenuation of the longitudinal wave in the liquid is:

$$\alpha_L \lambda_L = \pi\frac{\omega}{K}\left(\eta_v + \frac{4}{3}\eta_s\right) \qquad [1.327]$$

It is important to note that the transverse wave, also called the vorticity wave (Bruneau and Scelo 2006), is not purely propagative, since the real part of its wave number is equal to its imaginary part.

While the mass density, the adiabatic sound speed, and the shear viscosity can be accurately measured by different techniques, the measurements of bulk viscosity is still quite rare.

ORDER OF MAGNITUDE.– At 20°, the mass density of water is $\rho_0 = 1\,000$ kg/m^3, the sound speed is $c_0 = 1\,480$ m/s, its viscosities are $\eta_s = 1$ mPa.s and $\eta_v = 2.4$ mPa.s. The parameters for glycerol, at the same temperature, are $\rho_0 = 1\,260$ kg/m^3, $c_0 = 1\,920$ m/s and $\eta_s = 1.49$ Pa.s. At 1 MHz, the attenuation per longitudinal wavelength is $\alpha_L \lambda_L = 3.4 \times 10^{-5}$ for water and of the order of 10^{-2} for glycerol.

2

Reflection and Transmission at an Interface

When a monochromatic plane wave with a defined polarization encounters the boundary separating two solids, it gives rise to several waves that propagate on either side of the boundary. In the case of two isotropic materials, on the one hand, the incident wave generates at most two waves in each material, and, on the other hand, the polarizations are longitudinal or transverse. In the general case involving two anisotropic materials, an incident wave can generate three reflected waves in the crystal in which it propagates: a quasi-longitudinal wave and two quasi-transverse waves (one is slow and the other is fast) as well as three waves transmitted into the other crystal.

The problem of reflection and transmission is expressed as follows: assuming that the propagation direction, the polarization, the amplitude of the incident wave and the properties of the two materials are known, what are the directions of propagation, polarizations and amplitudes of the reflected and transmitted waves? In the first section, we specify the boundary conditions that apply on the surface of a solid, or at the interface between two media. In the second section, we start by showing that it is possible to predict the propagation direction of the waves and their polarization, without any calculations. The simple case of the transverse horizontal wave, which is reflected and transmitted without any change of polarization, is studied next to demonstrate the essential role played by the acoustic impedance and to illustrate the notions of critical angle and total reflection. The fourth section is dedicated to the case of an isotropic solid in contact with the vacuum or with a perfect fluid. The reflection and transmission coefficients are calculated by developing the propagation equations for each medium and the continuity equations on the interface. The calculations are relatively tedious and repetitive. In the last section, to study more complex configurations involving anisotropic solids, it is worthwhile to write the Christoffel equations in a form that satisfies the Snell–Descartes law and to use a matrix formulation. In addition to being elegant, this approach also offers the advantage of highlighting the waves propagating on a free surface (Rayleigh wave), at the interface between a solid and a liquid (Scholte wave) and, under restrictive conditions, at the interface between two solids (Stoneley wave). These surface and interface waves are studied in Chapter 3.

2.1. Boundary conditions

The solutions to the equation of motion [1.22] must satisfy *boundary conditions*, which connect the mechanical quantities at the interface between two media. They differ in nature depending on whether the surface of the solid is free, covered by a perfect or a viscous fluid, or whether the contact between the two solids is rigid or imperfect.

If the mechanical bond is *perfectly rigid*, there is no sliding of one of the solids with respect to the other, so that the displacement of matter is continuous at any point on the boundary Σ:

$$\underline{u}^{(1)} = \underline{u}^{(2)} \tag{2.1}$$

If the second medium is a non-viscous fluid, there is continuity only for the component normal to the interface.

For the traction vector $T_i(\underline{l}) = \sigma_{ij} l_j$, let us integrate equation [1.22] over a cylinder of height h, limited by two neighboring parallel discs, placed on either side of the interface Σ (Figure 2.1). Taking into account an external force with a density per unit area p_i, we obtain:

$$\left[\sigma_{ij}^{(2)} l_j - \sigma_{ij}^{(1)} l_j\right] dS + \rho F_i h\, dS + p_i\, dS = \frac{\rho_1 + \rho_2}{2} h\, dS \frac{dv_i}{dt} \tag{2.2}$$

When h tends to zero, only the terms due to the surface forces remain:

$$\sigma_{ij}^{(2)} l_j - \sigma_{ij}^{(1)} l_j = -p_i, \quad \text{hence} \quad T_i^{(2)} - T_i^{(1)} = -p_i \tag{2.3}$$

Thus, in the case of two solids separated by an interface of normal $\underline{l}$[1] and in the absence of any external force acting on the surface ($p_i = 0$), a perfect contact implies the continuity of displacements and traction vectors oriented along $\underline{l}$. These boundary conditions are formally written for two media (1) and (2), as:

$$\begin{cases} \underline{u}^{(1)}(\underline{x},t) = \underline{u}^{(2)}(\underline{x},t) \\ \underline{\underline{\sigma}}^{(1)}(\underline{x},t).\underline{l} = \underline{\underline{\sigma}}^{(2)}(\underline{x},t).\underline{l} \end{cases} \text{at} \quad \underline{x} \in \Sigma \tag{2.4}$$

---

[1] It does not matter whether the orientation is toward one material or the other.

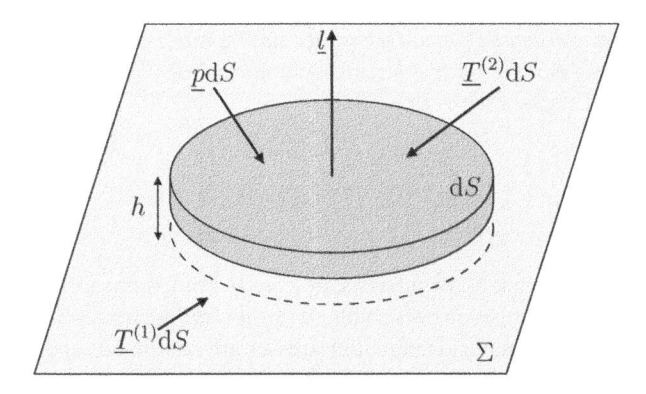

Figure 2.1. *If the density of the force p exerted on the surface Σ is zero, the traction vector $\underline{T}$ is continuous over Σ*

In the case of a *fluid* (superscript 2) in contact with a solid (superscript 1), the boundary conditions express the continuity of traction vectors and of normal components of the displacements:

$$\begin{cases} \underline{\underline{\sigma}}^{(1)}(\underline{x},t).\underline{l} = \underline{\underline{\sigma}}^{(2)}(\underline{x},t).\underline{l} \\ \underline{u}^{(1)}(\underline{x},t).\underline{l} = \underline{u}^{(2)}(\underline{x},t).\underline{l} \end{cases} \quad \text{at} \quad \underline{x} \in \Sigma \qquad [2.5]$$

In a *perfect fluid*, the stress tensor is diagonal and its components are the opposite of the acoustic pressure. Consequently, the normal stress is continuous at the interface Σ and the tangential stresses are zero since there is no viscosity. If the viscosity of the fluid is taken into account, the boundary conditions are the conditions at the interface between two solids, and not those between a solid and a perfect fluid. In addition, the particle velocity $\underline{v}(\underline{x},t) = \partial \underline{u}(\underline{x},t)/\partial t$ is used, rather than the mechanical displacement $\underline{u}(\underline{x},t)$.

There also exist two extreme boundary conditions: the *free surface* and the *clamped surface*. The first corresponds to an interface between a solid and the vacuum. The boundary conditions here result from the cancellation of the traction vector on the free surface: $\underline{\underline{\sigma}}(\underline{x},t).\underline{l} = \underline{0}$. These boundary conditions are also applicable to an interface between a solid and the air, which is a very light and highly compressible fluid, compared to usual solids. Conversely, a clamped surface corresponds to an interface between a material and a much more rigid solid. The boundary conditions are then the cancellation of the displacement vector on the surface: $\underline{u}(\underline{x},t) = \underline{0}$. This condition, which is quite rare for solids, may be useful when the material considered is highly deformable.

An *imperfect contact* between two solids can be modeled by a sliding boundary condition. In this case, the normal stresses and normal displacements are continuous and the tangential stresses are zero:

$$\begin{cases} \left[\underline{\underline{\sigma}}^{(1)}(\underline{x},t).\underline{l}\right].\underline{l} = \left[\underline{\underline{\sigma}}^{(2)}(\underline{x},t).\underline{l}\right].\underline{l} & \text{at} \quad \underline{x} \in \Sigma \\ \left[\underline{\underline{\sigma}}^{(1)}(\underline{x},t).\underline{l}\right].\underline{t} = \left[\underline{\underline{\sigma}}^{(2)}(\underline{x},t).\underline{l}\right].\underline{t} = 0 & \text{at} \quad \underline{x} \in \Sigma \\ \underline{u}^{(1)}(\underline{x},t).\underline{l} = \underline{u}^{(2)}(\underline{x},t).\underline{l} & \text{at} \quad \underline{x} \in \Sigma \end{cases} \quad [2.6]$$

where $\underline{t}$ is a vector tangent to the interface ($\underline{l}.\underline{t} = 0$). For a more accurate modeling of an imperfect interface between two solids, *interface stiffnesses* are often introduced. In these models, the normal and tangential stresses are continuous and proportional to the displacement jump on both sides of the interface:

$$\begin{cases} \underline{\underline{\sigma}}^{(1)}(\underline{x},t).\underline{l} = \underline{\underline{\sigma}}^{(2)}(\underline{x},t).\underline{l} & \text{at} \quad \underline{x} \in \Sigma \\ \left[\underline{\underline{\sigma}}^{(1)}(\underline{x},t).\underline{l}\right].\underline{l} = K_l \left[\underline{u}^{(1)}(\underline{x},t) - \underline{u}^{(2)}(\underline{x},t)\right].\underline{l} & \text{at} \quad \underline{x} \in \Sigma \\ \left[\underline{\underline{\sigma}}^{(1)}(\underline{x},t).\underline{l}\right].\underline{t} = K_t \left[\underline{u}^{(1)}(\underline{x},t) - \underline{u}^{(2)}(\underline{x},t)\right].\underline{t} & \text{at} \quad \underline{x} \in \Sigma \end{cases} \quad [2.7]$$

The interface stiffnesses K_l and K_t, homogeneous to an elasticity per unit length, are expressed in N/m^3. These parameters are used to model different phenomena: rough layer, lack of adhesion, thin layer compared to the wavelength, etc. When these two stiffnesses are zero, the boundary conditions are decoupled into two free surfaces. When these stiffnesses tend to infinity, with the stress remaining a finite quantity, the displacement jumps go to zero: the system tends toward that with a perfect contact. When K_t is zero and K_l tends to infinity, the system tends toward that with a sliding condition.

According to the atomic chain model, where the atoms are connected by springs with stiffnesses K_{1L} or K_{1T}, for longitudinal or transverse displacement, respectively, an interface stiffness represents the stiffness of a spring (in N/m) with respect to unit area. With a denoting the distance between two neighboring atomic chains, it is equal to:

$$K_m = \frac{K_{1M}}{a^2} \quad \text{with} \quad m = l, t \quad \text{and} \quad M = L, T \quad [2.8]$$

In this model, the bulk wave velocities are linked at low frequency to the elastic constants of the material, expressed in N/m^2 as K_{1L}/a for a traction force and K_{1T}/a for a shear force (Royer and Dieulesaint 1996):

$$V_M = \sqrt{\frac{K_{1M}}{\rho a}}, \quad \text{hence} \quad K_{1M} = \rho a V_M^2 \quad [2.9]$$

By introducing K_{1M} into formula [2.8], the interface stiffnesses are given by:

$$K_l = \frac{\rho V_L^2}{a} \quad \text{and} \quad K_t = \frac{\rho V_T^2}{a} \quad [2.10]$$

ORDER OF MAGNITUDE.– Considering the constants in Table 1.2 and $a = 0.4$ nm, we obtain for aluminum: $K_l = 2.7 \times 10^{20}$ N/m^3 and $K_t = 6.5 \times 10^{19}$ N/m^3. In the case of an epoxy resin, the distance between the polymer chains is not so well known. With a typical value of $a = 10$ nm, we obtain $K_l \approx 7.3 \times 10^{17}$ N/m^3 and $K_t \approx 1.4 \times 10^{17}$ N/m^3. These are the maximum values for the stiffnesses at an interface between an epoxy resin and a solid. To model this type of contact, it is reasonable to explore a domain from 10^{14} N/m^3 (weak bond) to 10^{17} N/m^3 (strong bond).

2.2. Direction and polarization of reflected and transmitted waves

In the following, the two solids in contact through a plane are assumed to be rigidly bonded (Figure 2.2). In a reference frame whose origin is located at the interface between the two solids, the equation for this plane of unit normal $\underline{l}$ is $\underline{l} \cdot \underline{x} = 0$. The continuity of the displacements u_j and of the traction vectors T_j at this interface is expressed by:

$$u_j^i + \sum_r u_j^r = \sum_t u_j^t \quad \text{and} \quad T_j^i + \sum_r T_j^r = \sum_t T_j^t \quad [2.11]$$

where the superscripts i, r, t denote the incident, reflected and transmitted waves, respectively. In the harmonic case, plane waves are described by expressions in the form:

$$e^{i(\underline{k}^m \cdot \underline{x} - \omega_m t)} = e^{i(k_1^m x_1 + k_2^m x_2 + k_3^m x_3 - \omega_m t)} \quad \text{with} \quad m = i, r, t \quad [2.12]$$

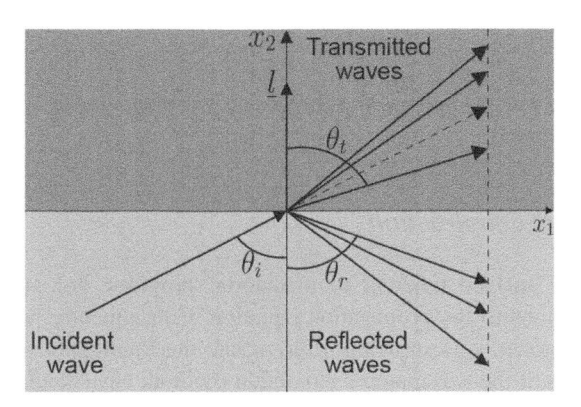

Figure 2.2. *Snell–Descartes law. The projections, on the interface between the two solids, of reflected and transmitted wave vectors are equal to the projection of the incident wave vector*

To be satisfied, the conditions for continuity impose:

– at each instant: $\omega_r = \omega_t = \omega_i = \omega$. The angular frequency of waves reflected and transmitted by a motionless surface is the same as that of the incident wave;

– at each point on the plane $\underline{l}.\underline{x} = 0$:

$$\underline{k}^r.\underline{x} = \underline{k}^t.\underline{x} = \underline{k}^i.\underline{x} \qquad [2.13]$$

so that, like $\underline{l}$, the vectors $\underline{k}^r - \underline{k}^i$ and $\underline{k}^t - \underline{k}^i$ are perpendicular to the interface.

The wave vectors of the reflected and transmitted waves are contained in the incident plane defined by the normal to the interface and the incident wave vector. Their projections onto the interface are equal to the projection k_1 of the incident wave vector. This result expresses the *Snell–Descartes law*:

$$k_r \sin \theta_r = k_t \sin \theta_t = k_i \sin \theta_i = k_1 \qquad [2.14]$$

with θ_i, θ_r and θ_t denoting the angles of incidence, reflection and transmission, respectively. Introducing the phase velocities, it comes:

$$\frac{\sin \theta_r}{V_r} = \frac{\sin \theta_t}{V_t} = \frac{\sin \theta_i}{V_i} = \frac{1}{V_s} = s_1 \qquad [2.15]$$

where s_1 and V_s are, respectively, the slowness and phase velocity along the interface, for all waves propagating in the two media. Relations [2.14] and [2.15] express the phase matching condition imposed by the incident wave: the planes of all the waves travel along the interface at the same velocity V_s. This one is larger than the phase velocity of the incident, reflected and transmitted waves. At normal incidence, V_s tends to infinity.

The two solids being known, the direction of propagation of the waves can be graphically determined.

2.2.1. *Graphical construction*

The slowness surface inherent to a material provides, for any direction, the plane-wave solutions to the propagation equation. Consequently, when the incident wave vector is known, it is sufficient to associate the Snell–Descartes law with the slowness surfaces of the two materials to obtain (without any calculation) the vectors of waves likely to propagate in one or the other solid, their polarization and the acoustic rays (along which the energy travels). In addition to the graphical construction, it is important to ensure that the displacement of the incident wave has one (or more) components that can give rise to the polarizations of the reflected or transmitted waves. This is illustrated by the constructions in Figure 2.3 in the case of two isotropic solids. The displacement of the transverse vertical wave contained in the plane of incidence is able to create a transverse motion and a longitudinal motion in each of the two media. Figure 2.3(b) reveals the existence of a critical angle of incidence: the longitudinal wave is no longer excited in medium (1), it is evanescent. This is the case when the projection of the incident wave vector does not intersect one of the sheets of the slowness surfaces: the corresponding wave is *evanescent or inhomogeneous*. Analytically, this is expressed by an imaginary or complex solution to the propagation equation: the amplitude of the wave decreases exponentially from the interface.

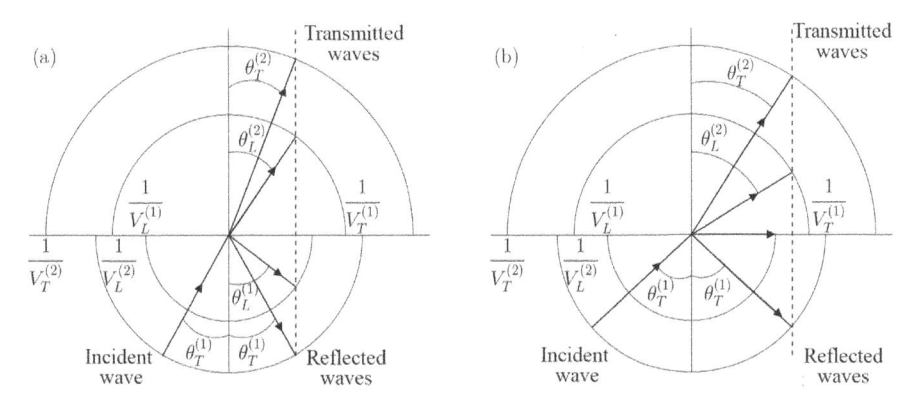

Figure 2.3. *Interface between two isotropic solids. The incident transverse vertical wave generates the following: a) two reflected waves and two transmitted waves (one transverse, the other longitudinal); b) two transmitted waves and a single reflected wave: the longitudinal reflected wave is evanescent*

2.2.2. *Wave decoupling*

In this section, we first study the general case of a crystalline solid where the decoupling conditions are linked to the presence of symmetry elements. We will then move to the case of an isotropic solid where these conditions are always satisfied.

2.2.2.1. *General case*

When the normal Ox_3 to the plane of incidence contains a direct or inverse binary axis of symmetry, the propagation equations for the elastic waves and the boundary conditions at the interface can be divided into two independent systems. The quasi-longitudinal and quasi-transverse waves polarized in the plane of incidence Ox_1x_2 are decoupled from the transverse horizontal (TH) wave polarized along Ox_3. Since $n_3 = 0$, the components [1.177] of the Christoffel tensor:

$$\Gamma_{il} = c_{i11l}n_1^2 + c_{i22l}n_2^2 + (c_{i12l} + c_{i21l})n_1n_2 \quad \text{with} \quad i, l = 1, 2, 3 \quad [2.16]$$

are simplified if the plane of incidence is perpendicular to a binary axis, parallel to Ox_3. Indeed, the elastic constants with a single index equal to 3 are zero (see section A2.5), so that $\Gamma_{13} = \Gamma_{23} = 0$. The *Christoffel equation* is divided into two independent sub-systems:

$$\begin{pmatrix} \Gamma_{11} - \rho V^2 & \Gamma_{12} \\ \Gamma_{12} & \Gamma_{22} - \rho V^2 \end{pmatrix} \begin{pmatrix} p_1 \\ p_2 \end{pmatrix} = \underline{0} \quad \text{and} \quad (\Gamma_{33} - \rho V^2)p_3 = 0 \qquad [2.17]$$

The solution to the first sub-system yields two waves polarized in the plane of incidence. The solution to the second one is a TH wave polarized along Ox_3.

In order for this decoupling to be total, it must also apply to the mechanical boundary conditions. On the surface with normal Ox_2, these conditions concern the normal stresses σ_{i2}. The development of the generalized Hooke's law [1.169]:

$$\sigma_{i2} = c_{i2kl}\frac{\partial u_l}{\partial x_k} = c_{i21l}\frac{\partial u_l}{\partial x_1} + c_{i22l}\frac{\partial u_l}{\partial x_2} = i\left(k_1 c_{i21l} + k_2 c_{i22l}\right) u_l \qquad [2.18]$$

leads to:

$$\sigma_{i2} = ik_1\left(c_{i211}u_1 + c_{i212}u_2 + c_{i213}u_3\right) + ik_2\left(c_{i221}u_1 + c_{i222}u_2 + c_{i223}u_3\right) \quad [2.19]$$

with $i = 1, 2, 3$. Since the elastic constants are zero when $i = 3$, because they have only one index equal to 3, the stress σ_{32} is identically zero for any wave polarized in the plane of incidence ($u_3 = 0$); the boundary conditions only apply to:

$$\begin{cases} \sigma_{12} = ik_1\left(C_{16}u_1 + C_{66}u_2\right) + ik_2\left(C_{66}u_1 + C_{26}u_2\right) \\ \sigma_{22} = ik_1\left(C_{12}u_1 + C_{26}u_2\right) + ik_2\left(C_{26}u_1 + C_{22}u_2\right) \end{cases} \quad \text{L and TV waves} \quad [2.20]$$

For the transverse horizontal wave ($u_1 = u_2 = 0$), since the elastic constants are zero when $i = 1$ or 2 (a single index equal to 3), the only non-identically zero stress is:

$$\sigma_{32} = i\left(k_1 C_{45} + k_2 C_{44}\right) u_3 \qquad \text{TH wave} \qquad [2.21]$$

In summary, the presence of a binary axis (direct or inverse) perpendicular to the plane of incidence implies the decoupling of the stresses of the waves polarized in this plane from those of the transverse horizontal wave.

2.2.2.2. *Isotropic case*

In an unbounded isotropic solid, the transverse wave is degenerate: its properties do not depend on its polarization. This is no longer true in the presence of a free surface. If the polarization of the incident wave is parallel to the interface, the TH wave gives rise to a reflected wave with no change in polarization (Figure 2.4(a)). If the polarization is contained in the plane of incidence, the transverse vertical (TV) wave may give rise to a TV wave and a longitudinal wave (Figure 2.4(b)). Similarly, a longitudinal (L) wave will never give rise to a reflected TH wave (Figure 2.4(c)).

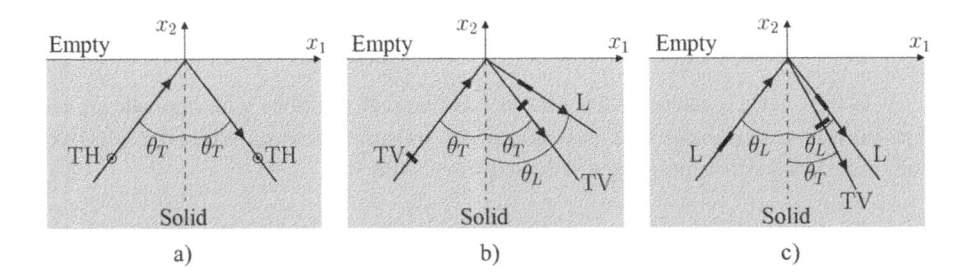

Figure 2.4. *Isotropic solid: Reflection of an incident wave that is a) transverse horizontal (TH), b) transverse vertical (TV) and c) longitudinal (L)*

At the interface between two isotropic media, the problems of the reflection and transmission of L and TV waves, on the one hand, and of TH waves, on the other hand, are resolved independently. Since the constants C_{61}, C_{62} and C_{45} are zero, expressions [2.20] and [2.21] for the stresses simplify:

$$\begin{cases} \sigma_{12} = ik_1 C_{66} u_2 + ik_2 C_{66} u_1 = ik_1 \mu u_2 + ik_2 \mu u_1 \\ \sigma_{22} = ik_1 C_{12} u_1 + ik_2 C_{22} u_2 = ik_1 \lambda u_1 + ik_2 (\lambda + 2\mu) u_2 \end{cases} \quad \text{L and TV waves}$$

[2.22]

and, since $C_{44} = C_{66}$:

$$\sigma_{32} = ik_2 C_{66} u_3 = ik_2 \mu u_3 \quad \text{TH wave} \quad [2.23]$$

2.2.3. *Critical angle, evanescent wave and total reflection*

When the projection of the incident wave vector does not intersect one of the sheets of the slowness surfaces, the corresponding wave is evanescent or inhomogeneous. This is analytically expressed by an imaginary or a complex solution to the propagation equation. The amplitude of the wave decreases exponentially from the interface. This situation appears beyond a *critical angle of incidence*. Let us examine the simple case of an isotropic solid, whose wave number $k = \omega/V$ does not depend on the direction of propagation:

$$k_1^2 + k_{2M}^2 + k_3^2 = \frac{\omega^2}{V_M^2} \quad \text{with} \quad M = L, T \quad [2.24]$$

according to the mode considered. In the $x_1 x_2$ plane, since $k_3 = 0$ and since the projection k_1 on the interface is imposed by the incident wave: $k_1 = (\omega/V_i) \sin \theta_i$, the normal component k_{2M} is given by:

$$k_{2M}^2 = \frac{\omega^2}{V_i^2} \left(\frac{V_i^2}{V_M^2} - \sin^2 \theta_i \right) \quad [2.25]$$

– If $V_M < V_i$, equation [2.25] always has two real solutions with opposite signs. The positive root indicates the direction of the transmitted wave, while the negative root indicates the direction of the reflected wave, with:

$$\sin \theta_M = \frac{V_M}{V_i} \sin \theta_i \quad [2.26]$$

– If $V_M > V_i$, there exists a critical angle of incidence θ_c, such that:

$$\sin \theta_c = \frac{V_i}{V_M} \quad \text{hence} \quad \theta_c = \arcsin \left(\frac{V_i}{V_M} \right) \quad [2.27]$$

beyond which, equation [2.25] no longer has any real solution. The component k_{2M} is imaginary:

$$k_{2M} = \pm i \alpha_M \quad \text{with} \quad \alpha_M = \frac{\omega}{V_i} \sqrt{\sin^2 \theta_i - \sin^2 \theta_c} \quad [2.28]$$

The root with a positive imaginary part corresponds to an evanescent transmitted wave ($x_2 > 0$), of displacement:

$$\underline{u}_M = a \underline{p}_M e^{-\alpha_M x_2} e^{i(k_1 x_1 - \omega t)} \quad [2.29]$$

The root with a negative imaginary part corresponds to an evanescent reflected wave ($x_2 < 0$). These waves, whose displacement decreases exponentially from the interface, do not transport any energy along x_2: there is total reflection.

For example, in the case of an isotropic solid occupying the half-space $x_2 < 0$, if the incident wave is longitudinal, there is no wave whose velocity is larger than V_L. Whatever the angle of incidence, the longitudinal wave is reflected into a wave of the same type and converted into a transverse (vertical) wave. If the incident wave is transverse vertical, of velocity $V_T < V_L$, there exists a critical angle $\theta_c^L = \arcsin(V_T/V_L)$ beyond which the reflected longitudinal wave is evanescent; the transverse wave is, therefore, totally reflected.

It is common to express the reflection and transmission coefficients as functions of the angle θ_i of the incident wave, and of the angles θ_r and θ_t of the reflected and transmitted waves; these last two angles being linked to θ_i by the Snell–Descartes law. When θ_i exceeds a critical angle, these expressions lead to erroneous results. To avoid this major drawback, we have chosen to express all the coefficients in terms of the normal components k_{2M} of the vector for the reflected and transmitted waves (the second index M refers to the mode, L or T). The formulae obtained are valid in the:

- subcritical case, by replacing k_{2M} with $(\omega/V_M)\cos\theta_M$;
- supercritical case, where at least one wave number is imaginary: $k_{2M} = \pm i\alpha_M$.

In the following, two types of coefficients will be used: the ratios r_{IJ} and t_{IJ} of the amplitudes of reflected and transmitted waves to the amplitude a of the incident wave (the first index denotes the incident mode) and the *intensity or energy* transmission and reflection coefficients (denoted by R_{IJ} and T_{IJ}). These latter coefficients satisfy a relationship associated with the conservation of energy.

2.2.4. Conservation of energy

In this section, we will express the conservation of the energy transported by the elastic waves. The objective is not to verify this law of physics, which cannot be challenged, but to indicate how to use it to ensure the accuracy of analytical or numerical calculations. The law of conservation of energy can be formulated as follows: the sum of energy fluxes of the reflected and transmitted waves at the interface $x_2 = 0$ must be equal to the energy flux of the incident wave. By reasoning on average values, the energy flux per unit area is $\langle \underline{J}.\underline{l}\rangle$ with $\underline{l} = \underline{e}_2$ for the incident and transmitted waves, and $\underline{l} = -\underline{e}_2$ for the reflected waves, so that:

$$\langle J_2^i\rangle = -\sum_r \langle J_2^r\rangle + \sum_t \langle J_2^t\rangle \qquad [2.30]$$

In the case of an *isotropic solid*, the Poynting vector is parallel to the propagation direction $\underline{n}$, for each traveling wave: $\langle J_2\rangle = \langle \underline{J}.\underline{n}\rangle n_2 = In_2$, where I is the acoustic intensity defined in section 1.2.5. In the expression deduced from [1.117] or [1.121], with $Z = \rho V$ and $k = \omega/V$:

$$\langle J_2\rangle = I\frac{k_2}{k} = \frac{1}{2}\rho\omega V^2 k_2|a|^2 \qquad [2.31]$$

k_2 is the algebraic value for the normal component of the wave vector. Its modulus $k_{2M} = \sqrt{k_M^2 - k_1^2}$ depends on the mode, M. Its sign depends on the type of wave: k_2 is positive for an incident wave ($M = I$) coming from the medium (1), and for any wave ($M = J$) transmitted in medium (2), with or without mode conversion; k_2 is negative for all reflected waves. In the following sections, in order to calculate the coefficients of reflection and transmission, it is useful to retain only the positive quantities, by writing:

$$k_2^i = k_{2I} > 0, \quad k_2^r = -k_{2I} \text{ or } -k_{2J} < 0 \text{ and } k_2^t = k_{2J} > 0 \qquad [2.32]$$

Of course, the quantities k_{2M} must be calculated with the velocity V_M of the medium in which the wave propagates, even if this is not specified, and the signs are opposite if the incident wave comes from medium (2).

The incident wave, of amplitude a, and the wave reflected with no change in mode, whose amplitude is $r_{II}a$, propagate in medium (1) of mass density ρ_1, at the same velocity V_I with opposite normal components k_2 for the wave vector, so that:

$$\langle J_2^i \rangle = \frac{1}{2}\rho_1 \omega V_I^2 k_{2I} |a|^2 \quad \text{and} \quad \langle J_2^r \rangle = -\frac{1}{2}\rho_1 \omega V_I^2 k_{2I} |r_{II}|^2 |a|^2 \quad [2.33]$$

For a reflected wave, of amplitude $r_{IJ}a$, converted from the incident mode I into a different mode J, only the mass density ρ_1 is unchanged:

$$\langle J_2^r \rangle = -\frac{1}{2}\rho_1 \omega V_J^2 k_{2J} |r_{IJ}|^2 |a|^2 \quad \text{with} \quad J \neq I \quad [2.34]$$

For a transmitted wave (mode J), of the amplitude $t_{IJ}a$, all the parameters are modified:

$$\langle J_2^t \rangle = \frac{1}{2}\rho_2 \omega V_J^2 k_{2J} |t_{IJ}|^2 |a|^2 \quad [2.35]$$

The intensity reflection coefficients (R_{II} and R_{IJ}), with the first index denoting the incident mode, are defined by the ratio of the energy fluxes: $-\langle J_2^r \rangle / \langle J_2^i \rangle$, that is:

$$R_{II} = |r_{II}|^2, \quad R_{IJ} = \frac{k_{2J} V_J^2}{k_{2I} V_I^2} |r_{IJ}|^2 \quad [2.36]$$

The intensity transmission coefficient T_{IJ} for the wave J is defined by the ratio $\langle J_2^t \rangle / \langle J_2^i \rangle$, that is:

$$T_{IJ} = \frac{\rho_2 k_{2J} V_J^2}{\rho_1 k_{2I} V_I^2} |t_{IJ}|^2 \quad [2.37]$$

Only the reflection coefficient without mode conversion is equal to the square of the modulus of the reflection coefficient in amplitude r_{II}. With these definitions, the law of conservation of energy [2.30] imposes, for isotropic solids:

$$R_{II} + R_{IJ} + \sum_J T_{IJ} = 1 \quad [2.38]$$

Evanescent waves, whose amplitude decreases exponentially from the interface, do not transport energy in the x_2 direction. They are essential to satisfy the boundary conditions beyond the critical angle, but they are not involved in the energy balance.

The *anisotropic* case is more complicated: for each wave, the average value of the component $J_2 = -\sigma_{2j}\dot{u}_j$ of the Poynting vector must be calculated, for instance, using the complex representation:

$$\langle J_2 \rangle = -\frac{1}{2}\operatorname{Re}\left[\sigma_{2j}\dot{u}_j^*\right] = -\frac{\omega}{2}\operatorname{Re}\left[i\sigma_{2j}u_j^*\right] \qquad [2.39]$$

for a time-harmonic variation in $e^{-i\omega t}$.

2.3. Isotropic solid: transverse horizontal wave

The case of the transverse horizontal wave, whose polarization is parallel to the interface between two isotropic solids is simple, because it is similar to the reflection and transmission of a pressure wave (scalar) between two perfect fluids (section 2.3.1). In section 2.3.2, we show how to choose the thickness and the acoustic impedance of a thin plate, rigidly attached to two solids, in order to optimize the transmission between the two media.

2.3.1. *Reflection and transmission between two solids*

Let us study the reflection and transmission of a transverse horizontal plane wave through a planar interface separating two semi-infinite, isotropic solids. The solid located at $x_2 \leq 0$ is characterized by its mass density ρ_1 and shear modulus μ_1, while the solid located at $x_2 \geq 0$ is characterized by the mass density ρ_2 and shear modulus μ_2. The transverse wave velocities in the two media are, therefore, $V_T^{(1)} = \sqrt{\mu_1/\rho_1}$ and $V_T^{(2)} = \sqrt{\mu_2/\rho_2}$. The transverse horizontal waves are polarized in the direction x_3. The incident wave, of amplitude a, propagates in medium (1) with the angle of incidence θ_1. The transmitted wave propagates in medium (2) with the angle θ_2, given by the Snell–Descartes law:

$$\sin \theta_2 = \frac{V_T^{(2)}}{V_T^{(1)}} \sin \theta_1 \qquad [2.40]$$

The expressions for the displacements in the two solids are given as:

$$\begin{cases} u_3^{(1)}(x_1, x_2, t) = a\left(e^{ik_2^{(1)}x_2} + r_{12}e^{-ik_2^{(1)}x_2}\right)e^{i(k_1 x_1 - \omega t)} \\ u_3^{(2)}(x_1, x_2, t) = t_{12}a e^{ik_2^{(2)}x_2}e^{i(k_1 x_1 - \omega t)} \end{cases} \qquad [2.41]$$

with:

$$\left(k_2^{(m)}\right)^2 = \left(\frac{\omega}{V_T^{(m)}}\right)^2 - k_1^2 \quad (m = 1, 2) \qquad [2.42]$$

and where r_{12} and t_{12} are the reflection and transmission coefficients for the displacement. The boundary conditions express the continuity of displacements u_3 and stresses $\sigma_{32} = \mu \partial u_3/\partial x_2$:

$$u_3^{(1)} = u_3^{(2)} \quad \text{and} \quad \mu_1 \frac{\partial u_3^{(1)}}{\partial x_2} = \mu_2 \frac{\partial u_3^{(2)}}{\partial x_2} \quad \text{at} \quad x_2 = 0 \quad [2.43]$$

Introducing the expressions for the displacements leads to the relations:

$$1 + r_{12} = t_{12} \quad \text{and} \quad \mu_1 k_2^{(1)}(1 - r_{12}) = \mu_2 k_2^{(2)} t_{12} \quad [2.44]$$

The solution to these two equations directly yields the reflection and transmission coefficients:

$$r_{12} = \frac{\mu_1 k_2^{(1)} - \mu_2 k_2^{(2)}}{\mu_1 k_2^{(1)} + \mu_2 k_2^{(2)}} \quad \text{and} \quad t_{12} = \frac{2\mu_1 k_2^{(1)}}{\mu_1 k_2^{(1)} + \mu_2 k_2^{(2)}} \quad [2.45]$$

By expressing the components $k_2^{(1)}$ and $k_2^{(2)}$ for the wave vectors as functions of the angles of incidence in each medium:

$$k_2^{(1)} = \frac{\omega}{V_T^{(1)}} \cos\theta_1 \quad \text{and} \quad k_2^{(2)} = \frac{\omega}{V_T^{(2)}} \cos\theta_2 \quad [2.46]$$

these coefficients can be written as:

$$r_{12} = \frac{Z_1 \cos\theta_1 - Z_2 \cos\theta_2}{Z_1 \cos\theta_1 + Z_2 \cos\theta_2} \quad \text{and} \quad t_{12} = \frac{2Z_1 \cos\theta_1}{Z_1 \cos\theta_1 + Z_2 \cos\theta_2} \quad [2.47]$$

where $Z_1 = \rho_1 V_T^{(1)}$ and $Z_2 = \rho_2 V_T^{(2)}$ are the acoustic impedances of the transverse waves in the two media. At normal incidence, the angles θ_1 and θ_2 are zero, so that:

$$r_{12} = \frac{Z_1 - Z_2}{Z_1 + Z_2} \quad \text{and} \quad t_{12} = \frac{2Z_1}{Z_1 + Z_2} \quad [2.48]$$

The acoustic impedances play a major role in the reflection and transmission phenomena. For example, if the impedance of medium (1) is very different from that of medium (2), reflection is high and transmission is low, and if the impedances of the two media are similar, reflection is weak and transmission is strong.

The intensity transmission and reflection coefficients are defined by relations [2.36] and [2.37], that is, since $\rho V_T^2 = \mu$ in each medium:

$$R_{12} = |r_{12}|^2 \quad \text{and} \quad T_{12} = \frac{\mu_2 k_2^{(2)}}{\mu_1 k_2^{(1)}} |t_{12}|^2 \quad [2.49]$$

A quick calculation shows that the sum of the coefficients R_{12} and T_{12} is indeed equal to unity.

The expressions of the reflection and transmission coefficients as functions of angles θ_1 and θ_2 must be handled with care. When the velocity $V_T^{(2)}$ is larger than $V_T^{(1)}$, then $\theta_2 > \theta_1$ and there exists a critical angle:

$$\theta_c = \arcsin\left(\frac{V_T^{(1)}}{V_T^{(2)}}\right) \qquad [2.50]$$

at which, the angle θ_2 of the transmitted wave in the second medium reaches the value $\pi/2$. Therefore, there is total reflection. According to formula [2.28], the component $k_2^{(2)}$ of the wave number is imaginary: $k_2^{(2)} = i\alpha^{(2)}$ and the coefficient of reflection is a unit complex number:

$$r_{12} = \frac{\mu_1 k_2^{(1)} - i\mu_2 \alpha^{(2)}}{\mu_1 k_2^{(1)} + i\mu_2 \alpha^{(2)}} \longrightarrow R_{12} = |r_{12}|^2 = 1 \qquad [2.51]$$

Beyond the critical angle, the incident energy is totally reflected.

REMARK.– Expressions [2.47] for the reflection and transmission coefficients for TH waves are similar to those at the interface between two perfect fluids.

If the contact between the two materials is not perfect, an interface stiffness K_t may be used to characterize the contact (section 2.1). In this case, the boundary conditions are modified:

$$\begin{cases} \sigma_{23}^{(1)}(x_2 = 0) = \sigma_{23}^{(2)}(x_2 = 0) \\ \sigma_{23}^{(1)}(x_2 = 0) = K_t \left[u_3^{(1)}(x_2 = 0) - u_3^{(2)}(x_2 = 0) \right] \end{cases} \qquad [2.52]$$

They lead to the system of equations:

$$\begin{cases} i\mu_1 k_2^{(1)}(1 - r_{12}) = i\mu_2 k_2^{(2)} t_{12} \\ i\mu_1 k_2^{(1)}(1 - r_{12}) = K_t \left[1 + r_{12} - t_{12}\right] \end{cases} \qquad [2.53]$$

whose solution provides the reflection and transmission coefficients:

$$\begin{cases} r_{12} = \dfrac{K_t \left[\mu_1 k_2^{(1)} - \mu_2 k_2^{(2)}\right] + i\mu_1 k_2^{(1)} \mu_2 k_2^{(2)}}{K_t \left[\mu_1 k_2^{(1)} + \mu_2 k_2^{(2)}\right] + i\mu_1 k_2^{(1)} \mu_2 k_2^{(2)}} \\ t_{12} = \dfrac{2 K_t \mu_1 k_2^{(1)}}{K_t \left[\mu_1 k_2^{(1)} + \mu_2 k_2^{(2)}\right] + i\mu_1 k_2^{(1)} \mu_2 k_2^{(2)}} \end{cases} \qquad [2.54]$$

If the interface stiffness K_t is zero, the reflection coefficient tends to unity, which corresponds to the case of a free surface. Conversely, if K_t tends to infinity, the expressions for the reflection and transmission coefficients tend to those of equation [2.45] between two rigidly bonded solids.

2.3.2. *Plate between two solids, impedance matching*

In this section, we will study the reflection and transmission of TH waves across a plate rigidly bound to two semi-infinite solids. The plate, limited by the planes $x_2 = 0$ and $x_2 = h$, is characterized by its mass density ρ_2 and its shear modulus μ_2. The first solid, located at $x_2 \leq 0$, is characterized by its mass density ρ_1 and its shear modulus μ_1; the second solid, located at $x_2 \geq h$, is characterized by ρ_3 and μ_3. The incident wave is a time-harmonic TH plane wave, polarized in the x_3 direction, which propagates in medium (1) with the angle of incidence θ_1. Since the interfaces are plane at the scale of the wavelength, there is no wave conversion. Then, all the displacement fields are of the type $\underline{u}^{(m)} = u_m \underline{e}_3$.

The reflection and transmission coefficients are calculated using an interferometric method that consists of analyzing the propagation of waves in the form of rays, and then studying the successive echoes in the plate (Figure 2.5). For each round trip in the intermediary plate, the amplitudes of the reflected and transmitted waves are multiplied by $r_{21}r_{23}$.

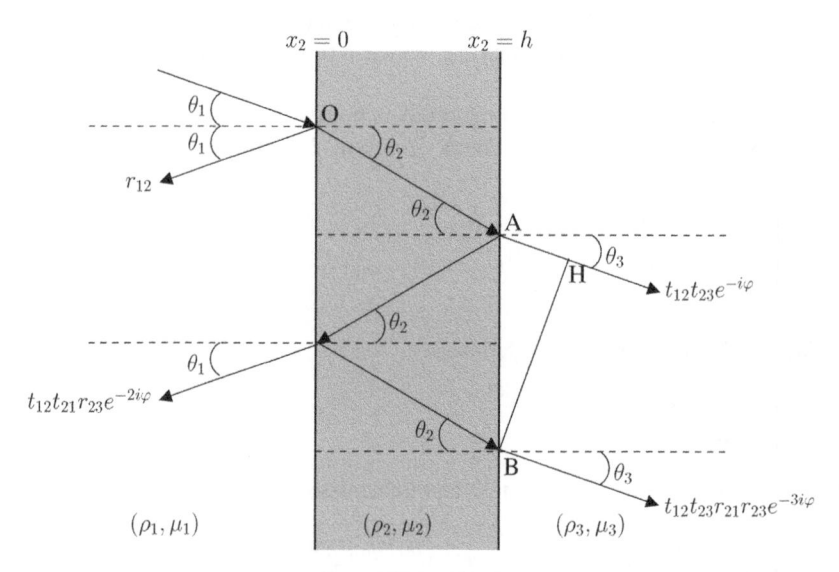

Figure 2.5. *Diagram of the successive echoes in the plate*

For a wave with complex representation $e^{i\omega t}$, the phase delay due to the path between two reflections is:

$$\varphi_2 = 2k^{(2)}\mathrm{OA} = \frac{2k^{(2)}h}{\cos\theta_2} \quad \text{with} \quad k^{(2)} = \frac{\omega}{V_T^{(2)}} \qquad [2.55]$$

The phase advance due to the path difference AH in medium (3) is:

$$\varphi_3 = k^{(3)} \text{AH} = k^{(3)} \text{AB} \sin \theta_3 = 2k^{(3)} h \tan \theta_2 \sin \theta_3 \quad \text{with} \quad k^{(3)} = \frac{\omega}{V_T^{(3)}} \quad [2.56]$$

Considering the Snell–Descartes law, that is, $k^{(3)} \sin \theta_3 = k^{(2)} \sin \theta_2$, we obtain:

$$\varphi_3 = \frac{2k^{(2)} h \sin^2 \theta_2}{\cos \theta_2} \quad [2.57]$$

The total phase shift is a phase delay:

$$\varphi_2 - \varphi_3 = \frac{2k^{(2)} h (1 - \sin^2 \theta_2)}{\cos \theta_2} = 2k^{(2)} h \cos \theta_2 = 2k_2^{(2)} h \quad [2.58]$$

At each round trip in the intermediary plate, the reflected and transmitted waves are phase-shifted by the quantity $2\varphi = 2k_2^{(2)} h$.

The summation of the backward waves for $x_2 \leq 0$ and of the forward waves for $x_2 \geq h$ lead to the expressions for the reflection and transmission coefficients:

$$\begin{cases} r = r_{12} + t_{12} t_{21} r_{23} e^{-2i\varphi} \sum_{n=0}^{\infty} \left(r_{21} r_{23} e^{-2i\varphi} \right)^n \\ t = t_{12} t_{23} e^{-i\varphi} \sum_{n=0}^{\infty} \left(r_{21} r_{23} e^{-2i\varphi} \right)^n \end{cases} \quad [2.59]$$

Knowing the sum of the geometric series

$$\sum_{n=0}^{\infty} q^n = \frac{1}{1-q}, \quad \text{for} \quad |q| < 1 \quad [2.60]$$

we obtain:

$$r = r_{12} + \frac{t_{12} t_{21} r_{23} e^{-2i\varphi}}{1 - r_{21} r_{23} e^{-2i\varphi}} \quad \text{and} \quad t = \frac{t_{12} t_{23} e^{-i\varphi}}{1 - r_{21} r_{23} e^{-2i\varphi}} \quad [2.61]$$

Given the expressions [2.47] for the reflection and transmission coefficients at a simple interface, and by writing $\tilde{Z}_p = Z_p \cos \theta_p$, the coefficients r and t can be expressed by:

$$r = \frac{\tilde{Z}_2 (\tilde{Z}_1 - \tilde{Z}_3) \cos \varphi - i(\tilde{Z}_2^2 - \tilde{Z}_1 \tilde{Z}_3) \sin \varphi}{\tilde{Z}_2 (\tilde{Z}_1 + \tilde{Z}_3) \cos \varphi + i(\tilde{Z}_2^2 + \tilde{Z}_1 \tilde{Z}_3) \sin \varphi} \quad [2.62]$$

and:

$$t = \frac{2 \tilde{Z}_1 \tilde{Z}_2}{\tilde{Z}_2 (\tilde{Z}_1 + \tilde{Z}_3) \cos \varphi + i(\tilde{Z}_2^2 + \tilde{Z}_1 \tilde{Z}_3) \sin \varphi} \quad [2.63]$$

REMARK.– If the impedances $\tilde{Z}_1$ and $\tilde{Z}_3$ are very small compared to the impedance $\tilde{Z}_2$, the transmission coefficient t tends to:

$$t = -\frac{2i\tilde{Z}_1}{\tilde{Z}_2 \sin \varphi} \quad \text{with} \quad \varphi = k_2^{(2)} h \qquad [2.64]$$

This coefficient has an infinity of poles corresponding to the cancellation of the function $\sin \varphi$, that is,

$$k_2^{(2)} = \frac{n\pi}{h} \quad \text{with} \quad n \in \mathbb{N} \qquad [2.65]$$

The denominator of the transmission coefficient is canceled when the angular frequency ω and the wave number $\beta = k_1^{(2)}$ along x_1 satisfy the dispersion relation $\omega(\beta)$ of the TH waves guided by the free plate (section 4.2.1):

$$\frac{\omega^2}{V_T^2} = \left(k_1^{(2)}\right)^2 + \left(k_2^{(2)}\right)^2 = \beta^2 + \frac{n^2 \pi^2}{h^2} \qquad [2.66]$$

When the thickness of the layer is comparable to the wavelengths of the waves which pass through it, Fabry–Pérot interferences appear. As an illustration of such interferences, the moduli of the reflection and transmission coefficients of a plexiglass plate in between two duralumin delay lines, acting as a half-space, are shown in Figure 2.6. Since the plexiglass is strongly viscoelastic at ultrasonic frequencies, the decrease in the average values of the reflection and transmission coefficients is the signature of an increase in the attenuation as a function of the frequency. The Fabry–Pérot interferences can be used to characterize the viscoelastic properties of the material (Lefebvre *et al.* 2018).

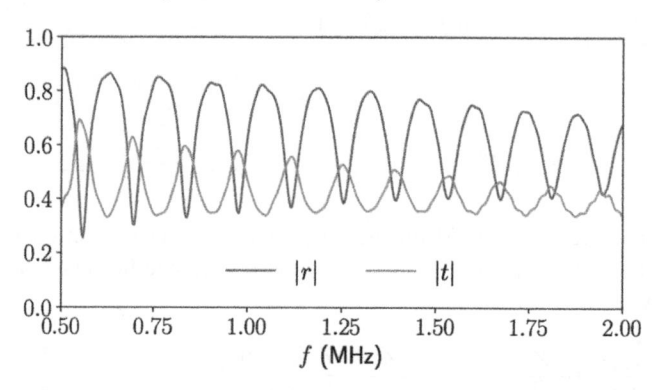

Figure 2.6. *Evolution of the moduli of the reflection (r) and transmission (t) coefficients of a plexiglass plate versus frequency (Simon et al. 2019). For a color version of this figure, see www.iste.co.uk/royer/waves1.zip*

At *normal incidence*, all angles of incidence are zero: $\tilde{Z}_p = Z_p$ and there is no wave conversion at the interfaces, whatever the nature of the incident wave. This modeling is thus valid for both longitudinal and transverse waves. Two situations, corresponding to the cancelation of the functions $\cos\varphi$ and $\sin\varphi$ in expressions [2.62] and [2.63] for r and t, are interesting to study.

– When the layer thickness is equal to a quarter of the wavelength, the phase shift φ is equal to $\pi/2$, so that the function $\cos\varphi$ is zero. The reflection and transmission coefficients are then given by:

$$r = \frac{Z_1 Z_3 - Z_2^2}{Z_2^2 + Z_1 Z_3} \quad \text{and} \quad t = -\frac{2i Z_1 Z_2}{Z_2^2 + Z_1 Z_3} \qquad [2.67]$$

and, according to relations [2.36] and [2.37], the intensity coefficients are:

$$R = |r|^2 = \left(\frac{Z_1 Z_3 - Z_2^2}{Z_2^2 + Z_1 Z_3}\right)^2 \quad \text{and} \quad T = \frac{Z_3}{Z_1}|t|^2 = \frac{4 Z_1 Z_3 Z_2^2}{(Z_2^2 + Z_1 Z_3)^2} \qquad [2.68]$$

If, in addition, the impedance of medium (2) is equal to $\sqrt{Z_1 Z_3}$, the reflection coefficient is zero and the transmission coefficient T is equal to unity: the transmission is perfect. This result is the basis of matching impedance layers deposited on immersion transducers in order to drastically increase the amplitude of the waves generated by a piezoelectric plate having an acoustic impedance much larger than that of water (Volume 2, section 3.1.1).

– When the thickness of the layer is equal to half the wavelength, the phase shift φ is equal to π, so that the function $\sin\varphi$ is zero. The reflection and transmission coefficients are then given by:

$$r = \frac{Z_1 - Z_3}{Z_1 + Z_3} \quad \text{and} \quad t = -\frac{2 Z_1}{Z_1 + Z_3} \qquad [2.69]$$

These expressions correspond to those obtained for a simple interface between media (1) and (3). This means that the layer is transparent from the point of view of elastic waves. It only introduces a phase shift equal to π during the transmission, which is expressed through the change in sign of the coefficient t.

When the thickness of the layer is very small compared to the wavelength, the reflection and transmission coefficients through the plate take the forms:

$$\begin{cases} r = \dfrac{\tilde{Z}_2(\tilde{Z}_1 - \tilde{Z}_3) - i(\tilde{Z}_2^2 - \tilde{Z}_1 \tilde{Z}_3)\varphi}{\tilde{Z}_2(\tilde{Z}_1 + \tilde{Z}_3) + i(\tilde{Z}_2^2 + \tilde{Z}_1 \tilde{Z}_3)\varphi} \\ t = \dfrac{2\tilde{Z}_1 \tilde{Z}_2}{\tilde{Z}_2(\tilde{Z}_1 + \tilde{Z}_3) + i(\tilde{Z}_2^2 + \tilde{Z}_1 \tilde{Z}_3)\varphi} \end{cases} \qquad [2.70]$$

If the material of this layer is soft compared to the other two, then $\tilde{Z}_2^2 \ll \tilde{Z}_1 \tilde{Z}_3$:

$$r = \frac{\tilde{Z}_2(\tilde{Z}_1 - \tilde{Z}_3) + i\tilde{Z}_1\tilde{Z}_3\varphi}{\tilde{Z}_2(\tilde{Z}_1 + \tilde{Z}_3) + i\tilde{Z}_1\tilde{Z}_3\varphi} \quad \text{and} \quad t = \frac{2\tilde{Z}_1\tilde{Z}_2}{\tilde{Z}_2(\tilde{Z}_1 + \tilde{Z}_3) + i\tilde{Z}_1\tilde{Z}_3\varphi} \qquad [2.71]$$

Considering the relations $\varphi = k_2^{(2)} h$ and $\tilde{Z}_p = \mu_p k_2^{(p)}/\omega$, these formulae lead to expressions [2.54], by writing:

$$K_t = \frac{\mu_2}{h} \qquad [2.72]$$

This interface stiffness can thus be used to model a thin and soft layer in contact between two solids. The stiffness of this layer expresses as an elasticity per unit length.

ORDER OF MAGNITUDE.– With a shear modulus in between 1 and 10 GPa and an intermediary layer whose thickness is between 10 nm and 1 µm, the interface stiffness K_t ranges from 10^{15} to 10^{18} N/m^3, that is, an order of magnitude comparable to that of the atomic chain model (section 2.1).

2.4. Isotropic media: longitudinal and transverse vertical waves

In this section, the propagation medium is either an isotropic solid, or a material with an isotropic axis perpendicular to the plane of incidence (which is the same as an isotropic solid), or a fluid. When it comes from the solid, the incident wave is either longitudinal or transverse vertical. In the fluid, assumed to be perfect, only a longitudinal wave can propagate. The reflection on a free surface is treated first, followed by the case of a solid–fluid interface. We finally study the interaction of a plane wave with an immersed plate, with the objective of introducing two methods for characterizing the plate material.

2.4.1. Reflection on a free surface

Let us study the reflection of a time-harmonic plane wave of angular frequency ω at the free surface of a homogeneous and isotropic solid half-space, located at $x_2 \leq 0$. The solid is characterized by its mass density ρ and its Lamé constants λ and μ. Since there is no transmitted wave, the only continuity condition to ensure is the cancelation of the traction vector on the free surface:

$$\sigma_{j2}^i + \sum_r \sigma_{j2}^r = 0 \quad \text{for} \quad x_2 = 0 \quad \text{and} \quad j = 1, 2 \qquad [2.73]$$

Two configurations are investigated, depending on whether the incident wave is longitudinal or transverse vertical.

2.4.1.1. *Incident longitudinal wave*

The incident wave propagates in the plane (x_1, x_2) with the angle of incidence θ_L and velocity V_L. At a point with coordinates $\underline{x} = (x_1, x_2, x_3)^t$, the expression for the displacement field is:

$$\underline{u}^i(\underline{x}, t) = a\underline{p}_L e^{i(k_1 x_1 + k_{2L} x_2 - \omega t)} \quad \text{with} \quad \underline{p}_L = \frac{V_L}{\omega}(k_1, k_{2L}, 0)^t \quad [2.74]$$

Since the polarization vector $\underline{p}_L$ for the incident wave is normalized ($k_L = \omega/V_L$), the amplitude a is homogeneous to a length. Introducing the displacement (equation [2.74]) into relations [2.22] leads to the following expressions for the stresses:

$$\sigma_{12} = \mu\left(\frac{\partial u_1}{\partial x_2} + \frac{\partial u_2}{\partial x_1}\right) = 2i\mu \frac{V_L}{\omega} k_1 k_{2L} a e^{i(k_1 x_1 + k_{2L} x_2 - \omega t)} \quad [2.75]$$

and:

$$\sigma_{22} = \lambda \frac{\partial u_1}{\partial x_1} + (\lambda + 2\mu)\frac{\partial u_2}{\partial x_2} = i\frac{V_L}{\omega}\left[\lambda k_1^2 + (\lambda + 2\mu)k_{2L}^2\right] a e^{i(k_1 x_1 + k_{2L} x_2 - \omega t)} \quad [2.76]$$

By noting that the term between square brackets for the stress σ_{22} is written as:

$$\lambda k_1^2 + (\lambda + 2\mu)k_{2L}^2 = (\lambda + 2\mu)\frac{\omega^2}{V_L^2} - 2\mu k_1^2 = \mu \frac{\omega^2}{V_T^2} - 2\mu k_1^2 = \mu\left(k_{2T}^2 - k_1^2\right) \quad [2.77]$$

the normal stress, like the tangential stress, is proportional to μ:

$$\sigma_{22} = i\mu \frac{V_L}{\omega}\left(k_{2T}^2 - k_1^2\right) a e^{i(k_1 x_1 + k_{2L} x_2 - \omega t)} \quad [2.78]$$

Considering the Snell–Descartes law, the reflected longitudinal wave propagates with the angle of incidence θ_L and the reflected transverse wave propagates with an angle:

$$\theta_T = \arcsin\left(\frac{V_T}{V_L}\sin\theta_L\right) \quad [2.79]$$

always smaller than the angle of incidence θ_L. The reflected transverse vertical wave is, therefore, always propagative. The normal components of the vectors for the reflected waves are always negative: $-k_{2L}$ and $-k_{2T}$, with

$$k_{2L} = \frac{\omega}{V_L}\cos\theta_L \quad \text{and} \quad k_{2T} = \frac{\omega}{V_T}\cos\theta_T \quad [2.80]$$

Given that the tangential components of the wave numbers are equal among themselves, the displacement field of the reflected waves is given by:

$$\underline{u}^r = \underline{u}_L + \underline{u}_T, \quad \text{with} \quad \begin{cases} \underline{u}_L(x_1, x_2, t) = ar_{LL}\underline{q}_L e^{i(k_1 x_1 - k_{2L} x_2 - \omega t)} \\ \underline{u}_T(x_1, x_2, t) = ar_{LT}\underline{q}_T e^{i(k_1 x_1 - k_{2T} x_2 - \omega t)} \end{cases} \quad [2.81]$$

where r_{LL} and r_{LT} are the reflection coefficients for the displacement (the first index recalls that the incident wave is longitudinal). The longitudinal displacement field is irrotationnal ($\underline{\text{rot}}\,\underline{u}_L = \underline{0}$), which means that the polarization vector $\underline{q}_L$ is collinear with the wave vector $\underline{k}_L = (k_1, -k_{2L}, 0)^t$. The transverse displacement field has zero divergence ($\text{div}\,\underline{u}_T = 0$), so that the polarization vector $\underline{q}_T$ is orthogonal to the wave vector $\underline{k}_T = (k_1, -k_{2T}, 0)^t$. Once normalized, the expressions for the polarization vectors, contained in the plane of incidence are:

$$\underline{q}_L = \frac{V_L}{\omega}\begin{pmatrix} k_1 \\ -k_{2L} \end{pmatrix} \quad \text{and} \quad \underline{q}_T = \frac{V_T}{\omega}\begin{pmatrix} k_{2T} \\ k_1 \end{pmatrix} \quad [2.82]$$

The stresses associated with the reflected longitudinal wave are derived from formulae [2.75] and [2.78], by replacing the amplitude a by ar_{LL} and k_{2L} by $-k_{2L}$. Similarly, the stresses associated with the transverse waves are derived from expressions [2.90] by replacing the amplitude a by ar_{LT} and k_{2T} by $-k_{2T}$. The total stresses for the reflected waves, have the expression, at $x_2 = 0$:

$$\begin{cases} \sigma_{12}^r = i\frac{\mu}{\omega}a\left[-2k_1 k_{2L} V_L r_{LL} + (k_1^2 - k_{2T}^2) V_T r_{LT}\right] e^{i(k_1 x_1 - \omega t)} \\ \sigma_{22}^r = i\frac{\mu}{\omega}a\left[(k_{2T}^2 - k_1^2) V_L r_{LL} - 2k_1 k_{2T} V_T r_{LT}\right] e^{i(k_1 x_1 - \omega t)} \end{cases} \quad [2.83]$$

Given the stresses [2.75] and [2.78] associated with the incident wave, the boundary conditions [2.73] lead to the system of equations written in the matrix form:

$$\begin{pmatrix} -2k_1 k_{2L} & k_1^2 - k_{2T}^2 \\ k_{2T}^2 - k_1^2 & -2k_1 k_{2T} \end{pmatrix} \begin{pmatrix} V_L r_{LL} \\ V_T r_{LT} \end{pmatrix} = -V_L \begin{pmatrix} 2k_1 k_{2L} \\ k_{2T}^2 - k_1^2 \end{pmatrix} \quad [2.84]$$

The solution to this system yields the reflection coefficients:

$$\begin{cases} r_{LL} = \frac{1}{d_R}\left[4k_1^2 k_{2L} k_{2T} - (k_{2T}^2 - k_1^2)^2\right] \\ r_{LT} = \frac{V_L}{V_T d_R} 4k_1 k_{2L}(k_{2T}^2 - k_1^2) \end{cases} \quad [2.85]$$

with:

$$d_R = 4k_1^2 k_{2L} k_{2T} + (k_{2T}^2 - k_1^2)^2 \quad [2.86]$$

At normal incidence ($k_1 = 0$), the reflection coefficient r_{LL} is equal to -1. Nonetheless, given the polarization vector [2.82] and since $k_{2L} = k_L = \omega/V_L$, the displacement of the reflected wave at $x_2 = 0$:

$$u_{2L} = a r_{LL} q_{2L} e^{i(k_1 x_1 - \omega t)} \qquad [2.87]$$

with $r_{LL} = -1 = q_{2L}$, is indeed added to the displacement of the incident wave, as it should be done on a surface that is free to move along x_2.

The *intensity reflection coefficients* are given by relations [2.36] with $k_{2I} = k_{2L}$ and $k_{2J} = k_{2T}$:

$$R_{LL} = |r_{LL}|^2 \quad \text{and} \quad R_{LT} = \frac{k_{2T}}{k_{2L}} \left| \frac{V_T}{V_L} r_{LT} \right|^2 \qquad [2.88]$$

Given the expressions [2.85] for the reflection coefficients r_{LL} and r_{LT}, the sum of the intensity reflection coefficients is equal to unity, signifying that energy is conserved in the x_2 direction.

The evolution of the moduli of the reflection coefficients r_{LL} and r_{LT} versus the angle of incidence θ_L is plotted in Figure 2.7 for different values of the ratio $\kappa = V_T/V_L$. The reflection coefficient r_{LL} is equal to unity and the coefficient r_{LT} is zero at normal ($\theta_L = 0°$) and grazing incidences ($\theta_L = 90°$). Thus, for these two angles of incidence, there is no mode conversion on the surface of the material. On the other hand, the coefficient r_{LT} presents a maximum in the interval $\theta_L \in [35, 50]°$ corresponding to a maximum, but not optimal, conversion, since the reflection coefficient r_{LL} is not necessarily zero. Moreover, it is interesting to note that the maximum for the coefficient r_{LT} exceeds the unity, unlike the maximum for the coefficient r_{LL}: an amplitude conversion coefficient may exceed unity, while still respecting the energy conservation (section 2.2.4). If the ratio $\kappa = V_T/V_L$ is larger than 0.565, the reflection coefficient cancels for two values of the angle of incidence: the incident longitudinal wave is thus entirely converted into a transverse vertical wave.

2.4.1.2. *Incident transverse vertical wave*

In a similar way, an incident TV wave is reflected into a transverse wave and converted into a longitudinal wave. The normalized displacement of the incident wave has the expression:

$$\underline{u}^i(x_1, x_2, t) = a \underline{p}_T e^{i(k_1 x_1 + k_{2T} x_2 - \omega t)} \quad \text{with} \quad \underline{p}_T = \frac{V_T}{\omega} (-k_{2T}, k_1)^t \qquad [2.89]$$

and the associated stresses are given by the relations:

$$\begin{cases} \sigma_{12}^i(x_1, x_2 = 0, t) = i \frac{\mu}{\omega} a \left(k_1^2 - k_{2T}^2 \right) V_T e^{i(k_1 x_1 - \omega_I t)} \\ \sigma_{22}^i(x_1, x_2 = 0, t) = 2i \frac{\mu}{\omega} a k_1 k_{2T} V_T e^{i(k_1 x_1 - \omega_I t)} \end{cases} \qquad [2.90]$$

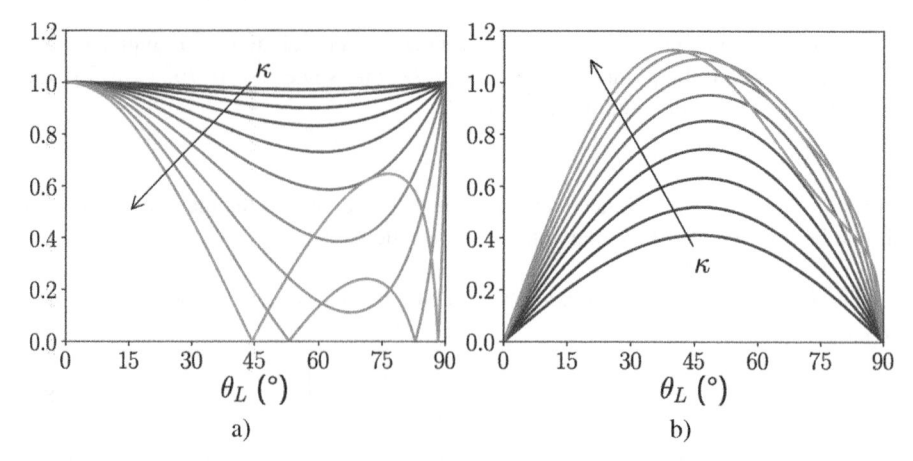

Figure 2.7. *Moduli of the reflection coefficients a) r_{LL} and b) r_{LT} as functions of the angle of incidence θ_L for different values of the ratio $\kappa = V_T/V_L$, lying between 0.2 and 0.65. For a color version of this figure, see www.iste.co.uk/royer/waves1.zip*

The application of the boundary conditions leads, after similar calculations to those of the longitudinal case, to the expression of the reflection coefficients for an incident transverse wave:

$$\begin{cases} r_{TT} = \dfrac{1}{d_R}\left[4k_1^2 k_{2L} k_{2T} - (k_{2T}^2 - k_1^2)^2\right] \\ r_{TL} = \dfrac{V_T}{V_L d_R} 4k_1 k_{2T} (k_1^2 - k_{2T}^2) \end{cases} \quad [2.91]$$

Below the critical angle, the components k_{2L} and k_{2T} can be expressed as functions of the angles θ_L and θ_T through [2.80]. By substituting these expressions into those of the reflection coefficients r_{TT} and r_{TL}, we obtain:

$$\begin{cases} r_{TT} = \dfrac{\kappa^2 \sin(2\theta_T)\sin(2\theta_L) - \cos^2(2\theta_T)}{\kappa^2 \sin(2\theta_T)\sin(2\theta_L) + \cos^2(2\theta_T)} \\ r_{TL} = -\dfrac{\kappa \sin(4\theta_T)}{\kappa^2 \sin(2\theta_T)\sin(2\theta_L) + \cos^2(2\theta_T)} \end{cases} \quad [2.92]$$

The intensity reflection coefficients are derived from the formulae [2.36], with $k_{2I} = k_{2T}$ and $k_{2J} = k_{2L}$:

$$R_{TT} = |r_{TT}|^2 \quad \text{and} \quad R_{TL} = \dfrac{k_{2L}}{k_{2T}}\left|\dfrac{V_L}{V_T} r_{TL}\right|^2 \quad [2.93]$$

and a quick calculation demonstrates that the law of conservation of energy is verified: $R_{TT} + R_{TL} = 1$.

The examination of the slowness surfaces corresponding to an incident transverse vertical wave (Figure 2.8) shows that the angle θ_L is always larger than the angle of incidence θ_T. θ_L reaches $90°$ as soon as the angle of incidence is equal to the critical angle $\theta_c^L = \arcsin(V_T/V_L)$, which separates the domain of propagative ($\theta_T < \theta_c^L$) from that of evanescent ($\theta_T \geq \theta_c^L$) reflected longitudinal waves. In the first case, the projections of the phase slowness are given by the expressions [2.80]; in the second case, the Snell–Descartes relations no longer apply for the longitudinal wave:

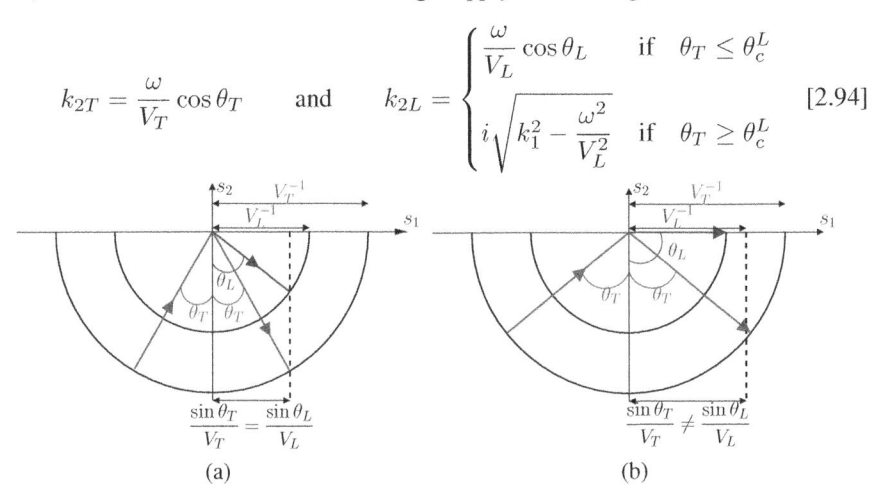

$$k_{2T} = \frac{\omega}{V_T}\cos\theta_T \quad \text{and} \quad k_{2L} = \begin{cases} \dfrac{\omega}{V_L}\cos\theta_L & \text{if } \theta_T \leq \theta_c^L \\ i\sqrt{k_1^2 - \dfrac{\omega^2}{V_L^2}} & \text{if } \theta_T \geq \theta_c^L \end{cases} \quad [2.94]$$

Figure 2.8. *Slowness curves for a free surface and an incident transverse vertical wave a) below the critical angle and b) beyond the critical angle. For a color version of this figure, see www.iste.co.uk/royer/waves1.zip*

The evolution of the moduli of the reflection coefficients r_{TT} and r_{TL} versus the angle of incidence θ_T is shown in Figure 2.9. The coefficient r_{TL} is zero at normal incidence ($\theta_T = 0°$) and grazing incidence ($\theta_T = 90°$), and also for $\theta_T = 45°$. Consequently, like in the case of an incident longitudinal wave, there is no mode conversion for this particular angle. For angles of incidence beyond the critical angle θ_c^L, the intensity reflection coefficient R_{TL} is zero and the coefficient R_{TT} is equal to unity, signifying that all the energy transported by the incident wave is transmitted to the reflected transverse vertical wave: the longitudinal wave, evanescent along x_2, does not transport energy in this direction.

2.4.2. Solid–fluid interface

The modeling of the reflection and transmission of waves at the interface between a perfect fluid and an isotropic solid is more complicated than in the case of a free surface. In the absence of any transverse wave in the fluid, only the mechanical displacement and acoustic pressure associated with the longitudinal wave must be taken into account.

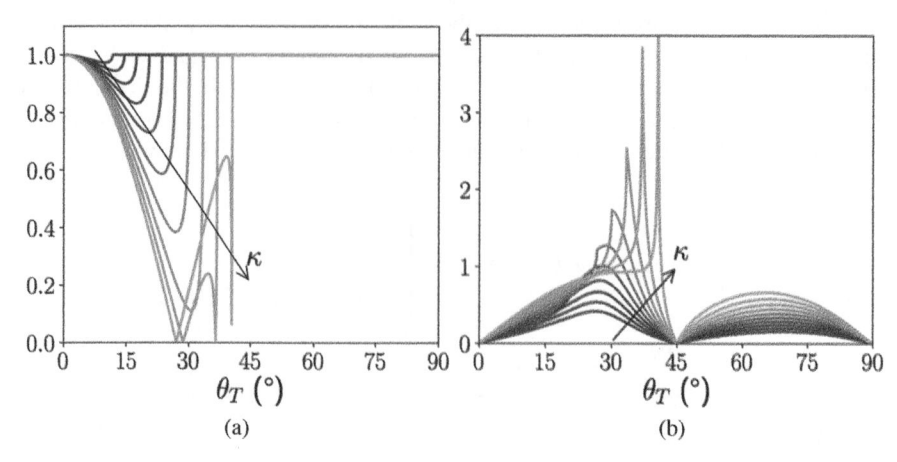

Figure 2.9. Moduli of reflection coefficients a) r_{TT} and b) r_{TL} versus the angle of incidence θ_T for different values of the ratio $\kappa = V_T/V_L$, lying between 0.2 and 0.65. For a color version of this figure, see www.iste.co.uk/royer/waves1.zip

2.4.2.1. *Incident longitudinal wave in the solid*

The incident longitudinal wave, with displacement amplitude a, originates from the solid which is in contact with a fluid of mass density ρ_f, occupying the half-space $x_2 > 0$. If this fluid is non-viscous, the wave transmitted with a coefficient t_{Lf} is necessarily longitudinal (velocity V_f). The normalized displacement field associated with this wave of amplitude at_{Lf} is given by:

$$\underline{u}^{(f)}(x_1, x_2, t) = at_{Lf}\frac{V_f}{\omega}\begin{pmatrix}k_1\\k_{2f}\end{pmatrix} e^{i(k_1 x_1 + k_{2f} x_2 - \omega t)} \quad \text{with} \quad k_{2f} > 0 \quad [2.95]$$

Considering equation [1.323], the acoustic pressure $p_a^{(f)}$ is proportional to the divergence of the particle displacement $\underline{u}^{(f)}$:

$$p_a^{(f)}(x_1, x_2, t) = -\rho_f V_f^2 \operatorname{div} \underline{u}^{(f)} = -i\omega\rho_f V_f at_{Lf} e^{i(k_1 x_1 + k_{2f} x_2 - \omega t)} \quad [2.96]$$

and the associated stresses are equal to $\sigma_{ij}^{(f)} = -p_a^{(f)}\delta_{ij}$, that is, $\sigma_{12}^{(f)} = 0$ and $\sigma_{22}^{(f)} = -p_a^{(f)}$. On the interface $x_2 = 0$, the boundary conditions require the cancelation of the tangential stress ($\sigma_{12} = 0$), the continuity of the normal stress ($\sigma_{22} = -p_a^{(f)}$) and the continuity of the normal displacement ($u_2 = u_2^{(f)}$).

According to the expressions for the displacement [2.74] and the stresses [2.75], [2.78] of the incident wave and the expression [2.83] for the stresses of the reflected waves in the isotropic solid, we obtain:

$$\begin{cases} 2k_1 k_{2L} V_L - 2k_1 k_{2L} V_L r_{LL} + \left(k_1^2 - k_{2T}^2\right) V_T r_{LT} = 0 \\ \left(k_{2T}^2 - k_1^2\right) V_L + \left(k_{2T}^2 - k_1^2\right) V_L r_{LL} - 2k_1 k_{2T} V_T r_{LT} = \dfrac{\rho_f V_f \omega^2}{\mu} t_{Lf} \\ k_{2L} V_L - k_{2L} V_L r_{LL} + k_1 V_T r_{LT} = k_{2f} V_f t_{Lf} \end{cases}$$

[2.97]

or, in matrix form:

$$\begin{pmatrix} -2k_1 k_{2L} & k_1^2 - k_{2T}^2 & 0 \\ k_{2T}^2 - k_1^2 & -2k_1 k_{2T} & -\dfrac{\rho_f \omega^2}{\mu} \\ -k_{2L} & k_1 & -k_{2f} \end{pmatrix} \begin{pmatrix} V_L r_{LL} \\ V_T r_{LT} \\ V_f t_{Lf} \end{pmatrix} = -V_L \begin{pmatrix} 2k_1 k_{2L} \\ k_{2T}^2 - k_1^2 \\ k_{2L} \end{pmatrix}$$ [2.98]

The resolution of this system of equations leads to the reflection and transmission coefficients for the displacement:

$$\begin{cases} r_{LL} = \dfrac{4k_1^2 k_{2L} k_{2T} - \left(k_{2T}^2 - k_1^2\right)^2 + \delta_f}{d_R + \delta_f} \\ r_{LT} = \dfrac{4k_1 k_{2L} \left(k_{2T}^2 - k_1^2\right)}{d_R + \delta_f} \dfrac{V_L}{V_T} \\ t_{Lf} = \dfrac{2k_T^2 \left(k_{2T}^2 - k_1^2\right)}{d_R + \delta_f} \dfrac{k_{2L} V_L}{k_{2f} V_f} \end{cases}$$ [2.99]

with:

$$d_R = 4k_1^2 k_{2L} k_{2T} + \left(k_{2T}^2 - k_1^2\right)^2 \quad \text{and} \quad \delta_f = \dfrac{\rho_f k_{2L}}{\rho k_{2f}} k_T^4$$ [2.100]

It must be noted that the reflection coefficient for the longitudinal wave is derived from the coefficient r_{LL} on a free surface [2.85] by adding, to the numerator and to the denominator, the quantity δ_f proportional to the mass density of the fluid ρ_f. The intensity reflection and transmission coefficients are given by relations similar to equations [2.36] and [2.37]:

$$R_{LL} = |r_{LL}|^2, \quad R_{LT} = \dfrac{k_{2T}}{k_{2L}} \left|\dfrac{V_T}{V_L} r_{LT}\right|^2 \quad \text{and} \quad T_{Lf} = \dfrac{\rho_f k_{2f}}{\rho k_{2L}} \left|\dfrac{V_f}{V_L} t_{Lf}\right|^2$$ [2.101]

The evolution of these coefficients as functions of the angle of incidence θ_L is shown in Figure 2.10 for a duralumin–water interface and a duralumin–air interface.

The reflection coefficient R_{LT} is always zero at normal incidence and grazing incidence and the evolution of the reflection coefficients is not very dependent on the fluid in contact with the solid. On the contrary, the transmitted acoustic intensity exceeds 20% in the case of water for less than 0.01% in the case of the air, due to its very small acoustic impedance.

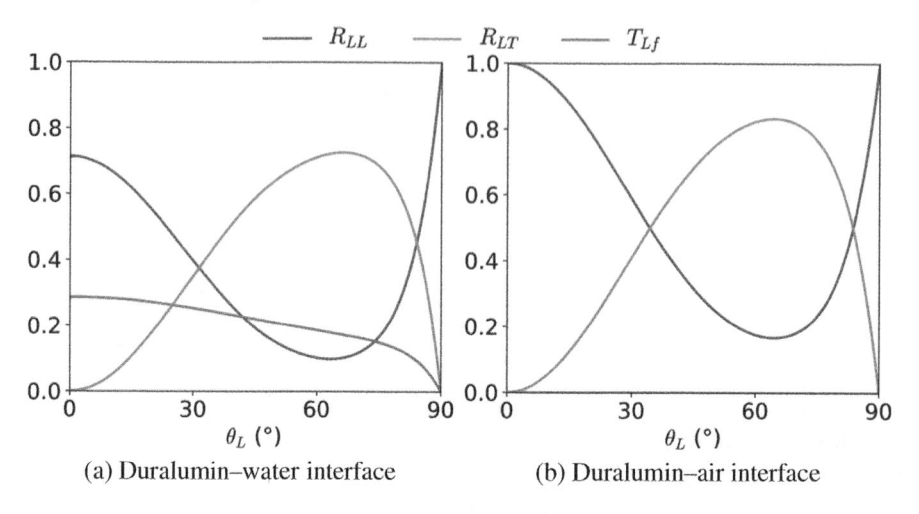

Figure 2.10. *Intensity reflection coefficients R_{LL} and R_{LT} and intensity transmission coefficient T_{Lf} versus the angle of incidence θ_L for a) a duralumin–water interface and b) a duralumin–air interface. For a color version of this figure, see www.iste.co.uk/royer/waves1.zip*

2.4.2.2. *Incident transverse vertical wave in the solid*

The reflection and transmission coefficients can be calculated in the same way as in the previous section:

$$\begin{cases} r_{TL} = \dfrac{4k_1 k_{2T}\left(k_{2T}^2 - k_1^2\right)}{d_R + \delta_f}\dfrac{V_T}{V_L} \\ r_{TT} = \dfrac{1}{d_R + \delta_f}\left[4k_1^2 k_{2L} k_{2T} - \left(k_{2T}^2 - k_1^2\right)^2 - \delta_f\right] \\ t_{Tf} = \dfrac{4k_1 k_{2L} k_T k_f}{d_R + \delta_f}\dfrac{k_{2T}}{k_{2f}} \end{cases} \quad [2.102]$$

The slowness curves corresponding to an incident transverse vertical wave are plotted in Figure 2.11. In the solid, the critical angle $\theta_c^L = \arcsin(V_T/V_L)$ expresses the boundary between the domain of propagative reflected longitudinal waves ($\theta_T < \theta_c^L$) and that of evanescent waves ($\theta_T > \theta_c^L$). In general, the speed of sound in the fluid V_f is lower than the velocity of the transverse wave V_T, thus the transmitted wave in the fluid is always propagative, regardless of the angle of incidence (Figure 2.11(a)). On the other hand, if V_T is smaller than V_f, there exists a critical angle $\theta_c^f = \arcsin(V_T/V_f)$ for the longitudinal wave in the fluid (Figure 2.11(b)) beyond which the incident transverse wave is totally reflected.

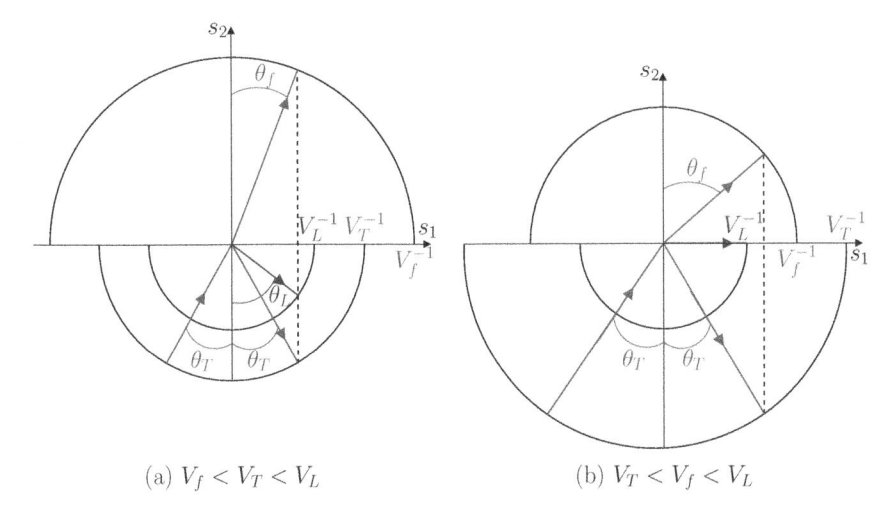

(a) $V_f < V_T < V_L$

(b) $V_T < V_f < V_L$

Figure 2.11. *Slowness curves for a solid–fluid interface and an incident transverse vertical wave according to the value of V_f with respect to V_T. For a color version of this figure, see www.iste.co.uk/royer/waves1.zip*

The evolution of the intensity reflection coefficients R_{TL} and R_{TT} and transmission coefficient T_{Tf} versus the angle of incidence θ_T is represented in Figure 2.12 for a duralumin–water interface and a plexiglass–water interface. The velocities V_L and V_T of the duralumin are larger than the velocity V_f of water. On the other hand, the velocity of the transverse wave in the plexiglass is smaller than V_f. The longitudinal wave in the solid becomes evanescent when the angle of incidence is larger than the critical angle $\theta_c^L = 29.6°$ for duralumin and $\theta_c^L = 29.5°$ for plexiglass. In the case of the plexiglass–water interface, the longitudinal wave in the fluid becomes evanescent when the angle of incidence is larger than the critical angle, $\theta_c^f = 62.5°$. Between the critical angles θ_c^L and θ_c^f, since the velocity of the transverse wave in the plexiglass ($V_T = 1\,330$ m/s) is very close to V_f (1500 m/s), the transmission in water is almost total, for example, $T_{Tf}(45°) = 0.997$.

2.4.2.3. *Incident longitudinal wave in the fluid*

The incident wave originating from the fluid is a longitudinal wave. This configuration is encountered in seismic exploration, in underwater acoustics (using sonar to inspect the ocean floor) and in non-destructive evaluation of materials. In this case, the sample, immersed in water, is insonified by an ultrasonic beam that generates, *a priori*, all types of elastic waves in the solid. The proportion of these waves depends on the angle of incidence. The mechanical behavior of the solid can be derived from the reflected acoustic field. The generation of a surface wave that radiates energy in the fluid when the angle of incidence is close to the characteristic angle of the Rayleigh wave leads to a significant distortion of the reflected field. This effect is exploited in acoustic microscopy (Briggs 1995).

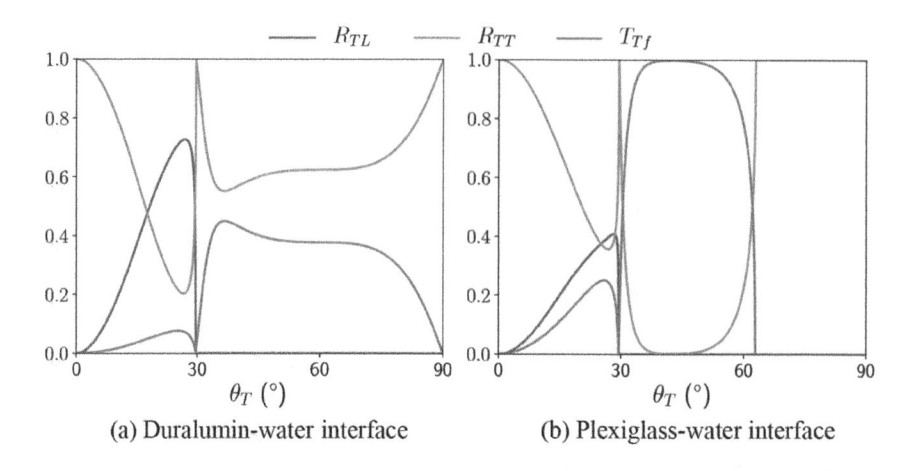

Figure 2.12. *Intensity reflection coefficients R_{TL} and R_{TT} and intensity transmission coefficient T_{Tf} versus the angle of incidence θ_T for a) a duralumin–water interface and b) a plexiglass–water interface. For a color version of this figure, see www.iste.co.uk/royer/waves1.zip*

Following the same approach as in the case of an incident wave in the solid, it is possible to model the reflection and transmission of an incident wave in the fluid. The expressions for the transmission coefficients t_{fL} and t_{fT} of the longitudinal and transverse waves in the solid, and for the reflection coefficient r_{ff} of the longitudinal wave in the fluid, are:

$$t_{fL} = -\frac{2k_T^2 \left(k_{2T}^2 - k_1^2\right)}{d_R + \delta_f} \frac{\rho_f}{\rho}, \quad t_{fT} = \frac{4k_1 k_{2L} k_T^2}{d_R + \delta_f} \frac{\rho_f}{\rho} \quad \text{and} \quad r_{ff} = -\frac{d_R - \delta_f}{d_R + \delta_f}$$

[2.103]

If the mass density of the fluid is very small compared to that of the solid ($\delta_f \approx 0$), the acoustic wave is totally reflected on the infinitely rigid surface of the solid, with a change of sign for the mechanical displacement ($r_{ff} = -1$). If $V_f < V_T < V_L$, there exist two critical angles: $\theta_c^L = \arcsin(V_f/V_L)$ and $\theta_c^T = \arcsin(V_f/V_T)$. When the angle of incidence is smaller than the angle θ_c^L, the longitudinal wave transmitted in the solid is propagative. When the angle of incidence is larger than θ_c^L, this wave is evanescent. Similarly, when the angle of incidence is smaller (larger) than θ_c^T, the transverse wave transmitted in the solid is propagative (evanescent).

Figure 2.13(a) illustrates the evolution of the intensity reflection and transmission coefficients *versus* the angle of incidence θ_f. The velocity in the liquid (water: $V_f = 1\,500$ m/s) is lower than that of the transverse wave in the solid (duralumin). The transmission coefficient in the longitudinal wave (T_{fL}) gradually decreases until it is canceled out for the first critical angle $\theta_c^L = 13.6°$. The transmission coefficient in the transverse wave (T_{fT}) then increases very rapidly. For $\theta_f = 19.7°$, the transverse wave is refracted under an angle $\theta_T = 45°$ with an efficiency close to 40%. In non-destructive evaluation, this configuration is suitable for the study of opening cracks. When the angle of incidence reaches and then exceeds the second critical value, $\theta_c^T = 28.4°$, no more waves are transmitted into the solid.

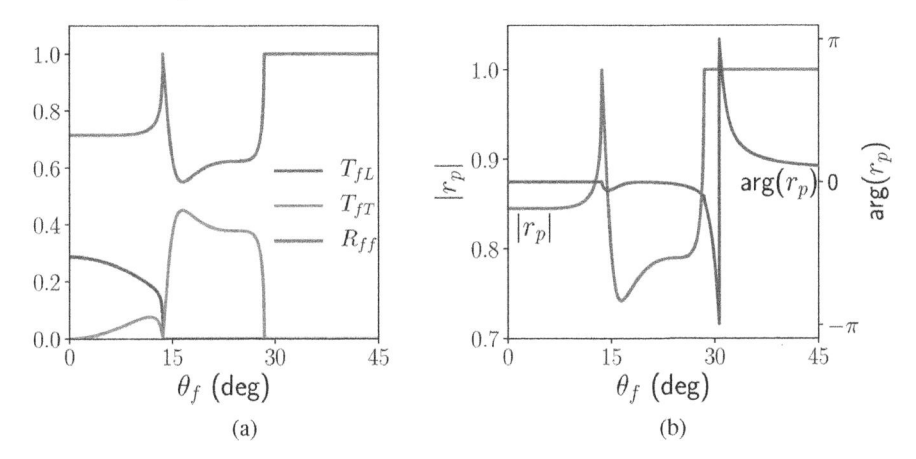

Figure 2.13. *Reflection and transmission at a liquid–solid interface. a) Intensity reflection and transmission coefficients and b) modulus and phase of the reflection coefficient of the acoustic pressure as functions of the angle of incidence θ_f. For a color version of this figure, see www.iste.co.uk/royer/waves1.zip*

Let us now examine, in parallel, the evolution of the amplitude and phase of the reflection coefficient (Figure 2.13(b)) as well as the behavior, depending on $k_1 = k_f \sin \theta_f$ of its analytical expression. This is deduced from the expression [2.100] of the factors d_R and δ_f; considering the reflection coefficient of the acoustic pressure $r_p = -r_{ff}$, we get:

$$r_p = \frac{\left(k_{2T}^2 - k_1^2\right)^2 + 4k_1^2 k_{2L} k_{2T} - \delta_f}{\left(k_{2T}^2 - k_1^2\right)^2 + 4k_1^2 k_{2L} k_{2T} + \delta_f} \quad \text{with} \quad \delta_f = \frac{\rho_f k_{2L}}{\rho k_{2f}} k_T^4 \quad [2.104]$$

When the angle of incidence θ_f is smaller than the first critical angle θ_c^L, all the modes are propagative and the reflection coefficient is real. When θ_f increases, the reflection coefficient is nearly constant and remains real up to the critical angle θ_c^L, for which $\theta_L = \pi/2$. For this precise value of the angle of incidence, the normal component k_{2L} and therefore the factor δ_f cancel out: $r_p = 1$, the reflection of the longitudinal wave coming from the fluid is total. Between the two critical angles ($\theta_c^L < \theta_f < \theta_c^T$), k_{2L} is imaginary: $k_{2L} = i\alpha_L$; since k_{2T} and k_{2f} are always real and positive, the reflection coefficient becomes complex:

$$r_p = \frac{(k_{2T}^2 - k_1^2)^2 + ia}{(k_{2T}^2 - k_1^2)^2 + ib} \quad \text{with} \quad \begin{cases} a = 4k_1^2 \alpha_L k_{2T} - |\delta_f| \\ b = 4k_1^2 \alpha_L k_{2T} + |\delta_f| \end{cases} \quad [2.105]$$

The reflection is accompanied by a small phase delay visible on the curve in Figure 2.13(b).

When the angle of incidence reaches and then exceeds the second critical angle $\theta_c^T = 28.4°$, k_{2T} also becomes imaginary: $k_{2T} = i\alpha_T$. By replacing k_{2L} by $i\alpha_L$ and k_{2T} by $i\alpha_T$ in equation [2.104], the reflection coefficient takes the form:

$$r_p = \frac{R(k_1) - iS(k_1)}{R(k_1) + iS(k_1)} \quad [2.106]$$

with:

$$R(k_1) = \left(k_1^2 + \alpha_T^2\right)^2 - 4k_1^2 \alpha_L \alpha_T \quad \text{and} \quad S(k_1) = \frac{\rho_f \alpha_L}{\rho\sqrt{k_f^2 - k_1^2}} k_T^4 \quad [2.107]$$

Since $k_1 < k_f$, the functions $R(k_1)$ and $S(k_1)$ are real. The modulus of r_p is equal to unity and the slope of its phase undergoes a small discontinuity.

In the first section of Chapter 3, we will show that the function $R(k_1)$ is canceled for a value $k_1 = k_R$ larger than k_T. This solution of the equation $R(k_1) = 0$ corresponds to a *Rayleigh wave*, which propagates without attenuation on the free surface of the isotropic solid, at a velocity V_R (2940 m/s for duralumin) lower than V_T. It consists of evanescent longitudinal and transverse displacements along x_2. The corresponding angle of incidence $\theta_R = 30.6°$, such that:

$$k_1 = k_f \sin\theta_R = k_R \quad \text{hence} \quad \theta_R = \arcsin\left(\frac{V_f}{V_R}\right) \quad [2.108]$$

is slightly larger than θ_c^T. The denominator of r_p cancels for a value k_{Rf} of k_1, whose real part is close to k_R: $k_{Rf} \approx k_R + i\alpha_R$. The numerator vanishes for the complex conjugated value $k_R - i\alpha_R$, with α_R being positive and small compared to k_R. In the vicinity of k_R, the reflection coefficient is approximately equal to:

$$r_p \approx \frac{k_1 - k_R + i\alpha_R}{k_1 - k_R - i\alpha_R} = e^{i\varphi} \quad \text{with} \quad \varphi = 2\arctan\left(\frac{\alpha_R}{k_1 - k_R}\right) \quad [2.109]$$

Its modulus is equal to unity, while its argument varies from $-\pi$ to $+\pi$ when k_1 moves from $k_R - \varepsilon$ to $k_R + \varepsilon$.

Although the incident longitudinal wave is converted into a Rayleigh wave, the reflection coefficient does not vanish at the angle θ_R because, in the presence of the fluid, the Rayleigh wave radiates energy at the angle θ_R. After traveling a few wavelengths along the interface, all the acoustic energy is returned to the fluid. The rapid variation of the phase around the critical angle θ_R reflects the generation of this *leaky Rayleigh wave*. The imaginary part α_R of the wave number expresses the attenuation due to this radiation (section 3.2.1.3). Another manifestation of this effect is a spatial shift of the order of λ_R in the reflected beam. This non-specular reflection can only be shown if the lateral dimensions of the incident beam are finite (Schoch 1950; Breazeale 1977).

2.4.2.4. Interaction of a plane wave with an immersed plate

The velocity and the attenuation of bulk waves in a solid are commonly measured by immersing a plate with parallel faces in water. Two specific techniques are used in ultrasonic experiments: measurements based on time-of-flight or on phase difference. From an experimental point of view, it must be ensured that the incident wave is plane (see Volume 2, section 1.1.4.2).

2.4.2.4.1. Velocity measurements by time-of-flight

Under oblique incidence, longitudinal and transverse waves are generated in the plate. The first two echoes transmitted through the sample correspond to the waves that have traveled a direct path through the plate, that is, without multiple reflections (Figure 2.14).

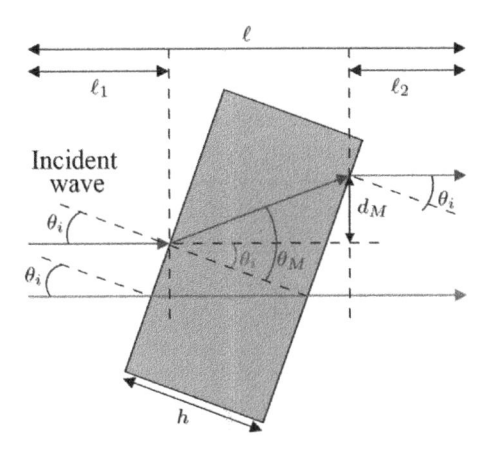

Figure 2.14. *Plate with parallel faces of thickness h immersed in a fluid*

When the plate is immersed, the time-of-flight for the first longitudinal echo ($M = L$) or transverse echo ($M = T$) is given by the relation:

$$t_M = \frac{\ell_1 + \ell_2}{V_f} + \frac{h}{V_M \cos\theta_M} \quad [2.110]$$

When the plate is removed, the expression of the time-of-flight for the acoustic wave in water is:

$$t_0 = \frac{\ell}{V_f} = \frac{\ell_1 + \ell_2}{V_f} + \frac{h \cos(\theta_M - \theta_i)}{V_f \cos\theta_M} \quad [2.111]$$

and the difference in propagation time is given by:

$$\tau_M = t_0 - t_M = \frac{h}{V_f \cos\theta_M} \left[\cos\theta_M \cos\theta_i + \sin\theta_M \sin\theta_i - \frac{V_f}{V_M} \right] \quad [2.112]$$

Using the Snell–Descartes law, we obtain:

$$\frac{\tau_M V_f}{h} = \cos\theta_i - \frac{V_f}{V_M} \cos\theta_M = \cos\theta_i - \sqrt{\frac{V_f^2}{V_M^2} - \sin^2\theta_i} \quad [2.113]$$

and the velocity V_M is thus given by the relation:

$$V_M = \frac{V_f}{\sqrt{1 - 2\frac{\tau_M V_f}{h} \cos\theta_i + \left(\frac{\tau_M V_f}{h}\right)^2}} \quad \text{with} \quad M = L \text{ or } T \quad [2.114]$$

Measurements can be performed using a pair of transducers. However, the receiver transducer must be shifted by a distance d_M with respect to the propagation direction of the incident wave generated by the emitter transducer. Pulse echo measurements can also be carried out, by replacing the receiver transducer by a plane reflector. Since the wave paths with or without the plate are identical both in the outward and return directions, the propagation distances are simply doubled. The velocity of the longitudinal and transverse waves in the material is given by formula [2.114], by replacing h with $2h$, where τ_M is always the difference in the time-of-flight with and without the plate.

The velocity of the longitudinal wave is often measured at normal incidence, while the velocity of transverse waves is measured at oblique incidence, beyond the critical angle for which the longitudinal wave becomes evanescent. Time-of-flight measurements that allow to access to wave surfaces are used to determine the elastic constants of anisotropic solids, such as composite materials (Castagnède 1990).

2.4.2.4.2. Phase velocity and attenuation measurements

At normal incidence and when the thickness of the sample is larger than the wavelength, the transmitted signal takes the form of a succession of pulses, spaced by a time $\tau = 2h/V_L$, where the velocity of the longitudinal wave in the plate is *a priori* unknown. The first echo corresponds to the directly transmitted wave; the following echoes correspond to successive reflections in the plate (section 2.3.2). Assuming that the incident wave is plane and by isolating the first echo by a windowing technique, the expression for the spectrum of the transmitted signal is:

$$S(\omega) = A(\omega) e^{ik_f \ell_1} t_{fL} e^{i\tilde{k}_L h} t_{Lf} e^{ik_f \ell_2} \qquad [2.115]$$

where $A(\omega)$ is an amplitude that takes into account the electro-mechanical transmittance of the emitter and receiver. If the solid material is weakly dissipative at the operating frequencies, the attenuation can be expressed by a complex wave number: $\tilde{k}_L = k_L + i\alpha_L$, where $k_L = \omega/V_L$. In the absence of the sample, the spectrum for the wave propagating over the same distance is:

$$S_0(\omega) = A(\omega) e^{ik_f \ell} \qquad \text{with} \qquad \ell = \ell_1 + h + \ell_2 \qquad [2.116]$$

The ratio of these two spectra is written as:

$$\frac{S(\omega)}{S_0(\omega)} = t_{fL} t_{Lf} e^{i(\tilde{k}_L - k_f)h} = t_{fL} t_{Lf} e^{i(k_L - k_f)h} e^{-\alpha_L h} \qquad [2.117]$$

When the imaginary part of the wave number is small compared to its real part, the product of the transmission coefficients is given by the approximate expression:

$$t_{Lf} t_{fL} = -\frac{4\tilde{\rho} \tilde{k}_L V_L \rho_f k_f}{V_f \left(\rho k_f + \rho_f \tilde{k}_L \right)^2} \approx -\frac{4\rho\rho_f V_L^2}{(Z_L + Z_f)^2} \qquad [2.118]$$

The modulus and phase of the spectrum ratio take the following form:

$$\left| \frac{S(\omega)}{S_0(\omega)} \right| = t_{fL} t_{Lf} e^{-\alpha_L h} \quad \text{and} \quad \varphi(\omega) = \arg\left[\frac{S(\omega)}{S_0(\omega)} \right] = (k_L - k_f)h \qquad [2.119]$$

The phase velocity and attenuation in the sample are deduced from these expressions (Sachse 1978):

$$V_L(\omega) = \frac{\omega h}{k_f h + \varphi(\omega)} \quad \text{and} \quad \alpha_L(\omega) = \frac{1}{h} \ln\left[-\frac{(Z_L + Z_f)^2}{4\rho\rho_f V_L^2} \left| \frac{S_0(\omega)}{S(\omega)} \right| \right] \qquad [2.120]$$

As an illustration, Figure 2.15 shows the variations *versus* frequency of the velocity and attenuation of longitudinal waves measured in a plexiglass plate and in a polyurethane plate. The attenuation curves are interpolated by the power law [1.314]. The attenuation in plexiglass varies almost linearly with frequency, while in polyurethane the attenuation follows a $f^{3/2}$ law.

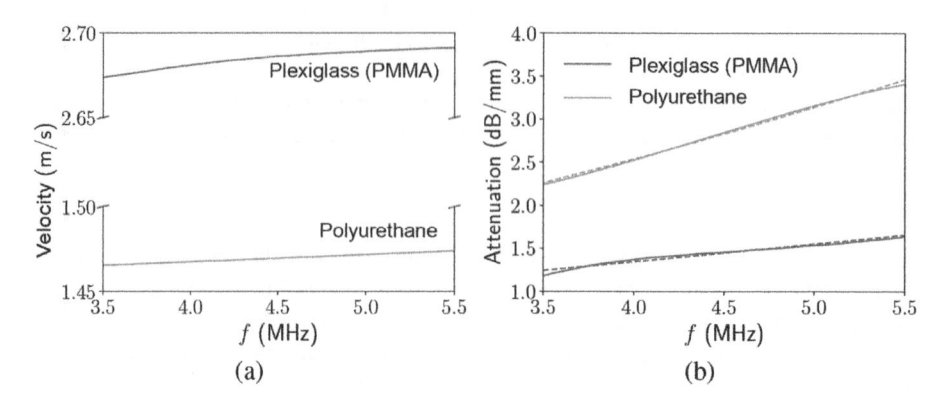

Figure 2.15. a) Velocity and b) attenuation of longitudinal waves measured in a plexiglass plate and in a polyurethane plate with thickness 10 mm

This modeling only remains valid when the attenuation in the material is low. If it is no longer the case, the transmission coefficients t_{fL} and t_{Lf} are complex and the velocity is no longer directly related to the phase difference φ. A way to overcome this problem is to measure the signals in transmission across two samples of the same material, but with different thicknesses. In this case, the phase of the spectrum ratio does not depend on the transmission coefficients and the velocity is directly linked to the phase shift resulting from the difference in the propagation distance, that is, the two thicknesses.

2.5. Anisotropic medium: diffraction matrix

In this section, we will express the transmission and reflection coefficients at the interface between two anisotropic solids. These coefficients are indeed the elements of a matrix that associates the amplitudes of all the reflected or transmitted waves to the amplitude of the incident wave. In the general case, this diffraction matrix is a (6×6) matrix. When the decoupling conditions, explained at the beginning of this chapter, are satisfied (plane of incidence $x_1 x_2$ perpendicular to a direct or inverse binary axis), it is only a (4×4) matrix. In all the cases examined here and in the absence of any incident wave, modes appear that propagate on the surface or at the interface. These surface and interface waves are studied in Chapter 3.

2.5.1. *Analytical resolution*

In an unbounded crystal, the phase velocity V of each plane wave is determined in terms of the propagation direction $\underline{n}$ by solving the Christoffel equation: $V = V(\underline{n})$. Here, it is the incident wave on the interface that is fixed and that creates different modes in each of the two media. The angle of incidence θ_i and the velocity V_i of this wave impose, for all modes, the component s_1 of the slowness vector $\underline{s} = \underline{n}/V$ on the interface $x_2 = 0$ (Snell–Descartes law). The slowness vectors of the incident, reflected and transmitted waves are contained in the plane of incidence (x_1, x_2):

$$\underline{s} = (s_1, s_2, 0)^t \quad \text{with} \quad s_1 = \frac{n_1^i}{V_i} = \frac{\sin\theta_i}{V_i} \qquad [2.121]$$

and they differ only in their projection s_2 on the normal to the interface. Let us introduce the dimensionless ratio m of the components along x_2 and x_1:

$$m = \frac{s_2}{s_1} = \frac{k_2}{k_1} \qquad [2.122]$$

By replacing ω by $k_1 V_s$, the mechanical displacement [1.171] of a wave with complex amplitude a and normalized polarization vector $\underline{p}$ takes a form:

$$u_l(\underline{x}, t) = a p_l e^{i(k_1 x_1 + k_2 x_2 - \omega t)} = a p_l e^{i k_1 (x_1 + m x_2 - V_s t)} \qquad [2.123]$$

similar to that of a time-harmonic plane wave that propagates in the direction $n_1 = 1$, $n_2 = m$ and $n_3 = 0$, and with the phase velocity V_s. This formal solution satisfies the phase matching condition imposed by the Snell–Descartes law. It reveals the factor $e^{i k_1 (x_1 - V_s t)}$ that expresses the propagation along the interface, with the velocity $V_s = 1/s_1$.

The presence of the interface is at the origin of non-propagative modes, described as *inhomogeneous* or *evanescent*, depending on the value of m. This unknown parameter and the polarization $\underline{p}$ are determined by the Christoffel equation [1.173]:

$$\left(\Gamma_{il} - \rho V_s^2 \delta_{il}\right) p_l = 0 \qquad [2.124]$$

Replacing n_1 by 1, n_2 by m and n_3 by 0, the components $\Gamma_{il} = c_{ijkl} n_j n_k$ are second-degree polynomials in m:

$$\Gamma_{il} = c_{i11l} + (c_{i12l} + c_{i21l}) m + c_{i22l} m^2 \qquad [2.125]$$

or using Voigt notation:

$$\begin{cases} \Gamma_{11} = C_{11} + 2C_{16}m + C_{66}m^2 \\ \Gamma_{22} = C_{66} + 2C_{26}m + C_{22}m^2 \\ \Gamma_{33} = C_{55} + 2C_{45}m + C_{44}m^2 \\ \Gamma_{12} = C_{16} + (C_{12} + C_{66}) m + C_{26}m^2 \\ \Gamma_{13} = C_{15} + (C_{14} + C_{56}) m + C_{46}m^2 \\ \Gamma_{23} = C_{56} + (C_{46} + C_{25}) m + C_{24}m^2 \end{cases} \qquad [2.126]$$

The characteristic equation obtained by replacing V by V_s in equation [1.175], which expresses the compatibility condition for system [2.124], is an equation of degree six in m:

$$\alpha_6 m^6 + \alpha_5 m^5 + \alpha_4 m^4 + \alpha_3 m^3 + \alpha_2 m^2 + \alpha_1 m + \alpha_0 = 0 \qquad [2.127]$$

The coefficients α_n of the polynomial are real and depend on the stiffnesses of the material, and on the velocity at the interface V_s. This is the same for the six roots, real or complex conjugated, m_K ($K = 1, 2... 6$), corresponding, *a priori*, to three forward waves and three backward waves, whose nature depends on the value of m. There are three different cases:

– if m is real, the slowness vector $\underline{s}$ is real. The propagation direction, defined by the angle θ, and the phase velocity V of the wave are, respectively:

$$\frac{\cos\theta}{\sin\theta} = \left|\frac{s_2}{s_1}\right| = |m| \quad \text{and} \quad V = \frac{1}{\sqrt{s_1^2 + s_2^2}} = \frac{V_s}{\sqrt{1+m^2}} \qquad [2.128]$$

This wave, called *homogeneous*, is easily identified by the intersection between the vertical line of abscissa $s_1 = V_s^{-1}$ and one of the slowness curves (Figure 2.3);

– if m is purely imaginary: $m = im''$, the wave is *evanescent*:

$$\underline{u}(x_1, x_2, t) = a\underline{p}\, e^{-k_1 m'' x_2} e^{i(k_1 x_1 - \omega t)} \qquad [2.129]$$

its amplitude decreases from the interface. The equi-phase (x_1 = Const) and equi-amplitude (x_2 = Const) planes are orthogonal;

– if m is complex: $m = m' + im''$ with $\underline{s} = \underline{s}' + i\underline{s}''$:

$$\underline{s}' = (s_1, m's_1, 0)^t \quad \text{and} \quad \underline{s}'' = (0, m''s_1, 0)^t \qquad [2.130]$$

and the corresponding plane wave:

$$\underline{u}(x_1, x_2, t) = a\underline{p}\, e^{-k_1 m'' x_2} e^{i\omega(\underline{s}'\cdot\underline{x} - t)} \qquad [2.131]$$

propagates in the direction $\underline{n} = \underline{s}'/|\underline{s}'|$ with the phase velocity $V = |\underline{s}'|^{-1}$ and undergoes an attenuation along x_2. The equi-phase ($s_i' x_i$ = Const) and equi-amplitude (x_2 = Const) planes are generally not orthogonal. This wave is said to be *inhomogeneous*.

For each root m_K, that is, for a given wave, the tensor Γ_{il} is known. The resolution of system [2.124] provides the components of the polarization p_l as the eigenvector associated with the eigenvalue ρV_s^2. In the case of an inhomogeneous or evanescent wave, the polarization p_l is complex and the normalization condition is: $p_l p_l^* = 1$. The displacement field on the interface $x_2 = 0$ takes the general form:

$$u_l(x_1, x_2 = 0, t) = \sum_{K=1}^{3} [p_{lK} a_K + q_{lK} b_K] e^{i(k_1 x_1 - \omega t)} \quad [2.132]$$

In this expression, the coefficients a_K and b_K correspond to the amplitudes of the forward and backward waves, respectively. The components p_{lK} and q_{lK} are the polarization vectors of these modes.

2.5.2. *Expression for the stresses*

To apply the boundary conditions at the surface $x_2 = 0$, the traction vector σ_{i2} must be calculated. By replacing k_2 by mk_1 in the developments [2.18], the stresses:

$$\sigma_{i2} = ik_1 (c_{i21l} + mc_{i22l}) p_l a e^{ik_1 m x_2} e^{i(k_1 x_1 - \omega t)} \quad \text{with} \quad i = 1, 2, 3 \quad [2.133]$$

take the form:

$$\sigma_{i2} = ik_1 A_i a e^{ik_1 m x_2} e^{i(k_1 x_1 - \omega t)} \quad [2.134]$$

by writing:

$$A_i(m) = (c_{i211} + mc_{i221}) p_1 + (c_{i212} + mc_{i222}) p_2 + (c_{i213} + mc_{i223}) p_3 \quad [2.135]$$

In the following developments, the propagation factor $e^{i(k_1 x_1 - \omega t)}$, common to all waves, is omitted. The vector of normal stresses is, therefore, written as:

$$(\sigma_{12}, \sigma_{22}, \sigma_{32})^t = ik_1 (A_1, A_2, A_3)^t a e^{ik_1 m x_2} \quad [2.136]$$

The proportionality coefficients A_1, A_2, A_3 between the normal stresses and the amplitude a of the incident displacement depend on the surface velocity V_s through the intermediary of the parameter m and the polarizations p_l:

$$\begin{cases} A_1(m) = (C_{16} + mC_{66}) p_1 + (C_{66} + mC_{26}) p_2 + (C_{56} + mC_{46}) p_3 \\ A_2(m) = (C_{12} + mC_{26}) p_1 + (C_{26} + mC_{22}) p_2 + (C_{25} + mC_{24}) p_3 \\ A_3(m) = (C_{14} + mC_{46}) p_1 + (C_{46} + mC_{24}) p_2 + (C_{45} + mC_{44}) p_3 \end{cases} \quad [2.137]$$

We have shown that in each of the two media, assumed to be semi-infinite, three of the six roots of the characteristic equation (m_I with $I = 1, 2, 3$) correspond to the traveling waves (incident), the three other roots (m_J with $J = 1, 2, 3$) correspond to the backward waves (reflected by the interface). The coefficients $A_{iI} = A_i(m_I)$ are, therefore, arranged in a (3×3) matrix A, in which each column corresponds to an incident wave of polarization p_{lI}. With $\underline{q}$ denoting the normalized polarization vector for a reflected wave in the first solid, that is, the solid situated in $x_2 < 0$, the expressions for the coefficients B_i that associate the normal stresses with the amplitude b of this wave are obtained by replacing p_l by q_l in equation [2.137]. These are also arranged in a (3×3) matrix $\boldsymbol{B}$: $B_{iJ} = B_i(m_J)$. It must be noted that:

– the coefficients A_i or B_i are homogeneous to a stiffness $(C = \rho V^2)$. They are proportional to the impedances $(Z = \rho V)$, defined by: $\sigma_{i2} = -Z_i v$, where v is the complex amplitude of the particle velocity. Since $v = -i\omega a$, it comes:

$$\sigma_{i2} = i\omega Z_i a = ik_1 A_i a \quad \text{with} \quad A_i = \frac{\omega}{k_1} Z_i = V_s Z_i \quad [2.138]$$

– in the components A_{iJ} and B_{iJ}, the first index (lowercase letter) denotes a coordinate ($i = 1, 2, 3$), the second index (capital letter) denotes a mode, by convention: $J = 1$ for the L or QL wave, $J = 2$ for the T or QT_1 wave, and $J = 3$ for the TH or QT_2 wave, depending on the conditions of symmetry.

As for the displacement given by relation [2.132], the traction vector σ_{i2} on the interface $x_2 = 0$ takes the form:

$$\sigma_{i2}(x_1, x_2 = 0, t) = ik_1 \sum_{K=1}^{3} [A_{iK} a_K + B_{iK} b_K] e^{i(k_1 x_1 - \omega t)} \quad [2.139]$$

where the matrices $\boldsymbol{A}$ and $\boldsymbol{B}$ represent the traction vectors associated with the forward and backward waves, respectively.

2.5.3. *Sorting the solutions*

The incident wave is assumed to be homogeneous and its amplitude is denoted by $a_I^{(1)}$ if it originates from the first medium (Figure 2.16). The normal component J_2 of the Poynting vector must, therefore, be positive for the energy transported by this wave to be directed toward the interface. In the harmonic case and considering equation [2.133], the average value of J_2 is given by relation [2.39], that is,

$$\langle J_2 \rangle = \frac{1}{2} \omega k_1 |a|^2 \left(c_{i21l} + m c_{i22l} \right) p_i p_l \quad [2.140]$$

for a homogeneous wave. Three waves ($I = 1, 2, 3$) at most fulfill the condition $\langle J_2 \rangle > 0$.

If the incident wave comes from the second medium, located on the positive side of the x_2 axis, its amplitude is denoted by $a_I^{(2)}$ and the condition for existence is $J_2^{(2)} < 0$. The components of the polarization of an incident wave are denoted by $p_{lI}^{(1)}$ or $p_{lI}^{(2)}$. Depending on the medium from which the incident wave comes, a wave whose energy moves away from the surface ($J_2^{(1)} < 0$ or $J_2^{(2)} > 0$) is a homogeneous reflected or transmitted wave, marked by the index J and of normalized polarization $q_{lJ}^{(1)}$ or $q_{lJ}^{(2)}$.

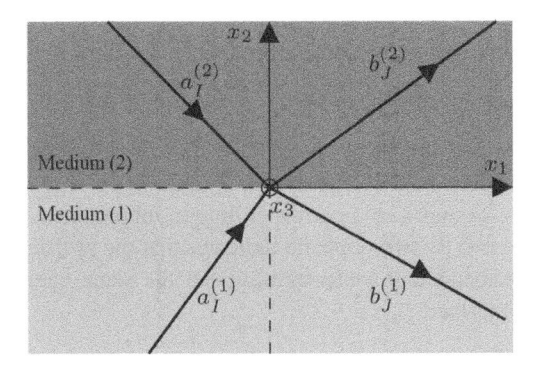

Figure 2.16. *Depending on the medium from which it comes, the amplitude of the incident wave is denoted by $a_I^{(1)}$ or $a_I^{(2)}$ ($I = 1, 2, 3$). Depending on the medium in which it propagates, the amplitude of a reflected or transmitted wave is denoted by $b_J^{(1)}$ or $b_J^{(2)}$ ($J = 1, 2, 3$)*

For an inhomogeneous wave, the component J_2 of the Poynting vector is always zero (Auld 1990). This type of wave, generated by the reflection-transmission of a homogeneous wave, transports the energy parallel to the interface. The previous selection criterion does not apply. Given the term $\exp(-k_1 m'' x_2)$ in the expression [2.131] for the mechanical displacement, a root with positive (negative) imaginary part m'' represents a wave whose amplitude decreases exponentially as x_2 increases (decreases). Since the roots of the characteristic equation are associated in complex conjugate pairs, only one inhomogeneous wave out of two fulfills the vanishing condition for the mechanical displacement at infinity.

2.5.4. *Considerations of symmetry*

The general expressions established in the previous section simplify if the medium possesses one or two binary axes of symmetry. Let us consider three cases according to the arrangement of the symmetry axes.

a) The *binary axis* (direct or inverse) *is parallel to the normal* Ox_2 to the interface (monoclinic symmetry). According to the results in section 1.3.1, the eight elastic constants c_{ijkl} with one or three indices equal to 2 are zero:

$$C_{14} = C_{24} = C_{34} = C_{16} = C_{26} = C_{36} = C_{45} = C_{56} = 0 \qquad [2.141]$$

The components [2.126] of the Christoffel tensor become:

$$\Gamma_{11} = C_{11} + C_{66}m^2, \quad \Gamma_{22} = C_{66} + C_{22}m^2, \quad \Gamma_{33} = C_{55} + C_{44}m^2$$

$$\Gamma_{12} = (C_{12} + C_{66})m, \quad \Gamma_{13} = C_{15} + C_{46}m^2, \quad \Gamma_{23} = (C_{46} + C_{25})m \quad [2.142]$$

and the characteristic equation is a third degree equation in m^2:

$$\alpha_6 m^6 + \alpha_4 m^4 + \alpha_2 m^2 + \alpha_0 = 0 \quad [2.143]$$

To each real positive (negative) root, $z = m^2$ corresponds two homogeneous (evanescent) waves. To a pair of complex conjugate roots z and z^* are associated four roots (m, m^*, $-m$ and $-m^*$) corresponding to inhomogeneous waves. The fact that the six roots are two by two opposite indicates that the propagation directions for the reflected and incident homogeneous waves of the same nature are symmetrical with respect to the interface.

The expressions [2.137] for the coefficients $A_i(m)$ giving the stresses σ_{i2} are also simplified:

$$\begin{cases} A_1(m) = C_{66}(mp_1 + p_2) + mC_{46}p_3 \\ A_2(m) = C_{12}p_1 + mC_{22}p_2 + C_{25}p_3 \\ A_3(m) = C_{46}(mp_1 + p_2) + mC_{44}p_3 \end{cases} \quad (A_2//Ox_2) \quad [2.144]$$

These considerations are valid for any orientation of the plane of incidence x_1x_2.

The polarization $\underline{q}_1$ of the reflected QL wave and the polarization $\underline{p}_1$ of the incident QL wave are antisymmetrical with respect to the x_1x_3 plane (Figure 2.17(a)). At normal incidence, the displacement is parallel to Ox_2, and in order that its reflection coefficient be equal to +1 on a free surface, we must choose: $q_{21} = p_{21}$, in the general case. This implies:

$$p_{11} = -q_{11}, \qquad p_{21} = q_{21} \quad \text{and} \quad p_{31} = -q_{31} \quad [2.145]$$

For the same reason, in the case of QT waves, polarized in the x_1x_3 plane at normal incidence, it is necessary to choose:

$$p_{1J} = q_{1J}, \quad p_{2J} = -q_{2J} \quad \text{and} \quad p_{3J} = q_{3J} \quad \text{for} \quad J = 2, 3 \quad [2.146]$$

The polarizations of the incident and reflected QT waves of the same nature are symmetrical with respect to the x_1x_3 plane (Figure 2.17(b)).

Considering the opposite sign of the ratio m for reflected waves, expressions [2.144] show that the coefficients B_{iJ} for the reflected waves, and the coefficients A_{iI} for the incident waves are linked:

$$\begin{pmatrix} A_{11} & A_{12} & A_{13} \\ A_{21} & A_{22} & A_{23} \\ A_{31} & A_{32} & A_{33} \end{pmatrix} = \begin{pmatrix} B_{11} & -B_{12} & -B_{13} \\ -B_{21} & B_{22} & B_{23} \\ B_{31} & -B_{32} & -B_{33} \end{pmatrix} \quad [2.147]$$

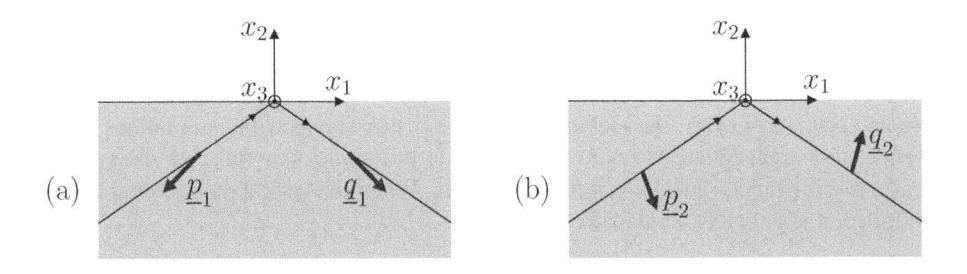

Figure 2.17. *Solid with monoclinic symmetry (binary axis parallel to Ox_2). Polarization of incident and reflected a) QL waves b) QT waves*

b) The *binary axis* is parallel to Ox_3, that is, perpendicular to the plane of incidence $x_1 x_2$. The elastic constants with one or three indices equal to 3: C_{14}, C_{24}, C_{15}, C_{25}, C_{46}, C_{56}, C_{34} and C_{35} are zero (section A2.5). This case has already been studied (section 2.2.2): the components Γ_{13} and Γ_{23} of the Christoffel tensor are canceled and the propagation of the QL and QT waves, polarized in the $x_1 x_2$ plane, is decoupled from the propagation of the TH wave, polarized along x_3. The normal stresses are also decoupled, since the coefficients [2.137] are separated into:

$$\begin{cases} A_{1J} = (C_{16} + m_J C_{66}) p_{1J} + (C_{66} + m_J C_{26}) p_{2J} \\ A_{2J} = (C_{12} + m_J C_{26}) p_{1J} + (C_{26} + m_J C_{22}) p_{2J} \end{cases} \quad [2.148]$$

for QL waves ($J = 1$) and QT waves ($J = 2$), and:

$$A_{33} = (C_{45} + m_3 C_{44}) p_3 \quad [2.149]$$

for the TH wave.

c) *Orthotropic symmetry*: if each of the Ox_2 and Ox_3 axes is parallel to a binary axis, the decoupling conditions given in section 2.2.2 are fulfilled. The coefficient α_6 is zero and the characteristic equation [2.143] is biquadratic. With respect to the stresses, only the terms that are common to developments [2.144] and [2.148] remain, that is, for the QL and QT waves:

$$\begin{cases} A_{1J} = C_{66}(m_J p_{1J} + p_{2J}) \\ A_{2J} = C_{12} p_{1J} + m_J C_{22} p_{2J} \end{cases} \quad J = 1, 2 \quad [2.150]$$

and for the TH wave, since $C_{45} = 0$:

$$A_{33} = m_3 C_{44} p_3 \quad [2.151]$$

2.5.5. *Reflection and transmission coefficients, interface waves*

In this section, we will express the reflection and transmission coefficients. These coefficients are, in fact, the elements of a matrix that connects the amplitudes of all reflected or transmitted waves to the amplitude of the incident wave. In the general case, this diffraction matrix is a (6 × 6) matrix. When the decoupling conditions are fulfilled, it is only a (4 × 4) matrix.

2.5.5.1. *Diffraction matrix*

In the case of two rigidly bonded solids, and considering the six possible waves in each medium, the boundary conditions at $x_2 = 0$ for the mechanical displacements u_l are written as:

$$\sum_{M=1}^{3} p_{lM}^{(1)} a_M^{(1)} + \sum_{M=1}^{3} q_{lM}^{(1)} b_M^{(1)} = \sum_{M=1}^{3} p_{lM}^{(2)} a_M^{(2)} + \sum_{M=1}^{3} q_{lM}^{(2)} b_M^{(2)} \quad l=1,2,3 \quad [2.152]$$

and for the normal stresses σ_{i2}:

$$\sum_{M=1}^{3} A_{iM}^{(1)} a_M^{(1)} + \sum_{M=1}^{3} B_{iM}^{(1)} b_M^{(1)} = \sum_{M=1}^{3} A_{iM}^{(2)} a_M^{(2)} + \sum_{M=1}^{3} B_{iM}^{(2)} b_M^{(2)} \quad i=1,2,3 \quad [2.153]$$

It must be recalled that the first index (i or l) indicates the coordinate; the second (M) specifies the mode. Each member of these equations represents the displacements or stresses on either side of the interface $x_2 = 0$. These relations are written in the matrix form as:

$$\begin{cases} \boldsymbol{P}^{(1)} \underline{a}^{(1)} + \boldsymbol{Q}^{(1)} \underline{b}^{(1)} = \boldsymbol{P}^{(2)} \underline{a}^{(2)} + \boldsymbol{Q}^{(2)} \underline{b}^{(2)} \\ \boldsymbol{A}^{(1)} \underline{a}^{(1)} + \boldsymbol{B}^{(1)} \underline{b}^{(1)} = \boldsymbol{A}^{(2)} \underline{a}^{(2)} + \boldsymbol{B}^{(2)} \underline{b}^{(2)} \end{cases} \quad [2.154]$$

where $\underline{a}^{(1)}$ and $\underline{a}^{(2)}$ are the vectors of amplitudes of the incident waves, $\underline{b}^{(1)}$ and $\underline{b}^{(2)}$ are the vectors of amplitudes of the reflected or transmitted waves. $\boldsymbol{P}^{(1)}$ and $\boldsymbol{P}^{(2)}$, on the one hand, and $\boldsymbol{Q}^{(1)}$ and $\boldsymbol{Q}^{(2)}$, on the other hand, represent the (3 × 3) polarization matrices for these waves. $\boldsymbol{A}^{(1)}$ and $\boldsymbol{B}^{(1)}$ are the characteristic matrices $[A_{iJ}]$ and $[B_{iJ}]$ of the medium occupying the half-space $x_2 < 0$ ($\boldsymbol{A}^{(2)}$ and $\boldsymbol{B}^{(2)}$ for the medium occupying the half-space $x_2 > 0$).

By introducing the (6 × 6) matrices:

$$\boldsymbol{N} = \begin{pmatrix} \boldsymbol{Q}^{(1)} & -\boldsymbol{Q}^{(2)} \\ \boldsymbol{B}^{(1)} & -\boldsymbol{B}^{(2)} \end{pmatrix} \quad \text{and} \quad \boldsymbol{M} = \begin{pmatrix} \boldsymbol{P}^{(1)} & -\boldsymbol{P}^{(2)} \\ \boldsymbol{A}^{(1)} & -\boldsymbol{A}^{(2)} \end{pmatrix} \quad [2.155]$$

the amplitudes $\underline{b}^{(1)}$ and $\underline{b}^{(2)}$ are expressed as functions of the amplitudes of the incident waves $\underline{a}^{(1)}$ and $\underline{a}^{(2)}$:

$$\boldsymbol{N} \begin{pmatrix} \underline{b}^{(1)} \\ \underline{b}^{(2)} \end{pmatrix} = -\boldsymbol{M} \begin{pmatrix} \underline{a}^{(1)} \\ \underline{a}^{(2)} \end{pmatrix} \quad \text{or} \quad \begin{pmatrix} \underline{b}^{(1)} \\ \underline{b}^{(2)} \end{pmatrix} = -\boldsymbol{N}^{-1} \boldsymbol{M} \begin{pmatrix} \underline{a}^{(1)} \\ \underline{a}^{(2)} \end{pmatrix} \quad [2.156]$$

by using the (6 × 6) diffraction matrix D:

$$D = -N^{-1}M = \begin{pmatrix} R^{(1)} & T^{(2)} \\ T^{(1)} & R^{(2)} \end{pmatrix} \quad [2.157]$$

In the general case, this matrix is composed of nine reflection coefficients, the elements of the (3 × 3) matrix $R^{(1)}$, when the incident wave originates from the first medium, and of nine other reflection coefficients, elements of the matrix $R^{(2)}$, when the incident wave comes from the second medium. In the same way, the elements of the matrices $T^{(1)}$ and $T^{(2)}$ represent the 18 transmission coefficients. Let us investigate different, increasingly simple configurations.

2.5.5.2. Reflection on a free surface

In the case of a solid in contact with the air or any other low-pressure fluid, there is no transmitted wave and the only continuity condition to satisfy is the cancelation of the traction vector at all points on the free surface. The field of normal stresses is the sum of that of the incident wave (index I) and that of the three reflected waves (there is only one medium, which need not be indexed):

$$\sigma_{i2} = ik_1 A_{iI} a_I e^{ik_1 m_I x_2} + ik_1 \sum_{J=1}^{3} B_{iJ} b_J e^{ik_1 m_J x_2}, \quad i = 1, 2, 3 \quad [2.158]$$

The relations imposed by the boundary conditions $\sigma_{i2} = 0$ at $x_2 = 0$ are written in matrix form as:

$$\begin{pmatrix} B_{11} & B_{12} & B_{13} \\ B_{21} & B_{22} & B_{23} \\ B_{31} & B_{32} & B_{33} \end{pmatrix} \begin{pmatrix} b_1 \\ b_2 \\ b_3 \end{pmatrix} = - \begin{pmatrix} A_{1I} \\ A_{2I} \\ A_{3I} \end{pmatrix} a_I \quad [2.159]$$

depending on whether the incident wave is quasi-longitudinal ($I = 1$), fast quasi-transverse ($I = 2$) or slow quasi-transverse ($I = 3$). This result can also be deduced from the general relations [2.154] with $\underline{b}^{(2)} = \underline{a}^{(2)} = \underline{0}$: $B\underline{b} = -A\underline{a}$. The resolution of this linear system provides the amplitudes of the reflected waves as functions of the amplitude of the incident wave. According to Cramer's rule, all reflection and conversion coefficients have the same denominator, the determinant Δ of the matrix B.

– *Crystal of monoclinic symmetry*: if the Ox_3 axis is parallel to a binary axis, the TH wave does not create any QL or QT wave polarized in the sagittal plane. By superimposing the two incident QL and QT waves, the conditions $\sigma_{12} = 0$ and $\sigma_{22} = 0$ over the free surface lead to a (2 × 2) system:

$$\begin{pmatrix} B_{11} & B_{12} \\ B_{21} & B_{22} \end{pmatrix} \begin{pmatrix} b_1 \\ b_2 \end{pmatrix} = - \begin{pmatrix} A_{11} & A_{12} \\ A_{21} & A_{22} \end{pmatrix} \begin{pmatrix} a_1 \\ a_2 \end{pmatrix} \quad [2.160]$$

– *Crystal of orthorhombic symmetry*: if the Ox_2 axis is also parallel to another binary axis, the relations [2.147] between the components B_{iJ} and A_{iJ} apply:

$$\begin{pmatrix} B_{11} & B_{12} \\ B_{21} & B_{22} \end{pmatrix} \begin{pmatrix} b_1 \\ b_2 \end{pmatrix} = - \begin{pmatrix} B_{11} & -B_{12} \\ -B_{21} & B_{22} \end{pmatrix} \begin{pmatrix} a_1 \\ a_2 \end{pmatrix} \qquad [2.161]$$

The resolution of this system leads to:

$$\begin{pmatrix} b_1 \\ b_2 \end{pmatrix} = -\frac{1}{\Delta} \begin{pmatrix} B_{11}B_{22} + B_{12}B_{21} & -2B_{22}B_{12} \\ -2B_{11}B_{21} & B_{11}B_{22} + B_{12}B_{21} \end{pmatrix} \begin{pmatrix} a_1 \\ a_2 \end{pmatrix} \qquad [2.162]$$

where the determinant Δ is equal to:

$$\Delta = B_{11}B_{22} - B_{12}B_{21} \qquad [2.163]$$

Formally, the reflection coefficient of the quasi-longitudinal wave ($r_{LL} = b_1/a_1$ when $a_2 = 0$) is the same as that of the quasi-transverse vertical wave ($r_{TT} = b_2/a_2$ when $a_1 = 0$):

$$r_{LL} = \frac{B_{12}B_{21} + B_{11}B_{22}}{B_{12}B_{21} - B_{11}B_{22}} = r_{TT} \qquad [2.164]$$

The conversion factors of the QL wave into a QT wave and *vice versa* (with the first index indicating the incident wave), are, respectively, equal to:

$$r_{LT} = \frac{b_2}{a_1} = \frac{2B_{11}B_{21}}{B_{11}B_{22} - B_{12}B_{21}} \quad \text{and} \quad r_{TL} = \frac{b_1}{a_2} = \frac{2B_{22}B_{12}}{B_{11}B_{22} - B_{12}B_{21}} \qquad [2.165]$$

2.5.5.3. *Interface and surface waves*

Waves are likely to propagate at the interface between two solids, or on the free surface of a solid; in this section, we establish their condition of existence.

– *Stoneley wave*: at the interface between two solids, and in the absence of an incident wave ($\underline{a}^{(1)} = \underline{a}^{(2)} = 0$), the homogeneous system [2.156] has a non-trivial solution ($\underline{b}^{(1)}$ and $\underline{b}^{(2)} \neq 0$) if the determinant of the matrix $\boldsymbol{N}$ cancels. Considering the identity:

$$\det \begin{pmatrix} \boldsymbol{E} & \boldsymbol{F} \\ \boldsymbol{G} & \boldsymbol{H} \end{pmatrix} = \det [\boldsymbol{E}] \det [\boldsymbol{H}\boldsymbol{F}^{-1} - \boldsymbol{G}\boldsymbol{E}^{-1}] \det [\boldsymbol{F}] \qquad [2.166]$$

and since the determinants of the polarization matrices, $\boldsymbol{Q}^{(1)}$ and $\boldsymbol{Q}^{(2)}$, are non-zero, a wave propagates on the interface between two solids if the characteristic equation:

$$\det \left[\boldsymbol{B}^{(2)} \left(\boldsymbol{Q}^{(2)} \right)^{-1} - \boldsymbol{B}^{(1)} \left(\boldsymbol{Q}^{(1)} \right)^{-1} \right] = 0 \qquad [2.167]$$

has a real root V_s. The conditions for the existence of this wave, discovered by Stoneley, are explained in the case of two isotropic solids in section 3.2.2.

– *Rayleigh wave*: on a free surface and in the absence of any incident wave ($a_I = 0$), the homogeneous system [2.159] has a non-trivial solution (b_1, b_2, $b_3 \neq 0$) if the determinant of the matrix B cancels. This determinant depends on the elastic constants of the material and on the surface velocity V_s. If the secular equation:

$$\det [B_{iJ}] = \Delta(V_s) = 0 \qquad [2.168]$$

has a real root $V_s = V_R$, a wave accompanied by a mechanical displacement propagates without attenuation on the free surface of the solid, at the velocity V_R. This surface wave was discovered by *Lord Raleigh* in the case of an isotropic solid. This root, which cancels the determinant Δ, is a pole for all the reflection coefficients [2.164] and [2.165].

2.5.6. *Interface between an orthotropic solid and an isotropic solid*

The medium (1), considered as isotropic, is characterized by the bulk wave velocities V_T and V_L. The *incident transverse vertical* wave, propagating in the direction θ_T, can give rise to a reflected longitudinal wave and a reflected transverse wave. In medium (2) of orthotropic symmetry, this wave can give rise to a quasi-longitudinal wave and a quasi-transverse wave (Figure 2.18(a)). Let us recall that there exists a critical angle $\theta_c^L = \arcsin(V_T/V_L)$ beyond which the longitudinal wave becomes evanescent in the isotropic solid.

It was shown in section 2.2.2.1 that the components Γ_{13} and Γ_{23} of the Christoffel tensor are zero in the XY plane of an orthotropic solid and that the other components are given by the relations [2.142]. The characteristic equation [2.124] is written as:

$$(\Gamma_{11} - \rho V_s^2)(\Gamma_{22} - \rho V_s^2) - \Gamma_{12}^2 = 0 \qquad [2.169]$$

where V_s is the surface velocity defined by $V_s = V_T/\sin\theta_T$. In the previous sections, the analysis was carried out with the velocity V_s as the variable and the dimensionless ratio $m = s_2/s_1$ as the unknown. In order to follow, with the help of the slowness curves, the evolution of the roots of the characteristic equation depending on the angle of incidence θ_T, it is preferable to choose the component $s_1 = (\sin\theta_T)/V_T$ of the slowness, as the variable, and the other component $s_2 = ms_1$ as the unknown. With these quantities and by multiplying by s_1^4, the characteristic equation [2.169] takes the form:

$$\left(C_{11}s_1^2 + C_{66}s_2^2 - \rho\right)\left(C_{66}s_1^2 + C_{22}s_2^2 - \rho\right) - \left(C_{12} + C_{66}\right)^2 s_1^2 s_2^2 = 0 \qquad [2.170]$$

This is a biquadratic polynomial equation in s_2^2:

$$\alpha_4 s_2^4 - \alpha_2 s_2^2 + \alpha_0 = 0 \qquad [2.171]$$

with:

$$\begin{cases} \alpha_4 = C_{22} C_{66} \\ \alpha_2 = C_{22}\left(\rho - s_1^2 C_{11}\right) + C_{66}\left(\rho - s_1^2 C_{66}\right) + (C_{12} + C_{66})^2 s_1^2 \\ \alpha_0 = \left(\rho - s_1^2 C_{11}\right)\left(\rho - s_1^2 C_{66}\right) \qquad \text{with} \qquad C_{11} > C_{66} \end{cases} \qquad [2.172]$$

The two roots s_{21}^2 and s_{22}^2 depend on the slowness s_1 through the intermediary of the coefficients α_2 and α_0:

$$2\alpha_4 s_{21}^2 = \alpha_2 - \sqrt{\alpha_2^2 - 4\alpha_0 \alpha_4} \quad \text{and} \quad 2\alpha_4 s_{22}^2 = \alpha_2 + \sqrt{\alpha_2^2 - 4\alpha_0 \alpha_4} \quad [2.173]$$

The coefficient α_0 cancels at a first time for the value $s_1 = \sqrt{\rho/C_{11}}$, corresponding to the critical angle θ_c^{QL} of the quasi-longitudinal wave in the orthotropic medium, and at a second time for the value $\sqrt{\rho/C_{66}}$ (indicated by an arrow in Figure 2.18(a)), corresponding to the angle θ_1 in Figure 2.18(b). Let us follow, on this figure, the evolution of the intensity reflection and transmission coefficients. For:

– $\theta_T = 0$, the incident transverse wave is partially reflected ($R_T \neq 0$) and partially transmitted ($T_{QT_1} \neq 0$) as a transverse wave;

– $0 < \theta_T < \theta_c^L$, an increasing part of the energy is reflected in the form of a longitudinal wave ($R_L \neq 0$);

– $\theta_c^L < \theta_T < \theta_c^{QL}$, the longitudinal wave is evanescent in the isotropic solid ($R_L = 0$) and the transmission as a quasi-longitudinal wave significantly increases;

– $\theta_c^{QL} < \theta_T < \theta_1$, that is, $\rho/C_{11} < s_1^2 < \rho/C_{66}$, the coefficient α_0 is negative and $\sqrt{\alpha_2^2 + 4|\alpha_0|\alpha_4}$ is larger than α_2, so that s_{21}^2 is negative. The acceptable root $s_{21} = i|s_{21}|$ corresponds to an evanescent QL wave, which does not transport energy ($T_{QL} = 0$). s_{22}^2 is positive and the real positive root s_{22} corresponds to the homogeneous quasi-transverse wave QT_1;

– $\theta_1 < \theta_T < \theta_c^{QT}$, that is, $s_1^2 > \rho/C_{66} > \rho/C_{11}$, the coefficient α_0 again becomes positive, as well as s_{21}^2 and s_{22}^2; the two real roots s_{21} and s_{22} correspond to the two quasi-transverse waves QT_1 and QT_2, one of which (QT_2) radiates toward the interface. Its wave vector is refracted, its energy vector is reflected. In this case, the root s_{22} must be taken to be negative;

$-\theta_T > \theta_c^{QT}$, the critical angle for the quasi-transverse wave corresponding to the maximum value of s_1 in the orthotropic solid, for which $s_{21} = s_{22}$, that is $\alpha_2^2 = 4\alpha_0\alpha_4$. The two roots:

$$2\alpha_4 s_{21}^2 = \alpha_2 - i\sqrt{4\alpha_0\alpha_4 - \alpha_2^2} \quad \text{and} \quad 2\alpha_4 s_{22}^2 = \alpha_2 + i\sqrt{4\alpha_0\alpha_4 - \alpha_2^2} \quad [2.174]$$

are complex. The slownesses s_{21} and s_{22} with positive imaginary parts correspond to two inhomogeneous waves, whose amplitude decreases with x_2 and whose polarizations are complex.

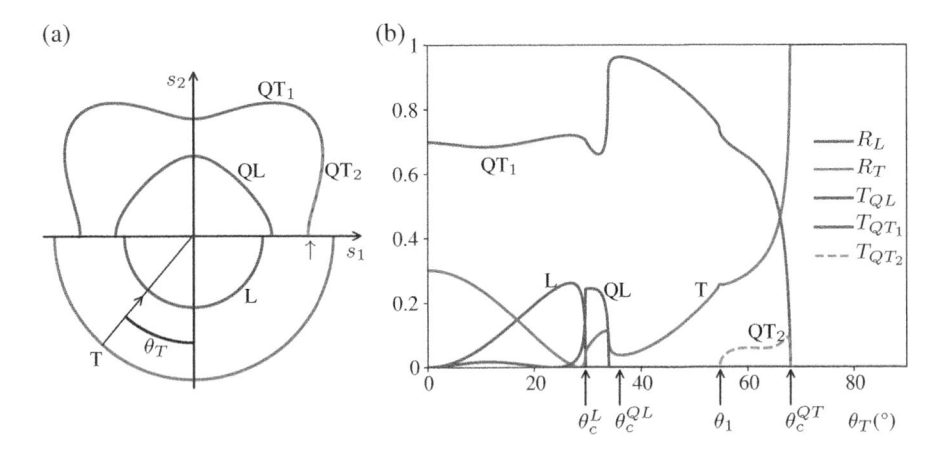

Figure 2.18. a) Slowness curves of the reflected and transmitted waves generated by a transverse vertical wave at the interface between a duralumin and an austenitic steel half-spaces. b) Intensity reflection coefficients for the longitudinal and transverse waves and intensity transmission coefficients for the quasi-longitudinal and quasi-transverse waves. For a color version of this figure, see www.iste.co.uk/royer/waves1.zip

3

Surface Waves and Interface Waves

Unlike bulk waves studied in the previous chapters, surface waves and interface waves propagate on the free surface or at the interface between two semi-infinite media. In both configurations, there is no finite dimension so that the velocity of the wave is a characteristic of one or both media, independent of the frequency: their propagation is not dispersive. Further, there exist a finite number of solutions. These two properties differentiate these waves from the waves that propagate in a medium where at least one of the lateral dimensions is finite, for which the successive reflections on the walls of the guide give rise to an infinite number of dispersive modes. These guided waves are the subject of Chapter 4.

In the last section of Chapter 2, at several instances, it appears that a wave could propagate on the surface of a solid. The most important of these waves is the one discovered by Lord Rayleigh in 1885 (Rayleigh 1885). In the middle of the 19th century, it was known that earthquakes propagate in the form of longitudinal and transverse waves. These vibrations explained the first and second echo recorded by seismographs. The late arrival of a third stronger echo remained unexplained for a long time, until Lord Rayleigh demonstrated the existence of a wave that propagates on the free surface of an isotropic solid more slowly than the transverse bulk wave. Subsequently, other propagation mechanisms were studied in specific structures: layer on a substrate, interface between two solids or between a solid and a fluid, free plate. Other geophysicist's names were thus added to Rayleigh: Love, Stoneley, Scholte and Lamb.

The first section in this chapter deals with waves that propagate on the free surface of a solid, first isotropic (Rayleigh wave) then anisotropic (generalized Rayleigh wave), and finally piezoelectric. While the isotropic case can be quickly resolved using the method of potentials, the developments in the anisotropic case can only be numerically solved, except for specific symmetries. The analysis method requires a formalism similar to that developed in section 2.5. It can be easily generalized to the case of a piezoelectric medium by adding a fourth component (the electric potential) to the mechanical displacement and by imposing electrical boundary conditions on the free surface.

When the surface of the solid is covered by a fluid, a Rayleigh wave can still propagate, but with an attenuation, due to the radiation of a part of the acoustic energy in the fluid. An interface wave (Scholte wave) also appears, which transports, without attenuation, a fraction of the acoustic energy in the fluid, and a part in the solid. In this section, we will also show that under certain conditions, a wave, discovered by Stoneley, can propagate without attenuation at the interface between two isotropic solids. When the decoupling conditions are fulfilled, these waves are polarized in the plane of incidence, called here the sagittal plane.

A transverse horizontal (TH) wave, which does not create any mechanical traction on the surface, is not retained if this surface is free. Bleustein and Gulyaev have shown, independently, that when the solid is piezoelectric (section 3.3), it is possible to confine a TH wave more or less close to the surface, depending on the electrical boundary conditions.

3.1. Surface waves

These waves obey the propagation equation and the mechanical conditions, as well as electrical conditions, if any, imposed by the free surface. In the case of a crystal, the solutions can generally only be obtained numerically; however, in the case of an isotropic solid (first studied) they can be derived analytically.

3.1.1. *Isotropic solid: Rayleigh wave*

The reference frame is shown in Figure 3.1. Ox_2 is the normal to the free surface, directed toward the bulk of the solid, and Ox_1 is the propagation direction.

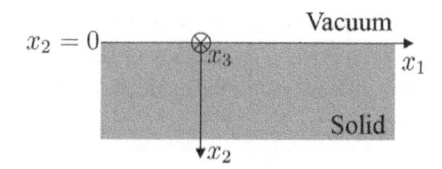

Figure 3.1. *Disposition of the axes*

3.1.1.1. *Method of potentials*

From the results of section 1.2.2.1, the mechanical displacement can be derived from a scalar potential ϕ_L and a vector potential $\underline{\psi}$. In the harmonic case, these potentials satisfy the Helmholtz equations:

$$\Delta \phi_L + k_L^2 \phi_L = 0 \quad \text{and} \quad \Delta \underline{\psi} + k_T^2 \underline{\psi} = 0 \quad [3.1]$$

with $k_L = \omega/V_L$ and $k_T = \omega/V_T$. Neglecting the diffraction along x_3 ($\partial/\partial x_3 = 0$), the components of the displacement of a wave propagating along x_1 ($\partial/\partial x_1 = ik$) are expressed as functions of the potentials using the Helmholtz decomposition [1.77]:

$$\begin{cases} u_1 = \dfrac{\partial \phi_L}{\partial x_1} + \dfrac{\partial \psi_3}{\partial x_2} = ik\phi_L + \dfrac{\partial \psi_3}{\partial x_2} \\[6pt] u_2 = \dfrac{\partial \phi_L}{\partial x_2} - \dfrac{\partial \psi_3}{\partial x_1} = \dfrac{\partial \phi_L}{\partial x_2} - ik\psi_3 \\[6pt] u_3 = \dfrac{\partial \psi_2}{\partial x_1} - \dfrac{\partial \psi_1}{\partial x_2} = ik\psi_2 - \dfrac{\partial \psi_1}{\partial x_2} \end{cases} \quad [3.2]$$

Since the components u_1 and u_2 only depend on the scalar potential ϕ_L and on the component $\psi_3 = \phi_T$ of the vector potential, the Rayleigh wave polarized in the sagittal plane $x_1 x_2$ (plane defined by the normal to the surface and the propagation direction) propagates independently from the transverse horizontal (TH) wave, polarized along the x_3 axis, whose displacement is expressed as a function of ψ_1 and ψ_2. Substituting the potentials:

$$\phi_M(x_1, x_2, t) = f_M(x_2) e^{i(kx_1 - \omega t)} \quad \text{with} \quad M = L, T \quad [3.3]$$

into equations [3.1], the functions $f_L(x_2)$ and $f_T(x_2)$ satisfy the equation:

$$\frac{d^2 f_M(x_2)}{dx_2^2} + \left(k_M^2 - k^2\right) f_M(x_2) = 0 \quad [3.4]$$

In order for the wave to be confined close to the surface, it is necessary that $k > k_T > k_L$: the phase velocity V of the Rayleigh wave is, therefore, less than V_T and V_L. By introducing the decay factors:

$$\alpha_M = \sqrt{k^2 - k_M^2} = k\chi_M \quad \text{or} \quad \chi_M = \sqrt{1 - \frac{V^2}{V_M^2}} \quad [3.5]$$

the solutions to equations [3.4] take the form:

$$f_M(x_2) = a_M e^{-\alpha_M x_2} + b_M e^{\alpha_M x_2} \quad \text{with} \quad M = L, T \quad [3.6]$$

The solutions associated with the amplitudes b_L and b_T are not physically admissible, because they diverge with depth. Consequently, the amplitudes b_L and b_T

are necessarily zero. Considering relations [3.2], the components of the displacement of the Rayleigh wave are:

$$\begin{cases} u_1(x_1, x_2, t) = [ika_L e^{-\alpha_L x_2} - \alpha_T a_T e^{-\alpha_T x_2}] e^{i(kx_1 - \omega t)} \\ u_2(x_1, x_2, t) = [-\alpha_L a_L e^{-\alpha_L x_2} - ika_T e^{-\alpha_T x_2}] e^{i(kx_1 - \omega t)} \end{cases} \quad [3.7]$$

The coefficients a_L and a_T are determined by the *boundary conditions* on the free surface $x_2 = 0$, which apply to the stresses:

$$\sigma_{12} = \mu \left(\frac{\partial u_1}{\partial x_2} + \frac{\partial u_2}{\partial x_1} \right) \quad [3.8]$$

that is:

$$\sigma_{12} = \mu \left[-2ik\alpha_L a_L e^{-\alpha_L x_2} + (k^2 + \alpha_T^2) a_T e^{-\alpha_T x_2} \right] e^{i(kx_1 - \omega t)} \quad [3.9]$$

and:

$$\sigma_{22} = \lambda \frac{\partial u_1}{\partial x_1} + (\lambda + 2\mu) \frac{\partial u_2}{\partial x_2} \quad [3.10]$$

that is:

$$\sigma_{22} = \left[\left(-\lambda k^2 + (\lambda + 2\mu) \alpha_L^2 \right) a_L e^{-\alpha_L x_2} + 2i\mu k \alpha_T a_T e^{-\alpha_T x_2} \right] e^{i(kx_1 - \omega t)}$$

$$[3.11]$$

Since $(\lambda + 2\mu) k_L^2 = \rho \omega^2 = \mu k_T^2$ and $\alpha_L^2 = k^2 - k_L^2$, the multiplicative factor of a_L:

$$(\lambda + 2\mu) \alpha_L^2 - \lambda k^2 = 2\mu k^2 - \mu k_T^2 = \mu(k^2 + \alpha_T^2) \quad [3.12]$$

is proportional to the Lamé constant μ, so that the stress [3.11] is given by:

$$\sigma_{22} = \mu \left[\left(k^2 + \alpha_T^2 \right) a_L e^{-\alpha_L x_2} + 2ik\alpha_T a_T e^{-\alpha_T x_2} \right] e^{i(kx_1 - \omega t)} \quad [3.13]$$

The boundary conditions $\sigma_{12} = \sigma_{22} = 0$ at $x_2 = 0$ are written, in matrix form, as:

$$\begin{pmatrix} -2ik\alpha_L & k^2 + \alpha_T^2 \\ k^2 + \alpha_T^2 & 2ik\alpha_T \end{pmatrix} \begin{pmatrix} a_L \\ a_T \end{pmatrix} = \begin{pmatrix} 0 \\ 0 \end{pmatrix} \quad [3.14]$$

The characteristic equation of the Rayleigh wave thus corresponds to the cancelation of the determinant of this matrix:

$$d_R = \left(k^2 + \alpha_T^2 \right)^2 - 4k^2 \alpha_L \alpha_T = 0 \quad [3.15]$$

which, considering relations [3.5], expresses in terms of the decay factors χ_L and χ_T as

$$\left(1 + \chi_T^2\right)^2 = 4\chi_L\chi_T \qquad [3.16]$$

Upon squaring this, the equation takes the form:

$$\left(2 - \frac{V^2}{V_T^2}\right)^4 = 16\left(1 - \frac{V^2}{V_L^2}\right)\left(1 - \frac{V^2}{V_T^2}\right) \qquad [3.17]$$

giving the velocity of the Rayleigh wave. Since the zero order term disappears, a simplification by V^2 shows that the *Rayleigh equation* is, indeed, a third degree equation in V^2. Defining the ratio $R = V^2/V_T^2$, it is written in a simpler form:

$$R^3 - 8(R-1)\left(R - 2 + 2\kappa^2\right) = 0 \quad \text{with} \quad \kappa = \frac{V_T}{V_L} \qquad [3.18]$$

This equation has only one positive root less than unity ($0 < V_R < V_T$) when the ratio κ varies between the limits 0 and $\sqrt{2}/2$ that are allowed for an isotropic solid. Since the root R lies between 0.912 and 0.764, the relative velocity $V_R/V_T = \sqrt{R}$ varies from 0.955 to 0.874 (Figure 3.2). The velocity V_R being independent of the frequency, the propagation of the Rayleigh wave *is not dispersive*. The fact that V_R is always smaller than V_T can be interpreted by the absence of matter above the free surface, which is equivalent to reduce the stiffness constants. A useful approximation was proposed by Bergmann and cited by Viktorov (1967):

$$\frac{V_R}{V_T} \approx \frac{0.718 - \kappa^2}{0.75 - \kappa^2} \qquad [3.19]$$

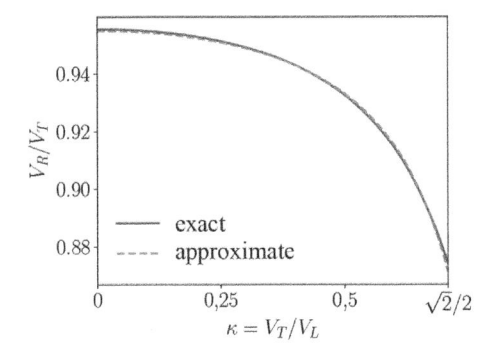

Figure 3.2. *Variation of the Rayleigh wave velocity V_R with the ratio of the bulk wave velocities $\kappa = V_T/V_L$. For a color version of this figure, see www.iste.co.uk/royer/waves1.zip*

For the particular value of the Poisson's ratio $\nu = 0.25$, corresponding to $\lambda = \mu$, it is easy to verify that with $\kappa^2 = 1/3$, equation [3.18] admits the solution:

$$R = \frac{4}{3+\sqrt{3}} \quad \text{hence} \quad V_R = \frac{2V_T}{\sqrt{3+\sqrt{3}}} = 0.9194 V_T \qquad [3.20]$$

It must be noted that by defining:

$$k_1 = k, \quad k_{2L} = \sqrt{k_L^2 - k_1^2} = i\alpha_L \quad \text{and} \quad k_{2T} = \sqrt{k_T^2 - k_1^2} = i\alpha_T \qquad [3.21]$$

the determinant [3.15], which is written as:

$$d_R = \left(k_1^2 - k_{2T}^2\right)^2 + 4k_1^2 k_{2L} k_{2T} \qquad [3.22]$$

is the same as the denominator [2.86] of the reflection coefficients [2.85] and [2.91] of longitudinal and transverse waves. The root $k_1 = \omega/V_R$ of the characteristic equation $d_R = 0$, corresponding to the Rayleigh wave velocity V_R, is a pole for all the reflection coefficients on the free surface of an isotropic solid. This property was demonstrated, in a general way, in section 2.5.5.

3.1.1.2. *Mechanical displacement and stresses*

Considering equations [3.14] and [3.15], the amplitudes a_L and a_T are proportional:

$$a_T = i\sqrt{\frac{\alpha_L}{\alpha_T}} a_L \qquad [3.23]$$

and the mechanical displacement [3.7] can be written as:

$$\underline{u}(x_1, x_2, t) = \left[\begin{pmatrix} ik \\ -\alpha_L \end{pmatrix} e^{-\alpha_L x_2} + \begin{pmatrix} -i\alpha_T \\ k \end{pmatrix} \sqrt{\frac{\alpha_L}{\alpha_T}} e^{-\alpha_T x_2} \right] a_L e^{i(kx_1 - \omega t)}$$

$$[3.24]$$

The expressions for the real parts of the displacement components and of the stresses of the Rayleigh wave are as follows:

$$\begin{cases} \text{Re}[u_1] = U_1(x_2) \sin(kx_1 - \omega t) \\ \text{Re}[u_2] = U_2(x_2) \cos(kx_1 - \omega t) \end{cases} \text{and} \begin{cases} \text{Re}[\sigma_{12}] = \Sigma_{12}(x_2) \sin(kx_1 - \omega t) \\ \text{Re}[\sigma_{22}] = \Sigma_{22}(x_2) \cos(kx_1 - \omega t) \end{cases}$$

$$[3.25]$$

with:

$$\begin{cases} U_1(x_2) = a_L \left[-ke^{-\alpha_L x_2} + \sqrt{\alpha_L \alpha_T} e^{-\alpha_T x_2} \right] \\ U_2(x_2) = a_L \sqrt{\dfrac{\alpha_L}{\alpha_T}} \left[-\sqrt{\alpha_L \alpha_T} e^{-\alpha_L x_2} + ke^{-\alpha_T x_2} \right] \\ \Sigma_{12}(x_2) = 2\mu k \alpha_L a_L \left[e^{-\alpha_L x_2} - e^{-\alpha_T x_2} \right] \\ \Sigma_{22}(x_2) = 2\mu k \sqrt{\alpha_L \alpha_T} a_L \left[e^{-\alpha_L x_2} - e^{-\alpha_T x_2} \right] \end{cases}$$

[3.26]

The displacement components are in phase quadrature and, since the decay factors α_L and α_T are proportional to the wave number $k = k_R = \omega/V_R$, they depend on the dimensionless variable $X_2 = k_R x_2$, that is, on the ratio of the depth to the Rayleigh wavelength: x_2/λ_R. The particles describe an ellipse that is deformed below the surface, because the amplitudes vary differently with the depth x_2. From retrograde, at the surface, the polarization becomes direct at a depth of about $0.2\lambda_R$, at which the sign of the longitudinal component changes. In Figure 3.3(a), the normal component U_2 and tangential component U_1 of the mechanical displacement, calculated at a fixed position x_1 and at a given time t for a duralumin half-space, are represented as a function of the ratio of depth to wavelength x_2/λ_R.

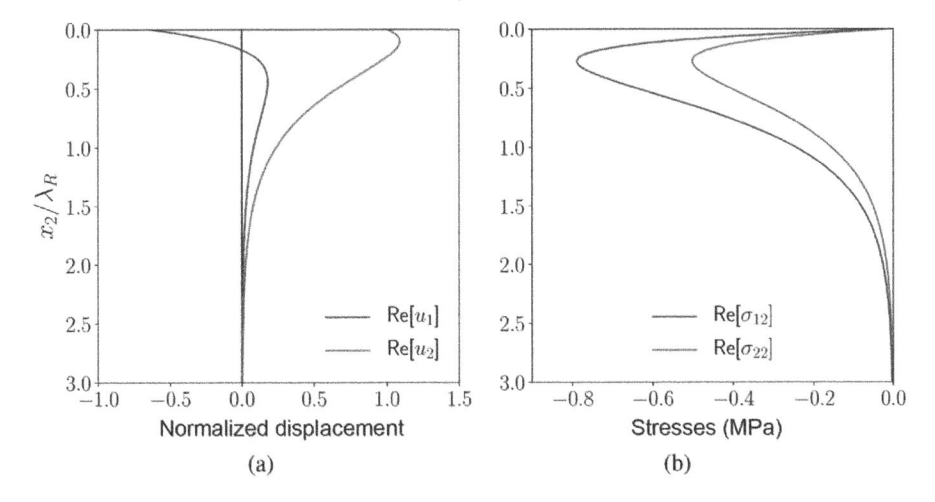

Figure 3.3. a) Displacement and b) stresses of the Rayleigh wave in the depth of a duralumin half-space. The stresses are drawn for an amplitude $U_2(0) = 1$ nm and an angular frequency $\omega = 10^8$ rad/s ($f = 16$ MHz). For a color version of this figure, see www.iste.co.uk/royer/waves1.zip

The stresses σ_{12} and σ_{22} are also in phase quadrature, but they vary in the same way with the depth (Figure 3.3(b)). Their ratio is, therefore, independent of x_2; it is equal to the ratio of the displacement components on the free surface:

$$\left.\frac{\Sigma_{12}}{\Sigma_{22}}\right|_{x_2>0} = \sqrt{\frac{\alpha_L}{\alpha_T}} = \left.\frac{U_2}{U_1}\right|_{x_2=0} \qquad [3.27]$$

This similarity is a consequence of the cancellation of the two normal stresses σ_{12} and σ_{22} on the surface $x_2 = 0$. For many isotropic materials, the ratio $\sqrt{\alpha_L/\alpha_T}$ is of the order of 1.5.

For a depth larger than two wavelengths, the two displacement components are very small and, therefore, hard to measure (Figure 3.4(a)). The elliptical polarization of the Rayleigh wave is shown in Figure 3.4(b). The radii of the ellipses decreases as the depth increases.

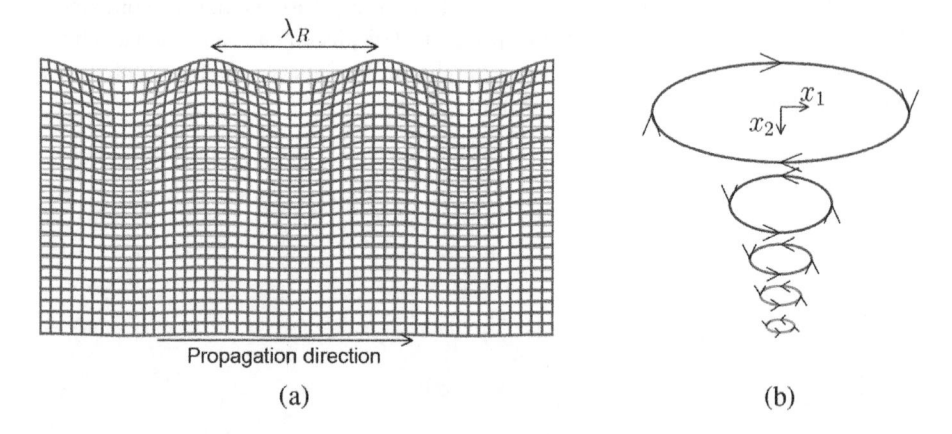

Figure 3.4. *Rayleigh wave: a) Deformation of the material as the wave passes through it. b) Elliptical motion of the particles in the depth of the solid*

3.1.1.3. *Displacement amplitude and power*

The absolute amplitude of the Rayleigh wave displacement is linked to the density of the transported power. The average value of the elastic power is expressed as a function of the displacement components by formula [A3.36]. In the absence of dispersion and for a beam of width w (along x_3), we obtain:

$$\langle P \rangle = \frac{1}{2}\rho\omega^2 V_R w \int_0^\infty \left[|u_1(x_2)|^2 + |u_2(x_2)|^2\right] \, \mathrm{d}x_2 \qquad [3.28]$$

By introducing the dimensionless decay factors χ_L and χ_T, defined by $\alpha_L = k_R\chi_L$ and $\alpha_T = k_R\chi_T$, the squares of the moduli of the displacement components [3.24] are written as functions of $X_2 = k_R x_2$:

$$\begin{cases} |u_1(X_2)|^2 = (k_R a_L)^2 \left[e^{-\chi_L X_2} - \sqrt{\chi_L \chi_T} e^{-\chi_T X_2}\right]^2 \\ |u_2(X_2)|^2 = (k_R a_L)^2 \dfrac{\chi_L}{\chi_T} \left[e^{-\chi_T X_2} - \sqrt{\chi_L \chi_T} e^{-\chi_L X_2}\right]^2 \end{cases} \quad [3.29]$$

The power density per unit width of the beam w is expressed by:

$$P_R = \frac{\langle P \rangle}{w} = \frac{1}{2}\rho\omega^2 V_R k_R a_L^2 (I_1 + I_2) \quad [3.30]$$

where the integrals:

$$\begin{cases} I_1 = \displaystyle\int_0^\infty \left[e^{-2\chi_L X_2} - 2\sqrt{\chi_L \chi_T} e^{-(\chi_L+\chi_T)X_2} + \chi_L \chi_T e^{-2\chi_T X_2}\right] dX_2 \\ I_2 = \dfrac{\chi_L}{\chi_T}\displaystyle\int_0^\infty \left[e^{-2\chi_T X_2} - 2\sqrt{\chi_L \chi_T} e^{-(\chi_L+\chi_T)X_2} + \chi_L \chi_T e^{-2\chi_L X_2}\right] dX_2 \end{cases}$$

$$[3.31]$$

can be easily calculated ($\int_0^\infty e^{-\chi X_2} dX_2 = 1/\chi$):

$$I_1 = \frac{\chi_L}{2} + \frac{1}{2\chi_L} - \frac{2\sqrt{\chi_L \chi_T}}{\chi_L + \chi_T} \quad \text{and} \quad I_2 = \frac{\chi_L}{\chi_T}\left[\frac{\chi_T}{2} + \frac{1}{2\chi_T} - \frac{2\sqrt{\chi_L \chi_T}}{\chi_L + \chi_T}\right] \quad [3.32]$$

Since the tangential displacement amplitude on the free surface is $|u_1(0)| = k_R a_L \left(1 - \sqrt{\chi_L \chi_T}\right)$, we get:

$$P_R = \frac{1}{2}\rho\omega V_R^2 |u_1(0)|^2 \frac{I_1 + I_2}{\left(1 - \sqrt{\chi_L \chi_T}\right)^2} \quad [3.33]$$

According to equation [3.16], written as $2\sqrt{\chi_L \chi_T} = 1 + \chi_T^2$, and by introducing the factor:

$$E = \frac{I_1 + I_2}{\chi_L - \chi_T}, \quad \text{hence} \quad P_R = \frac{1}{2}\rho\omega V_R^2 |u_1(0)|^2 \frac{E}{\chi_T} \quad [3.34]$$

the displacement amplitudes on the surface for the Rayleigh wave are given by:

$$|u_1(0)| = \sqrt{\frac{2P_R}{\rho\omega V_R^2}}\sqrt{\frac{\chi_T}{E}} \quad \text{and} \quad |u_2(0)| = \sqrt{\frac{2P_R}{\rho\omega V_R^2}}\sqrt{\frac{\chi_L}{E}} \quad [3.35]$$

A simple calculation of the sum $I_1 + I_2$ shows that the factor E is always larger than unity (Figure 3.5):

$$E = 1 + \frac{\chi_L - \chi_T}{2\chi_L \chi_T^2} \qquad [3.36]$$

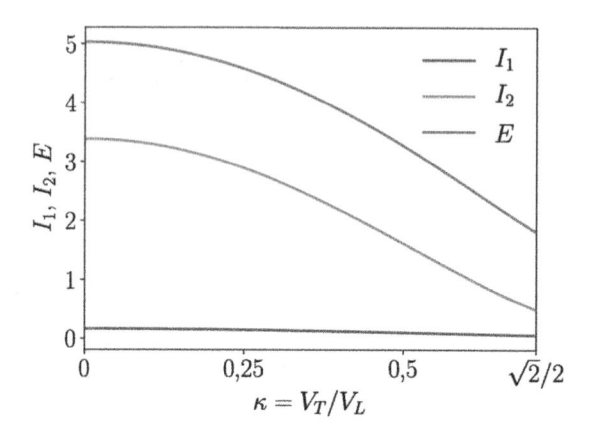

Figure 3.5. *Evolution of parameters E, I_1 and I_2 depending on the ratio of the bulk wave velocities κ. For a color version of this figure, see www.iste.co.uk/royer/waves1.zip*

ORDER OF MAGNITUDE.– For duralumin ($\kappa = 0.495$): $I_1 = 0.10$, $I_2 = 1.66$, $\chi_L = 0.89$, $\chi_T = 0.36$ and $E = 3.32$, at a frequency $f = 16$ MHz ($\omega = 10^8$ rad/s) and for a power density $P_R = 1$ W/cm, the amplitude of the displacement normal to the free surface is $|u_2(0)| = 4.7$ nm.

This typical example of a material with Poisson's ratio $\nu = 0.34$ shows that 93% of the energy is transported by the transverse displacement. This displacement, which is larger on the surface than the longitudinal one, and which penetrates more deeply into the material ($\chi_T < \chi_L$) without changing of sign, is the main component of the Rayleigh wave. To cancel the stresses σ_{12} and σ_{22} on the free surface, a small longitudinal component must be added to the transverse displacement.

Physically, there is nothing to prevent the vertical motion of the free face of the cube A in Figure 3.6; on the contrary, the motion of the five other faces is restricted by the adjacent material. The removal of the matter located on the side $x_2 < 0$ decreases the stiffness in the vicinity of the surface: the stiffness C_A of cube A is about 5/6ths of the stiffness C_B of cube B located in the depth. The velocity of an elastic wave being of the form $\sqrt{C/\rho}$, the velocity V_S of the surface wave is of the order of:

$$V_S = \sqrt{\frac{C_A}{\rho}} \approx \sqrt{\frac{5C_B}{6\rho}} = \sqrt{\frac{5}{6}}V_T = 0.91V_T \qquad [3.37]$$

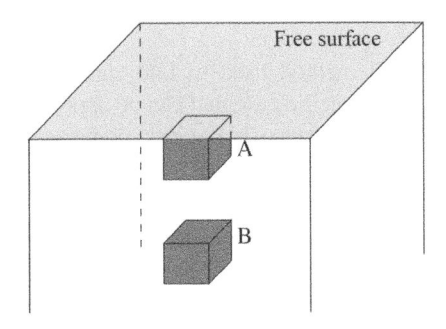

Figure 3.6. *The stiffness of cube A, which has one of its six faces free, is reduced compared to that of cube B located in the core of the solid*

This result is in good agreement with the complete calculation, which leads to a ratio V_R/V_T between 0.89 and 0.94 for an isotropic solid. This is no longer valid in the case of an anisotropic solid because the stiffness depends on the orientation of the faces: the ratio V_R/V_T can be much smaller than unity (section 3.1.2.3).

3.1.2. *Anisotropic solid*

The anisotropy of the propagation medium significantly complicates the analysis of the propagation of surface waves. Several types of waves appear depending on the orientation of the free surface and of the propagation direction. The terminology and classification of these waves vary among authors (Favretto-Cristini *et al.* 2011). Let us summarize here the main results (more or less expected):

– the Rayleigh wave velocity depends on the propagation direction on the free surface, if this plane is not perpendicular to an A_6 axis, which is equivalent to an axis of isotropy from the point of view of elasticity;

– the energy vector is no longer parallel to the propagation direction;

– the mechanical displacement has three components and the particles of matter describe an ellipse in a plane that is inclined with respect to the sagittal plane, if this one is not a symmetry plane;

– the decay factors are complex: the amplitudes of the displacement components of this *generalized Rayleigh wave* tend toward zero while oscillating;

– in isolated directions, solutions of the Rayleigh wave type exist with a phase velocity higher than that of the slow transverse wave; however, as soon as we move away from these particular directions, a component appears and causes a radiation of energy toward the interior of the crystal. The displacement of these *pseudo surface waves* is no longer zero at an infinite depth.

To illustrate the general method used to find surface waves, explained in the following section, we treat a specific case that can be approached analytically: that of an orthotropic solid whose free surface and sagittal plane are symmetry planes.

3.1.2.1. *Method of partial waves*

This method of analysis calls for a formalism very close to that developed in section 2.5. It consists of looking for a solution analogous to [2.123]:

$$u_l(x_1, x_2, t) = a p_l e^{imkx_2} e^{i(kx_1 - \omega t)} \quad \text{with} \quad \text{Im}[mk] > 0 \qquad [3.38]$$

This solution represents an inhomogeneous wave of amplitude a and wave number k along x_1, whose mechanical displacement components u_1, u_2, u_3 decrease with depth.

Substituting this expression in the propagation equations lead to the Christoffel equations [1.173] in which the components Γ_{ik} are polynomials of the second-degree in m. The velocity V being considered as a parameter to be determined, the compatibility condition of the Christoffel equations is a sixth-degree equation in m. The coefficients of the powers of m are real, and there exist three pairs of complex conjugate roots. The only acceptable roots are the three roots m_J ($J = 1, 2, 3$), such that $\text{Im}[m_J k] > 0$. The others, with $\text{Im}[m_J k] < 0$, lead to a wave whose amplitude increases with x_2 and are, therefore, excluded. For each root m_J, the system [2.124] provides the polarization vector p_{lJ}. The general solution is a linear combination of these three partial waves with the same phase velocity V:

$$u_l = \left(\sum_{J=1}^{3} p_{lJ} a_J e^{im_J kx_2} \right) e^{i(kx_1 - \omega t)} \qquad [3.39]$$

The mechanical boundary conditions concern the normal stresses: $\sigma_{i2} = 0$ on the free surface $x_2 = 0$. For each partial wave, their expression is similar to equations [2.133] and [2.134] by replacing k_1 by k:

$$\sigma_{i2} = ik A_i a e^{imkx_2} e^{i(kx_1 - \omega t)} \quad \text{with} \quad A_i = (c_{i21l} + m c_{i22l}) p_l \qquad [3.40]$$

The stresses specific to the general solution are written as:

$$\sigma_{i2} = ik \left(\sum_{J=1}^{3} A_{iJ} a_J e^{im_J kx_2} \right) e^{i(kx_1 - \omega t)} \qquad i = 1, 2, 3 \qquad [3.41]$$

where the coefficients $A_{iJ} = A_i(m_J)$ depend on the phase velocity V through the intermediary of the roots m_J. The stresses σ_{i2} are zero on the free surface $x_2 = 0$ and the velocity V results from the compatibility condition of the linear and homogeneous system:

$$\sum_{J=1}^{3} A_{iJ} a_J = 0 \quad \text{hence} \quad \det(A_{iJ}) = 0, \quad i, J = 1, 2, 3 \quad [3.42]$$

Generally, a single root $V = V_R$ is acceptable, that is, such that $\text{Im}[m_J k] > 0$ (Barnett and Lothe 1974). The mechanical displacement of this surface wave has three components. When the $x_1 x_2$ plane, called the sagittal plane, is perpendicular to a binary axis or parallel to a mirror plane, the propagation equations and mechanical boundary conditions are decoupled (section 2.2.2). A surface wave with two components, u_1 and u_2, and a TH wave, polarized along x_3, propagate independently. This is the case for the Rayleigh wave propagating on the free surface of an orthotropic solid.

3.1.2.2. *Solid with an orthotropic symmetry*

The normal Ox_2 to the free surface and the normal Ox_3 to the sagittal plane are parallel to two of the three binary axes of symmetry (direct or inverse). It has been shown, in section 2.2.2, that the propagation equations and boundary conditions at the interface were divided into two independent systems. The waves polarized in the sagittal plane $x_1 x_2$ satisfy the Christoffel equation [2.17]. By taking into account the expression of the components [2.142] and denoting $\zeta = \rho V^2$, this equation is written as:

$$\begin{pmatrix} C_{11} + m^2 C_{66} - \zeta & m(C_{12} + C_{66}) \\ m(C_{12} + C_{66}) & C_{66} + m^2 C_{22} - \zeta \end{pmatrix} \begin{pmatrix} p_1 \\ p_2 \end{pmatrix} = \begin{pmatrix} 0 \\ 0 \end{pmatrix} \quad [3.43]$$

The characteristic equation is biquadratic in m:

$$C_{22} C_{66} m^4 - \left[\zeta(C_{22} + C_{66}) + C_{12}^2 + 2 C_{12} C_{66} - C_{11} C_{22} \right] m^2 \\ + (C_{11} - \zeta)(C_{66} - \zeta) = 0 \quad [3.44]$$

that is:

$$m^4 - S m^2 + P = 0 \quad [3.45]$$

P and S being the product and the sum of the roots m_1^2 and m_2^2:

$$\begin{cases} P = m_1^2 m_2^2 = \dfrac{(C_{11} - \zeta)(C_{66} - \zeta)}{C_{22} C_{66}} \\ S = m_1^2 + m_2^2 = \dfrac{2 C_{12} C_{66} + C_{66} \zeta - C_{22}(C - \zeta)}{C_{22} C_{66}} \end{cases} \quad [3.46]$$

where:

$$C = C_{11} - \frac{C_{12}^2}{C_{22}} \qquad [3.47]$$

For the two acceptable roots m_1 and m_2 having a positive imaginary part, the polarizations are the eigenvectors of the system [3.43]. On taking $p_{1J} = 1$, it comes:

$$p_{2J} = -\frac{C_{11} - \zeta + m_J^2 C_{66}}{m_J(C_{12} + C_{66})}, \qquad J = 1, 2 \qquad [3.48]$$

The general solution is a linear combination of these two displacements, propagating at the same velocity V:

$$\begin{cases} u_1(x_1, x_2, t) = \left[a_1 e^{im_1 k x_2} + a_2 e^{im_2 k x_2}\right] e^{i(kx_1 - \omega t)} \\ u_2(x_1, x_2, t) = \left[a_1 p_{21} e^{im_2 k x_2} + a_2 p_{22} e^{im_1 k x_2}\right] e^{i(kx_1 - \omega t)} \end{cases} \qquad [3.49]$$

The coefficients a_1 and a_2 and the velocity V are deduced from system [3.42] with only i and $J = 1, 2$. Considering the expressions [2.150] giving the factors A_{iJ}, with $p_{1J} = 1$ we obtain:

$$\begin{cases} (m_1 + p_{21})a_1 + (m_2 + p_{22})a_2 = 0 \\ (C_{12} + m_1 p_{21} C_{22})a_1 + (C_{12} + m_2 p_{22} C_{22})a_2 = 0 \end{cases} \qquad [3.50]$$

These two equations are compatible if the determinant is zero:

$$(p_{21} - p_{22})(C_{12} - m_1 m_2 C_{22}) + (m_1 - m_2)(C_{12} - p_{21} p_{22} C_{22}) = 0 \qquad [3.51]$$

By substituting the value of $p_{21} - p_{22}$, deduced from [3.48] and noting that the factor $m_1 m_2 p_{21} p_{22} C_{22}$ is expressed using the sum S and the product P of the squares of the roots, by:

$$m_1 m_2 p_{21} p_{22} C_{22} = C_{11} - \zeta \qquad [3.52]$$

equation [3.51] is reduced to:

$$m_1 m_2 = -\frac{\zeta(C_{11} - \zeta)}{C_{22}(C - \zeta)} \qquad [3.53]$$

Introducing the product $m_1 m_2$ in equations [3.46] leads to an equation for velocity:

$$f(\zeta) = C_{66}(C_{11} - \zeta)\zeta^2 - C_{22}(C_{66} - \zeta)(C - \zeta)^2 = 0 \qquad [3.54]$$

similar to the equation established by Stoneley (1955) in the specific case of a Rayleigh wave propagating in the basal plane of a hexagonal crystal. However, this equation is more general: it applies to 16 configurations belonging to the orthorhombic, tetragonal, hexagonal and cubic systems (Royer and Dieulesaint 1984). When $\zeta = 0$: $f(\zeta)$ is negative, and when $\zeta = \min[C_{66}, C]$: $f(\zeta)$ is positive, because $C < C_{11}$, so that there is one and only one root ζ_R, lying between zero and the minimum value of C_{66} and C.

Since the Rayleigh wave velocity V_R is known ($\zeta_R = \rho V_R^2$), it is possible to derive the components of the mechanical displacement. Substituting $(C_{11} - \zeta)$, taken from equation [3.52], into equation [3.48] gives:

$$\begin{cases} p_{21}(C_{12} + m_2 p_{22} C_{22}) = -C_{66}(m_1 + p_{21}) \\ p_{22}(C_{12} + m_1 p_{21} C_{22}) = -C_{66}(m_2 + p_{22}) \end{cases} \quad [3.55]$$

By forming the product of the ratio a_2/a_1, deduced from each of the equations [3.50], we obtain:

$$\frac{a_2^2}{a_1^2} = \frac{p_{21}}{p_{22}} \quad \text{or} \quad p_{22} a_2^2 = p_{21} a_1^2 \quad [3.56]$$

The longitudinal and transverse components of the displacement [3.49] take the form:

$$\begin{cases} u_1(x_1, x_2, t) = a_1 \left[e^{im_1 k x_2} + \dfrac{a_2}{a_1} e^{im_2 k x_2} \right] e^{i(kx_1 - \omega t)} \\ u_2(x_1, x_2, t) = p_{22} a_2 \left[e^{im_2 k x_2} + \dfrac{a_2}{a_1} e^{im_1 k x_2} \right] e^{i(kx_1 - \omega t)} \end{cases} \quad [3.57]$$

We must distinguish between two cases, according to the sign of the discriminant $\Delta = S^2 - 4P$ in the characteristic equation [3.45].

– If Δ is positive, the roots m_1^2 and m_2^2 are real and negative, and the two physically acceptable solutions are $m_J = i\chi_J$, with $\chi_J > 0$. The components p_{2J}, given by equation [3.48], are purely imaginary and the product $p_{21} p_{22} = -p^2$ is negative, so that:

$$(p_{22} a_2)^2 = p_{22} p_{21} a_1^2 = -p^2 a_1^2 \quad \text{or} \quad p_{22} a_2 = ip a_1 \quad [3.58]$$

The longitudinal and transverse components of the mechanical displacement [3.57] thus undergo a phase shift of $\pi/2$. According to [3.48], the factors ip_{21} and ip_{22} are real negative and equation [3.50] shows that it is the same for the ratio:

$$\frac{a_2}{a_1} = -\frac{m_1 + p_{21}}{m_2 + p_{22}} = -\frac{\chi_1 - ip_{21}}{\chi_2 - ip_{22}} \quad \text{hence} \quad \frac{a_2}{a_1} = -\left|\frac{a_2}{a_1}\right| \quad [3.59]$$

The decay with depth is similar to that of the components [3.24] of the Rayleigh wave propagating in an isotropic solid:

$$\begin{cases} u_1(x_1, x_2, t) = a_1 \left[e^{-\chi_1 k x_2} - \left|\dfrac{a_2}{a_1}\right| e^{-\chi_2 k x_2} \right] e^{i(kx_1 - \omega t)} \\ u_2(x_1, x_2, t) = ip a_1 \left[e^{-\chi_2 k x_2} - \left|\dfrac{a_2}{a_1}\right| e^{-\chi_1 k x_2} \right] e^{i(kx_1 - \omega t)} \end{cases} \quad [3.60]$$

It is a Rayleigh wave whose components are evanescent.

– If Δ is negative, the roots m_1^2 and m_2^2 of equation [3.45] are complex conjugates and the two acceptable solutions having a positive imaginary part are:

$$m_1 = g + ih \quad \text{and} \quad m_2 = -m_1^* = -g + ih \quad [3.61]$$

with:

$$g = \frac{1}{2}\sqrt{2\sqrt{P} + S} \quad \text{and} \quad h = \frac{1}{2}\sqrt{2\sqrt{P} - S} \quad [3.62]$$

By denoting, as before, $p^2 = -p_{21}p_{22}$, equation [3.48] shows that $p_{21} = ipe^{-2i\varphi}$ and $p_{22} = -p_{21}^* = ipe^{2i\varphi}$ and equation [3.56] shows that $a_2 = ae^{-i\varphi}$ if we write $a_1 = ae^{i\varphi}$. The two components [3.49] are still in phase quadrature:

$$\begin{cases} u_1(x_1, x_2, t) = 2ae^{-hkx_2} \cos(gkx_2 + \varphi)e^{i(kx_1 - \omega t)} \\ u_2(x_1, x_2, t) = 2ipae^{-hkx_2} \cos(gkx_2 - \varphi)e^{i(kx_1 - \omega t)} \end{cases} \quad [3.63]$$

However, their amplitudes oscillate as a function of the depth, with a spatial period λ_R/g equal to the ratio of the wavelength λ_R to the real part g of the roots. This is a *generalized Rayleigh* wave, for which the decay of the mechanical displacement with the depth, linked to the imaginary part h of these roots, can be much slower than for a Rayleigh wave. Indeed, this difference arises from the anisotropy of the solid in the sagittal plane.

3.1.2.3. *Influence of the anisotropy factor*

For crystals belonging to cubic and quadratic systems, $C_{22} = C_{11}$, and if the fourfold axis of symmetry is parallel to the x_3 axis, the anisotropy in the sagittal plane is characterized by the factor [1.215]: $A = 2C_{66}/(C_{11} - C_{12})$. If A is large with respect to unity, that is, when C_{12} is close to C_{11}, the solution to equation [3.54], necessarily less than $C = C_{11} - C_{12}^2/C_{11}$, is small compared to C_{11} and C_{66}: $\zeta_R \approx C/2$. The development, to the second degree in ζ_R, of the sum and the product of the roots provides an approximate value of their imaginary part (equation [3.62]):

$$h \approx \frac{(1 + C_{66}/C_{11})}{2A} \quad \text{for} \quad A \gg 1 \quad [3.64]$$

Consequently, the damping of these oscillations gets smaller as the anisotropy factor increases. This result is illustrated by the curves in Figure 3.7, for a propagation along the [100]-binary axis of symmetry in crystals of silicon (Si, $m3m$), copper (Cu, $m3m$), rutile (TiO$_2$, $4/mmm$) and tellurium dioxide (TeO$_2$, 422). The stiffness constants used are from Table 1.6. Tellurium dioxide is piezoelectric; however, this effect has not been taken into account because the sagittal plane is perpendicular to a direct binary axis (section 3.1.3). It is interesting to note that as A

increases, the ratio of the Rayleigh wave velocity V_R to the transverse bulk wave velocity $V_T = \sqrt{C_{66}/\rho}$ decreases as can be predicted for $A \gg 1$:

$$\zeta_R = \rho V_R^2 \approx \frac{C}{2} = \frac{C_{11}^2 - C_{12}^2}{2C_{11}} \quad \text{or} \quad \frac{V_R}{V_T} \approx \sqrt{\frac{2}{A}\left(1 - \frac{C_{66}}{AC_{11}}\right)} \qquad [3.65]$$

For instance, in tellurium dioxide, V_R is almost four times smaller than V_T. Let us recall that for an isotropic solid, V_R lies between $0.874V_T$ and $0.955V_T$. In practice, equations [3.64] and [3.65] are good approximations if $A > 4$.

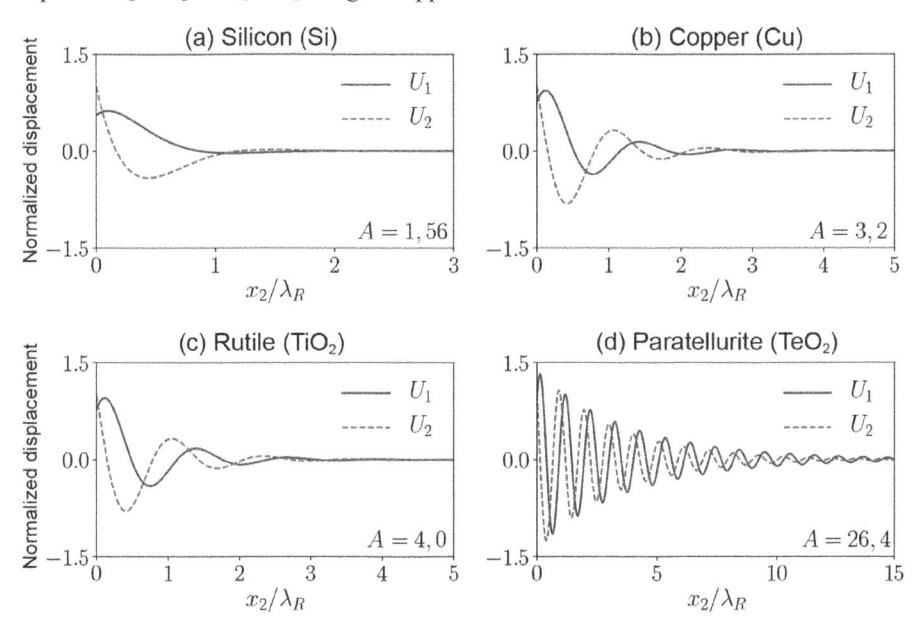

Figure 3.7. *Variation, with depth, of the longitudinal (L) and transverse (T) components of the mechanical displacement for a Rayleigh wave, propagating along the [100] axis in the (010) plane of Si, Cu, TiO$_2$ and TeO$_2$ crystals. The damping of the oscillations is reduced as the anisotropy factor A increases. For a color version of this figure, see www.iste.co.uk/royer/waves1.zip*

This generalized Rayleigh wave, whose polarization is elliptical in the sagittal plane, propagates without attenuation at a velocity that is lower than that of the bulk waves. In the next paragraph, we will describe a "supersonic" surface wave, whose velocity is higher than that of the TH wave, and which is attenuated during its propagation by the radiation of part of its energy toward the bulk of the material.

3.1.2.4. *Pseudo-surface acoustic wave*

A *pseudo-surface wave* appears in some crystals, as a result of anisotropy, when the velocity of the Rayleigh wave is greater than that of one of the transverse bulk waves. For example, in a crystal with cubic symmetry, described by three constants C_{11}, C_{12} and C_{66}, a transverse wave, polarized vertically (TV) and a quasi-transverse wave, polarized horizontally (QTH) propagate in a plane such as (001) (Figure 3.8). The transverse vertical wave propagates at the velocity $V_{TV} = \sqrt{C_{66}/\rho}$, independent of the direction. On the contrary, the velocity $V_{TH}(\varphi)$ of the quasi-transverse wave depends on the angle φ with the [100] direction and $V_{TH}(0) = V_{TV}$. For $\varphi = 0$ and $\pm\pi/4$, the sagittal plane is a symmetry plane. Consequently, there exists a Rayleigh wave whose components contained in this plane are not coupled with the TH wave. The velocity of this Rayleigh wave, propagating in the directions [100] and [110], may be higher than V_{TH}. If we move away from these two specific directions, two cases can be distinguished depending on whether the anisotropy factor:

$$A = \frac{2C_{66}}{C_{11} - C_{12}} = \left[\frac{V_{TV}}{V_{TH}(\pi/4)}\right]^2 \qquad [3.66]$$

is smaller than or greater than unity.

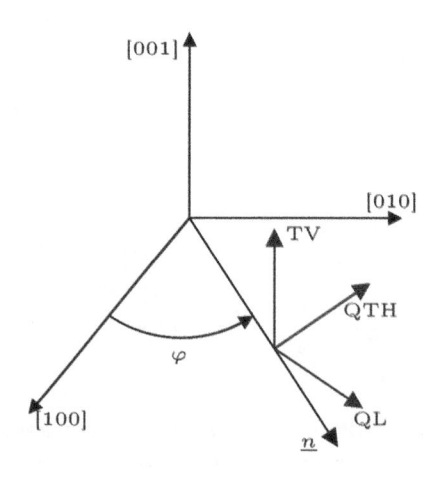

Figure 3.8. *TV and QTH waves propagating in the (001) plane of a cubic crystal*

The first case ($A < 1 \longrightarrow V_{TH}(\pi/4) > V_{TV}$) is illustrated in Figure 3.10(a): the velocity of the TH wave increases with φ and is maximum in the [110] direction. The coupling between this wave and the Rayleigh wave is weak, so that the Rayleigh wave velocity V_R remains almost constant and slightly lower than V_{TV}.

The second case ($A > 1 \longrightarrow V_{TH}(\pi/4) < V_{TV}$) is represented in Figure 3.10(b). The velocity $V_{TH}(\varphi)$, which decreases continuously with φ, starting from its value along [100], is minimal along the [110] direction: its value is lower than the Rayleigh wave velocity $V_R(\pi/4)$. As φ increases, the coupling between the TH wave and the

Rayleigh wave increases. When the curves meet together, the Rayleigh wave degenerates into a TH wave. In the reverse direction, starting with the velocity $V_R(\pi/4)$ along [110], the coupling between the Rayleigh wave and the QTH wave appears also naturally. The surface wave with velocity $V_R > V_{QTH}$ excites an increasing amount of QTH wave at an angle β that satisfies the phase matching condition for the wave planes (Figure 3.9).

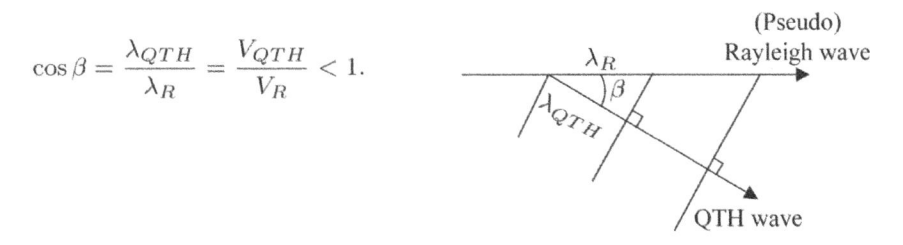

$$\cos\beta = \frac{\lambda_{QTH}}{\lambda_R} = \frac{V_{QTH}}{V_R} < 1.$$

Figure 3.9. *Phase matching condition between the Rayleigh wave and the QTH wave when $V_R > V_{QTH}$*

The Rayleigh wave with two components propagating in the direction of symmetry [110] becomes a pseudo-surface acoustic wave (PSAW) or *leaky surface acoustic wave* (leaky SAW) with three components, which is no longer confined close to the surface. The additional solution m_3 of the characteristic equation does not fulfill the cancelation condition of displacement at infinity: $\text{Im}[m_3] > 0$. The leakage of energy leads to an attenuation (Figure 3.10(b)), as in the case of the leaky Rayleigh wave propagating at the interface between a solid and a liquid (section 3.2.1.3). For an angle less than 24°, the root m_3 becomes real and the solution no longer corresponds to a surface wave.

These results for $A > 1$ and a (001) plane may be valid for $A < 1$ and a plane different from (001) if the relative values of the characteristic velocities are similar (Farnell 1970).

These different surface waves result from the choice of partial waves, required to satisfy the boundary conditions on the free surface. According to their behavior as a function of the depth x_2:

– for a SAW, the three modes are decreasing; the wave propagates, without attenuation, at a velocity lower than the smallest velocity of bulk waves;

– for a PSAW, a decreasing partial wave is replaced by an increasing mode close to the QTH bulk wave. This PSAW propagates at a velocity slightly lower than that of the fastest transverse wave;

– for a high-velocity pseudo-surface wave (HVPSAW), two decreasing partial waves are replaced by two increasing modes, one of which resembles a longitudinal bulk wave. This HVPSAW propagates at a velocity slightly lower than that of the longitudinal wave, that is, a velocity much greater (twice as fast, for example) than that of a surface wave.

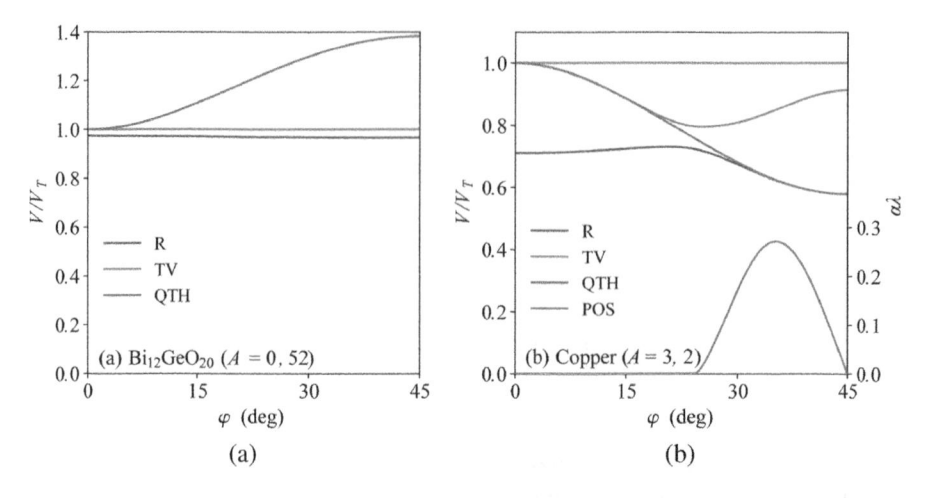

Figure 3.10. *Propagation of Rayleigh waves (R), and pseudo-surface acoustic waves (PSAW) in the (001) plane of a cubic crystal whose anisotropy factor A is smaller a) or greater b) than unity. Along the symmetry axes ($\varphi = 0$ or $45°$), the Rayleigh wave with two components is not coupled with the bulk TH wave, so that its velocity can be greater than V_{TH} For a color version of this figure, see www.iste.co.uk/royer/waves1.zip*

For certain crystalline cuts, and in well-chosen directions, the attenuation by radiation of a PSAW or HVPSAW can be small enough for this wave to be observed over distances of a few hundred wavelengths (Adler 1998).

The phase slowness of the transverse waves and of the Rayleigh wave in the (X, Y) plane for a copper crystal of cubic symmetry are represented in Figure 3.11(a). The wave surface is plotted for the Rayleigh wave in Figure 3.11(b). This one is obtained by calculating the energy velocity V^e and the deviation angle ψ defined by relations [1.211] and [1.210]. Starting from the angle $\varphi = 24°$, the slowness curve of the Rayleigh wave merges with that of the QTH bulk wave and a pseudo-surface wave appears (Figure 3.10(b)). Due to the double inflection of the slowness curve of the surface waves, there exist along the X and Y axes two cusps and three distinct energy velocities.

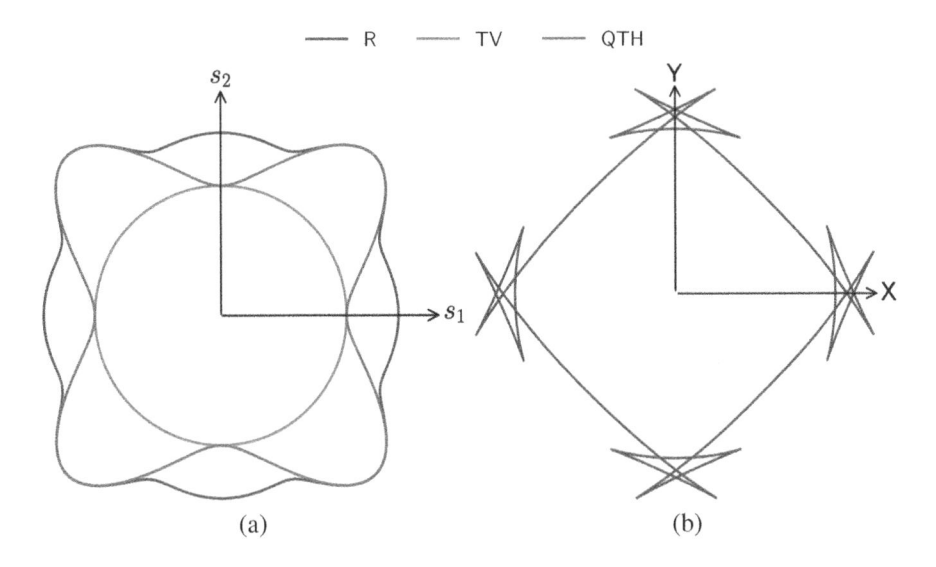

Figure 3.11. *Characteristic curves in the XY plane of crystallographic axes for a copper crystal with cubic symmetry. (a) Slowness of transverse bulk waves and Rayleigh wave, and (b) energy velocity of the Rayleigh wave. For a color version of this figure, see www.iste.co.uk/royer/waves1.zip*

3.1.3. *Piezoelectric crystal*

The search for surface waves in a piezoelectric solid necessitates the addition of the Poisson's equation to the local equation of motion and the addition of the electrical boundary conditions to the mechanical boundary conditions. However, numerical solutions show that the character of the Rayleigh wave is not modified very much by the piezoelectricity, with respect to the phase velocity, the decay of its components, or the penetration depth. Two independent solutions exist in the two cases of decoupling of the propagation equations and of the mechanical and electrical boundary conditions. The interest in piezoelectricity lies in the electrical coupling of the surface wave with the external medium, which facilitates its generation by a transducer with interdigitated comb-shaped electrodes (Volume 2, section 3.2). The method chosen to express this interaction consists of introducing a quantity, called the surface piezoelectric permittivity, which associates the normal electric induction and the potential at the surface, once all mechanical phenomena are taken into account.

3.1.3.1. *Search procedure, wave decoupling*

The propagation equations in an insulating piezoelectric material were established in section 1.4.3. For a time-harmonic plane wave with phase velocity $V = \omega/k$ and polarization $\underline{p}$, accompanied by an electric potential Φ_0, these equations take the form [1.266]:

$$\Gamma_{il}p_l + \gamma_i\Phi_0 = \rho V^2 p_i \quad \text{and} \quad \gamma_l p_l - \varepsilon^S \Phi_0 = 0, \qquad i,l = 1,2,3 \qquad [3.67]$$

In matrix form and considering Φ_0 as the fourth component p_4 of a generalized polarization vector, the Christoffel equations are written as:

$$\begin{pmatrix} \Gamma_{11} - \rho V^2 & \Gamma_{12} & \Gamma_{13} & \gamma_1 \\ \Gamma_{12} & \Gamma_{22} - \rho V^2 & \Gamma_{23} & \gamma_2 \\ \Gamma_{13} & \Gamma_{23} & \Gamma_{33} - \rho V^2 & \gamma_3 \\ \gamma_1 & \gamma_2 & \gamma_3 & -\varepsilon^S \end{pmatrix} \begin{pmatrix} p_1 \\ p_2 \\ p_3 \\ p_4 \end{pmatrix} = \begin{pmatrix} 0 \\ 0 \\ 0 \\ 0 \end{pmatrix} \qquad [3.68]$$

The search procedure for surface waves is similar to the method used in section 3.1.2. A solution of the form:

$$u_l = a p_l e^{imkx_2} e^{i(kx_1 - \omega t)} \quad \text{with} \quad \text{Im}[m] > 0 \quad \text{and} \quad l = 1,2,3,4 \qquad [3.69]$$

represents an inhomogeneous wave of amplitude a. Assuming the wave number k to be positive, the mechanical displacement components u_1, u_2, u_3 and the electric potential $\Phi = u_4$ decrease with depth x_2. With $n_1 = 1$, $n_2 = m$ and $n_3 = 0$, the components of the Christoffel tensor [1.177] and [1.267] are given by:

$$\begin{cases} \Gamma_{il} = c^E_{i11l} + (c^E_{i12l} + c^E_{i21l})m + c^E_{i22l}m^2 \\ \gamma_i = e_{1i1} + (e_{1i2} + e_{2i1})m + e_{2i2}m^2 \\ \varepsilon^S = \epsilon^S_{11} + 2\epsilon^S_{12}m + \epsilon^S_{22}m^2 \end{cases} \qquad [3.70]$$

The velocity V being a parameter to be determined, the compatibility condition for the homogeneous system [3.68] is an eight-degree equation in m. With the coefficients of the powers of m being real, there exist four pairs of complex conjugate roots. The only acceptable ones are the four roots m_J ($J = 1,2,3,4$) with a positive imaginary part. The others, with $\text{Im}[m_J] < 0$, lead to a wave whose amplitude increases with x_2. For each of the acceptable values, the system of equations [3.68] gives the polarization vector p_{lJ} ($l = 1,2,3$) and the amplitude $\Phi_{0J} = p_{4J}$ of the potential, apart from a proportionality coefficient, which makes it possible to normalize this potential to unity. The general solution is a linear combination of these four partial waves having the same phase velocity V:

$$u_l = \left(\sum_{J=1}^{4} p_{lJ} a_J e^{im_J k x_2}\right) e^{i(kx_1 - \omega t)} \qquad [3.71]$$

On the free surface $x_2 = 0$, the mechanical boundary conditions impose the cancelation of the stresses σ_{i2}. From the general expression [1.263], and omitting the propagation factor $e^{i(kx_1 - \omega t)}$ and the superscript E for the elastic constants c_{ijkl}, we obtain:

$$\sigma_{i2} = \frac{\partial}{\partial x_k}(c_{i2kl}u_2 + e_{ki2}\phi) = ika\left[c_{i21l}p_l + mc_{i22l}p_l + (e_{1i2} + me_{2i2})p_4\right]e^{ikmx_2}$$

[3.72]

with $i = 1, 2, 3$. By generalizing the definition [2.136] for the coefficients A_{iJ} in section 2.5.2 to the piezoelectric case:

$$\tilde{A}_{iJ} = \tilde{A}_i(m_J) = A_{iJ} + (e_{1i2} + e_{2i2}m_J)p_{4J}$$

[3.73]

the cancelation of the stresses σ_{i2} at $x_2 = 0$ leads to the system:

$$\sum_{J=1}^{3} \tilde{A}_{iJ}a_J = -\tilde{A}_{i4}a_4, \qquad i = 1, 2, 3$$

[3.74]

If the solid is not piezoelectric, $\tilde{A}_{iJ} = A_{iJ}$ and $a_4 = 0$: the velocity V results from the compatibility condition of the linear and homogeneous system [3.42]: $\det[A_{iJ}] = 0$. If the solid is piezoelectric, equations [3.74] provide the amplitudes a_1, a_2, a_3 of the partial waves as functions of a_4. The velocity is, therefore, determined by the electrical boundary conditions (see section 3.1.3.2). In general, a single root $V = V_R$ is acceptable, that is, such that $\text{Im}[m_J] > 0$. It corresponds to a Rayleigh wave with three components of mechanical displacement, accompanied by an electric field contained in the sagittal plane x_1x_2, since $E_3 = -\partial\Phi/\partial x_3 = 0$.

Nonetheless, two solutions exist in the two cases of decoupling of the propagation equations and of the mechanical boundary conditions examined in section 2.2.2. The propagation equations [3.68] simplify if the sagittal plane is perpendicular to a binary axis parallel to Ox_3. Indeed, the elastic constants with a single index equal to 3 are zero, so that $\Gamma_{13} = \Gamma_{23} = 0$. The piezoelectricity imposes to distinguish the case of a direct axis from the case of an inverse axis (equivalent to a mirror parallel to the sagittal plane).

a) *Sagittal plane x_1x_2 perpendicular to a direct axis of even order* (which includes a binary axis): the piezoelectric constants with no index equal to 3 are zero [1.255], so that $\gamma_1 = \gamma_2 = 0$. The system [3.68] is divided into two independent sub-systems. The first:

$$\begin{pmatrix} \Gamma_{11} - \rho V^2 & \Gamma_{12} \\ \Gamma_{12} & \Gamma_{22} - \rho V^2 \end{pmatrix}\begin{pmatrix} p_1 \\ p_2 \end{pmatrix} = \begin{pmatrix} 0 \\ 0 \end{pmatrix}$$

[3.75]

corresponds to a *non-piezoelectric Rayleigh wave* polarized in the sagittal plane x_1x_2 (mode R_2). The second sub-system:

$$\begin{pmatrix} \Gamma_{33} - \rho V^2 & \gamma_3 \\ \gamma_3 & -\varepsilon^S \end{pmatrix}\begin{pmatrix} p_3 \\ p_4 \end{pmatrix} = \begin{pmatrix} 0 \\ 0 \end{pmatrix}$$

[3.76]

corresponds to a *piezoelectric transverse horizontal wave* polarized along x_3 ($\tilde{\text{TH}}$), discovered by Bleustein and Gulyaev in 1968.

b) *Sagittal plane $x_1 x_2$ parallel to a mirror*: the piezoelectric constants with a single index equal to 3 are zero [1.254], so that $\gamma_3 = 0$. The system splits again into two parts. The first:

$$\begin{pmatrix} \Gamma_{11} - \rho V^2 & \Gamma_{12} & \gamma_1 \\ \Gamma_{12} & \Gamma_{22} - \rho V^2 & \gamma_2 \\ \gamma_1 & \gamma_2 & -\varepsilon^S \end{pmatrix} \begin{pmatrix} p_1 \\ p_2 \\ p_4 \end{pmatrix} = \begin{pmatrix} 0 \\ 0 \\ 0 \end{pmatrix} \qquad [3.77]$$

gives a Rayleigh wave polarized in the sagittal plane, accompanied by an electric field (denoted $\tilde{R}_2$). The second:

$$(\Gamma_{33} - \rho V^2) p_3 = 0 \qquad [3.78]$$

corresponds to a non-piezoelectric TH wave.

In summary, a binary axis (direct or inverse) normal to the sagittal plane decouples the propagation of waves polarized in the sagittal plane from the propagation of TH waves. This is also the case for the *boundary conditions*. At the beginning of Chapter 2 (section 2.2.2), it was established that the coefficient A_{3J} is zero for any wave polarized in the sagittal plane ($p_3 = 0$). The piezoelectric term $(e_{132} + m_J e_{232}) p_{4J}$ in equation [3.73] is also zero in both cases:

a) $A_2 // x_3$, because the wave polarized in the sagittal plane is not piezoelectric: $p_4 = 0$;

b) $M \perp x_3$, because the piezoelectric constants with only one index equal to 3: e_{132} and e_{232} are zero.

The stress σ_{32} is, therefore, identically zero. Furthermore, for any TH wave, the coefficients A_{1J} and A_{2J} are zero (section 2.2.2). The additional terms $(e_{112} + m_J e_{212}) p_{4J}$ and $(e_{122} + m_J e_{222}) p_{4J}$ are also zero in the two cases, where:

a) $A_2 // x_3$, since the piezoelectric constants with no index equal to 3 are zero;

b) $M \perp x_3$, since the TH wave is not piezoelectric: $p_4 = 0$.

The stresses σ_{12} and σ_{22} are, therefore, identically zero, and the results obtained are summarized below:

Sagittal plane $\perp$ axis $A_2 \longrightarrow$ Modes R_2 and $\tilde{\tilde{TH}}$
Sagittal plane $//$ mirror M $\longrightarrow$ Modes $\tilde{R}_2$ and TH

3.1.3.2. *Electrical boundary conditions, surface permittivity*

The piezoelectric solid is in contact with a (non-piezoelectric) dielectric occupying the half-space $x_2 < 0$. In the quasi-static approximation framework, the electrical boundary conditions involve the electric potential $\Phi = u_4$ and the normal component of the electric induction D_2, which are continuous at the interface, in the absence of any free charges. Let us start by defining the surface permittivity for the dielectric. The electric potential that accompanies the surface wave:

$$\Phi(x_1, x_2, t) = \Phi_0(x_2) e^{i(kx_1 - \omega t)} \quad \text{for} \quad x_2 < 0 \quad [3.79]$$

satisfies the Laplace equation:

$$\varepsilon_{11} \frac{\partial^2 \Phi}{\partial x_1^2} + \varepsilon_{22} \frac{\partial^2 \Phi}{\partial x_2^2} + 2\varepsilon_{12} \frac{\partial^2 \Phi}{\partial x_1 \partial x_2} = 0 \quad [3.80]$$

that is:

$$\varepsilon_{22} \frac{d^2 \Phi_0}{dx_2^2} + 2ik\varepsilon_{12} \frac{d\Phi_0}{dx_2} - \varepsilon_{11} k^2 \Phi_0 = 0 \quad [3.81]$$

The physical solution, vanishing at $x_2 = -\infty$, takes the form $e^{\alpha x_2}$ with $\text{Re}[\alpha] > 0$:

$$\varepsilon_{22} \alpha^2 + 2ik\alpha \varepsilon_{12} - \varepsilon_{11} k^2 = 0 \quad [3.82]$$

Since $\varepsilon_{11} \varepsilon_{22} > \varepsilon_{12}^2$, the acceptable root is:

$$\alpha = \frac{1}{\varepsilon_{22}} \left[-ik\varepsilon_{12} + |k| \sqrt{\varepsilon_{11}\varepsilon_{22} - \varepsilon_{12}^2} \right] \quad [3.83]$$

The normal component of the induction D_2 is proportional to the potential Φ:

$$D_2 = -\varepsilon_{12} \frac{\partial \Phi}{\partial x_1} - \varepsilon_{22} \frac{\partial \Phi}{\partial x_2} = -(ik\varepsilon_{12} + \alpha\varepsilon_{22})\Phi = -|k|\sqrt{\varepsilon_{11}\varepsilon_{22} - \varepsilon_{12}^2}\,\Phi \quad [3.84]$$

The surface permittivity of the dielectric is defined by the ratio between the two electric quantities D_2 and Φ that are conserved at the interface:

$$\varepsilon_d = -\frac{D_2}{|k|\Phi} = \sqrt{\varepsilon_{11}\varepsilon_{22} - \varepsilon_{12}^2} \quad [3.85]$$

This constant is independent of ω and k. The sign is negative because the dielectric is located on the $x_2 < 0$ side, and the division by $|k|$ ensures that ε_d does not change sign with the direction of wave propagation.

In a *piezoelectric material*, the electric induction D_2 and the electric potential Φ accompanying the elastic surface wave are no longer proportional, for any x_2, because the wave is a linear combination of partial waves with different factors m_J (equation [3.71]). However, according to the mechanical boundary conditions [3.74] on the free surface, the amplitudes a_1, a_2, a_3 of the partial waves are proportional to a_4, so that on this surface, the electric potential and the normal component of the electric induction are also proportional to a_4. The *surface piezoelectric permittivity* $\tilde{\varepsilon}$ is therefore defined by the relation:

$$\tilde{\varepsilon} = \frac{D_2(0_+)}{|k|\Phi(0_+)} \quad [3.86]$$

with a positive sign, because the piezoelectric solid is located on the $x_2 > 0$ side. According to the relation [1.229]:

$$D_2(0_+) = \frac{\partial}{\partial x_j}\left(e_{2jl}u_l - \varepsilon_{2j}\Phi\right)\bigg|_{0_+} = ik\left[e_{21l}p_l - \varepsilon_{12}\Phi_0 + m\left(e_{22l}p_l - \varepsilon_{22}\Phi_0\right)\right] \quad [3.87]$$

and by taking $\Phi_{0J} = p_{4J} = 1$, the summation of all partial waves yields:

$$\tilde{\varepsilon} = \frac{ik}{|k|\sum_{J=1}^{4} a_J} \sum_{J=1}^{4} a_J\left[\left(e_{21l} + m_J e_{22l}\right)p_{lJ} - \left(\varepsilon_{12} + m_J\varepsilon_{22}\right)\right], \quad l = 1, 2, 3 \quad [3.88]$$

For a given crystal and orientation, the surface piezoelectric permittivity only depends on the phase velocity $V = \omega/k$: $\tilde{\varepsilon} = \tilde{\varepsilon}(V)$.

Considering the definitions [3.85] and [3.86] of the surface permittivities and the continuity of electrical quantities D_2 and Φ at the interface $x_2 = 0$, the relation:

$$\tilde{\varepsilon}(V) + \varepsilon_d = 0 \quad [3.89]$$

shows that the phase velocity V of the Rayleigh wave depends on the adjacent medium through its permittivity ε_d. Let us examine three cases:

– if the surface is covered by a very thin metallic film (short-circuit condition: $\Phi = 0 \longrightarrow \varepsilon_d = \infty$):

$$V = V_\infty \quad \text{with} \quad \tilde{\varepsilon}(V_\infty) = \infty \quad [3.90]$$

– if the surface is covered by a hypothetical medium with zero-permittivity (open-circuit condition: $D_2(0) = 0$):

$$V = V_0 \quad \text{with} \quad \tilde{\varepsilon}(V_0) = 0 \quad [3.91]$$

– if the adjacent "medium" is the vacuum ($\varepsilon_d = \varepsilon_0$), the velocity V_R of the Rayleigh wave lies between the two extreme values V_∞ and V_0:

$$V_\infty < V_R < V_0 \quad \text{with} \quad \tilde{\varepsilon}(V_R) = -\varepsilon_0 \quad [3.92]$$

The velocity V_∞ is lower than the velocity V_0 since a metallic film deposited on the surface partially removes the piezoelectricity of the material, by canceling the tangential component of the electric field. For materials with a high relative permittivity, like lithium niobate, the velocities V_R and V_0 are very close. The relative variation in velocity:

$$\frac{\Delta V}{V_0} = \frac{V_0 - V_\infty}{V_0} \qquad [3.93]$$

is always small (at most a few percent).

Starting from relations [3.91] and [3.90], it is possible to postulate an approximate expression for $\tilde{\varepsilon}$ as a function of the phase velocity $V = \omega/k$:

$$\tilde{\varepsilon}(\omega, k) = \varepsilon_p^T \frac{V^2 - V_0^2}{V^2 - V_\infty^2} \qquad [3.94]$$

valid in the vicinity of V_R. For very high velocities, that is, at high frequency, the limiting value $\tilde{\varepsilon}_{HF}$ is equal to the effective permittivity [3.85] of the piezoelectric medium:

$$\tilde{\varepsilon}_{HF} = \varepsilon_p^T = \sqrt{\varepsilon_{11}^T \varepsilon_{22}^T - \left(\varepsilon_{12}^T\right)^2} \qquad [3.95]$$

because the mechanical phenomena, which are much slower, are no longer coupled to the electrical phenomena. The constants ε_{ij}^T must be chosen, since the mechanical traction is zero on the surface. At low frequency, the permittivity $\tilde{\varepsilon}_{BF} = \varepsilon_p^T (V_0/V_\infty)^2$ is greater than the high-frequency value ε_p^T. If the direction of propagation of the wave is specified, as $x_1 > 0$, for example, the previous formula can be simplified to:

$$\tilde{\varepsilon}(\omega, k) = \varepsilon_p^T \frac{V - V_0}{V - V_\infty} = \varepsilon_p^T \frac{\omega - kV_0}{\omega - kV_\infty}, \qquad k > 0 \qquad [3.96]$$

3.1.3.3. *Electromechanical coupling coefficient*

As for bulk waves, it is useful to define a factor of merit K_S, called the electromechanical coupling coefficient. This factor expresses the fact that a piezoelectric material transforms a fraction K_S^2 of the electrical energy supplied to it into surface mechanical energy ($0 < K_S < 1$). It is related to the existence of the surface piezoelectric permittivity, whose value at low frequencies is greater than the high-frequency value. Let us assume that a continuous voltage v_0 is applied between two electrodes deposited on the free surface. With the solid, these electrodes constitute a capacitor of capacitance C. The energy supplied by the electrical source and stored in this capacitor is:

$$U_s = \frac{1}{2} C_{LF} v_0^2 \qquad [3.97]$$

where the static capacitance C_{LF} is proportional to $\tilde{\varepsilon}_{LF} + \varepsilon_d$. If, at time $t = 0$, the electrodes are short-circuited (Figure 3.12(a)), the capacitor returns the electrical energy:

$$U_r = \frac{1}{2} C_{HF} v_0^2 \qquad [3.98]$$

because the spectrum of the discontinuity at $t = 0$, the negative step of voltage, consists of very high frequencies. The difference $U_s - U_r$ represents the acoustic energy U_{ac} generated:

$$U_{ac} = \frac{1}{2}(C_{LF} - C_{HF}) v_0^2 = \frac{C_{LF} - C_{HF}}{C_{LF}} U_s \qquad [3.99]$$

By definition, the ratio U_{ac}/U_s is equal to the square of the electromechanical coupling coefficient:

$$K_S^2 = \frac{U_{ac}}{U_s} = \frac{\tilde{\varepsilon}_{LF} - \tilde{\varepsilon}_{HF}}{\tilde{\varepsilon}_{LF} + \varepsilon_d} \qquad [3.100]$$

The same result is obtained by assuming that at time $t = 0$ a charge $\pm Q_0$ is injected on the electrodes (Figure 3.12(b)). The energy supplied by the generator in the form of a current impulse $[i(t) = Q_0 \delta(t)]$ is $Q_0^2/2C_{HF}$ since the material reacts to the discontinuity with its high-frequency permittivity. In the steady-state regime, the stored energy $Q_0^2/2C_{LF}$ is smaller. The difference, which represents the acoustic energy generated:

$$U_{ac} = \left(\frac{1}{C_{HF}} - \frac{1}{C_{LF}} \right) \frac{Q_0^2}{2} = \left(1 - \frac{C_{HF}}{C_{LF}} \right) U_s \qquad [3.101]$$

is identical to the previous value [3.99]. The two reasonings show that regardless of the excitation procedure, by a voltage or a current source, the coupling coefficient K_S is very useful to compare the efficiency of piezoelectric materials for the generation of SAW by transducers with interdigitated comb-shaped electrodes (Volume 2, Chapter 3).

When $\tilde{\varepsilon}(\omega, k)$ is approximated by formula [3.94], it comes:

$$\tilde{\varepsilon}_{LF} = \varepsilon_p^T \frac{V_0^2}{V_\infty^2} \quad \text{and} \quad \tilde{\varepsilon}_{HF} = \varepsilon_p^T \qquad [3.102]$$

and then:

$$K_S^2 = \frac{V_0^2 - V_\infty^2}{V_0^2 + V_\infty^2 \varepsilon_d/\varepsilon_p} = \frac{V_0 - V_\infty}{V_0} \frac{V_0 + V_\infty}{V_0} \left(1 + \frac{\varepsilon_d V_\infty^2}{\varepsilon_p V_0^2} \right)^{-1} \qquad [3.103]$$

Since $V_0 \approx V_\infty$, we obtain:

$$K_S^2 \approx \frac{2}{1 + \varepsilon_d/\varepsilon_p} \frac{\Delta V}{V} \qquad [3.104]$$

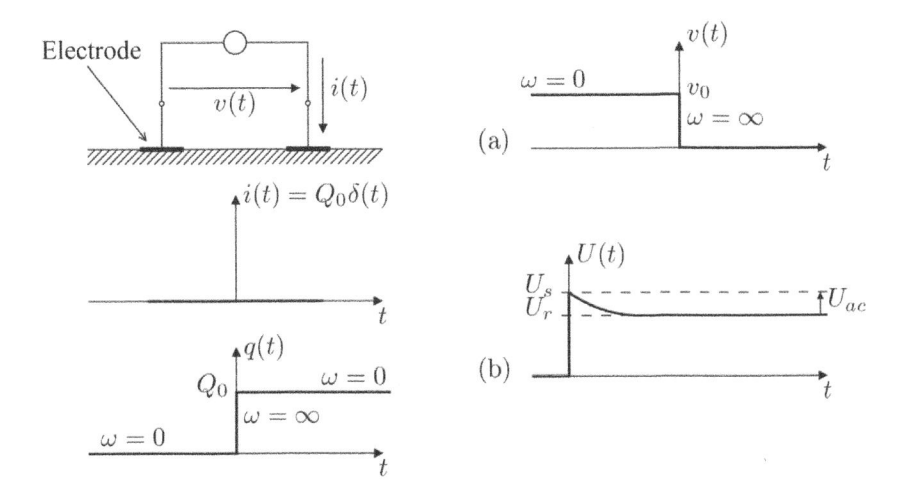

Figure 3.12. *Capacitor formed of two electrodes in contact with the piezoelectric solid. a) "Instantaneous" discharge of the capacitor previously charged with a voltage v_0. b) Charge by a current pulse $Q_0 \delta(t)$. In both cases, the mechanical energy generated, U_{ac}, is the fraction K_S^2 of the supplied electrical energy U_s*

If the dielectric is a vacuum and if the relative effective permittivity of the piezoelectric solid is much larger than unity (as for lithium niobate or lithium tantalate), the approximate formula can be written as:

$$K_S^2 \approx 2\frac{\Delta V}{V} = 2\frac{V_0 - V_\infty}{V_0} \qquad [3.105]$$

3.1.3.4. *Amplitude of electric potential and power*

The amplitude of mechanical displacement and the electric potential Φ_0 on the surface are related to the power density, $P_R = \langle P \rangle / w$, of the Rayleigh wave, where $\langle P \rangle$ is the average flux of acoustic power (Appendix 3). According to formula [A3.27], we have:

$$\langle P \rangle = P_R w = -\frac{1}{4}\frac{\partial B}{\partial k}|\Phi_0|^2 \qquad [3.106]$$

where the susceptance $B(\omega, k)$, defined by relation [A3.25], is related to the surface piezoelectric permittivity $\tilde{\varepsilon}(\omega/k)$. Indeed, in the case of a planar waveguide made of an insulating medium ($j_n = 0$) and since the surface density of electric charges $\dot{\sigma}_e = j_n - j'_n = -j'_n$ is constant over the width w (along x_3) of the Rayleigh wave beam, and zero outside it, we obtain:

$$j = -\int_C j'_n \, dc = \int_C \dot{\sigma}_e \, dc = i\omega w \sigma_e \qquad [3.107]$$

Considering the definitions for the surface permittivities of the dielectric [3.85], the piezoelectric medium [3.86] and the electrical boundary conditions $\Phi(0_+) = \Phi(0_-)$:

$$\sigma_e = D_2(0_+) - D_2(0_-) = (\tilde{\varepsilon} + \varepsilon_d)|k|\Phi(0) \qquad [3.108]$$

the susceptance is equal to:

$$B(\omega, k) = ww k [\tilde{\varepsilon}(\omega/k) + \varepsilon_d] \quad \text{for} \quad k > 0 \qquad [3.109]$$

The first term of the derivative:

$$\frac{\partial B}{\partial k} = \frac{B(\omega, k)}{k} + ww k \frac{\partial \tilde{\varepsilon}}{\partial k} \qquad [3.110]$$

is zero for a free mode $[B(\omega, k) = 0]$, therefore:

$$P_R = -\frac{\omega}{4} k_R \frac{\partial \tilde{\varepsilon}}{\partial k}\bigg|_{k_R} |\Phi_0|^2 = \frac{\omega}{4} V_R \frac{\partial \tilde{\varepsilon}}{\partial V}\bigg|_{V_R} |\Phi_0|^2 \qquad [3.111]$$

According to equation [3.96], the derivative of the surface permittivity can be expressed as a function of the electromechanical coupling coefficient K_R of the Rayleigh wave:

$$K_R^2 = 2\frac{V_R - V_\infty}{V_R} \quad \text{hence} \quad V_R \frac{\partial \tilde{\varepsilon}}{\partial V}\bigg|_{V_R} = \frac{2(\varepsilon_p + \varepsilon_d)}{K_R^2} \qquad [3.112]$$

On the free surface, the amplitudes of the potential and of the tangential component of the electric field $E_1 = -ik_R \Phi$ that accompany the Rayleigh wave are given by:

$$|\Phi_0| = K_R \sqrt{\frac{2}{\varepsilon_p + \varepsilon_d} \frac{P_R}{\omega}} \quad \text{and} \quad |E_1| = \frac{K_R}{V_R} \sqrt{\frac{2\omega P_R}{\varepsilon_p + \varepsilon_d}} \qquad [3.113]$$

ORDER OF MAGNITUDE.– For lithium niobate: $K_R = 0.24$, $V_R = 4\,000$ m/s, $\varepsilon_p + \varepsilon_0 = 6 \times 10^{-10}$ F/m. Let $P_R = 1$ W/cm and $\omega = 10^9$ rad/s, that is, $f = 160$ MHz: $|\Phi_0| = 4.3$ V. The value of the electric field: $|E_1| = 1.1 \times 10^6$ V/m is inversely proportional to the Rayleigh wavelength ($\lambda_R = 25$ μm).

3.1.3.5. *Characteristics of the main materials*

Once the velocity V is known, the polarizations $\underline{p}_J$ of the four partial waves result from the system [3.68] for each value of the decay factors satisfying the characteristic equation. The amplitudes a_J are deduced from the mechanical and electrical boundary conditions. Equation [3.71] then gives the components of the mechanical displacement and the electric potential. In practice, the equations can only be solved numerically. The complete exploration of a crystal through a rotation of the plane of the cut and the propagation direction requires a very large number of operations. It is therefore recommended to exploit, as much as possible, the symmetry of the crystal.

The most natural method of investigation consists of rotating the propagation direction $\underline{n}$ in the plane P of the free surface, which is fixed. This method is not very interesting, considering that in practice only pure modes, that is, modes whose energy velocity $\underline{V}^e$ is parallel to $\underline{n}$, are usable. Indeed, except for a few directions for which the angle of deviation $(\underline{V}^e, \underline{n})$ is accidentally zero, the pure mode directions are the symmetry axes.

The second method consists of rotating the free surface P around the fixed propagation direction $\underline{n}$ (Figure 3.13(a)). Regardless of the orientation of this surface, the mode is pure if $\underline{n}$ is along a symmetry axis that conserves the plane P, that is:

– an A_2, A_4, $\overline{A}_4$ or A_6 axis, which contains a binary axis;

– an $\overline{A}_2$ or $\overline{A}_6$ axis that includes a perpendicular mirror plane.

In these conditions, it is possible to optimize a characteristic by choosing the angle of the cut, for instance, to maximize the coupling coefficient ($Y + 131.5°$ cut of lithium niobate – propagation along the X axis which is perpendicular to a mirror) or to cancel the temperature coefficient (ST-cut of quartz – propagation along the binary axis X).

In the third method, the propagation direction rotates in the sagittal plane, which is fixed (Figure 3.13(b)). Therefore, there is a simultaneous rotation of the propagation direction and of the cut. Whatever the direction $\underline{n}$, the mode is pure if the sagittal plane is a mirror plane because the energy velocity, contained in the plane P of the free surface and in the mirror, must be directed along $\underline{n}$.

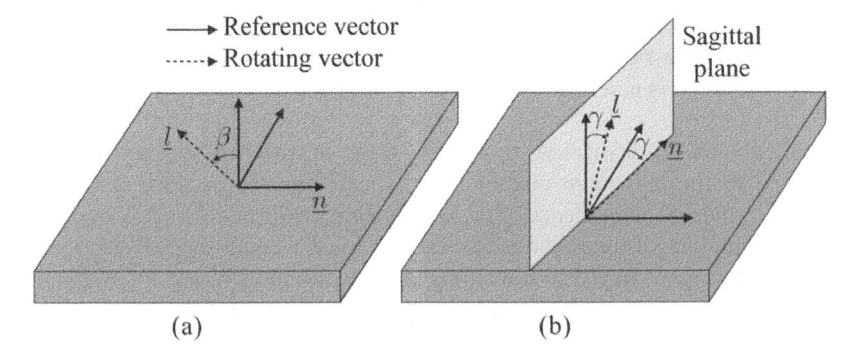

Figure 3.13. *Methods for investigating surface acoustic wave properties of a crystal. Rotation a) of the cut around the propagation direction, b) of the propagation direction and the cut, simultaneously*

Among the main piezoelectric solids that have been studied, we can mention zinc oxide (ZnO), bismuth germanium oxide ($Bi_{12}GeO_{20}$), berlinite ($AlPO_4$), lithium niobate ($LiNbO_3$), lithium tantalate ($LiTaO_3$) and quartz (SiO_2). The last three are the most commonly used.

– *Lithium niobate* (class $3m$) is a strongly piezoelectric material, available in large crystals (cylinders of diameter 20 cm and length more than 20 cm). Since the YZ plane is a mirror, the Rayleigh waves are pure modes for any free surface containing the X axis if the propagation direction is perpendicular or parallel to this axis. The coupling coefficient is maximum ($K_R^2 = 5.6$ %) for propagation along the X axis and an angle $\beta = 131.5°$ between the cut and the Y axis (Slobodnik 1973). It is equal to 4.8% when the propagation is along Z in the XZ plane (Y-cut).

– *Quartz* (class 32) has no mirror plane and does not lend itself to the propagation of $\tilde{R}_2$ type wave. However, waves of $\tilde{R}_3$ type propagating along the binary axis X are pure, for any orientation of the free surface. Although weakly piezoelectric ($K_R^2 = 0.22$% for the Y-cut and propagation along X), quartz is widely used, as its properties vary little with temperature. The temperature coefficient of the delay τ is zero at 25°C for the ST-cut (Y-cut rotated by 42.5° in the YZ plane) and propagation along X.

– *Lithium tantalate* is a piezoelectric material whose properties (electromechanical coupling factor and temperature coefficient) are intermediate between those of lithium niobate and quartz.

Table 3.1 summarizes the characteristics of these crystals. ZnO/glass refers to a layer of zinc oxide deposited on two comb-shaped aluminum electrodes, themselves engraved on a glass substrate. Such interdigitated electrodes are used to generate Rayleigh waves (see Volume 2, section 3.2). When the substrate is not piezoelectric, the electrodes must be covered with a piezoelectric layer (ZnO, AlN). This layer, whose Z axis is perpendicular to the substrate, is generally deposited by sputtering.

The cuts and propagation directions used in practice do not necessarily correspond to the maximum coupling coefficient, because in some devices, it is important to reduce the influence of the temperature, while in others it is important to reduce the amount of spurious waves excited by the comb-shaped electrodes, in addition to the useful Rayleigh waves. From this point of view, the $Y + 128°$ cut of lithium niobate is more advantageous than the $Y + 131.5°$ cut. Finally, the number of available piezoelectric crystals is restricted. Consequently, some laboratories elaborate others, such as gallium phosphate ($GaPO_4$) or langasite ($La_2Ga_5SiO_{14}$). However, to be usable, a material must satisfy requirements related to purity, coupling coefficient, thermal stability, dimensions and cost.

Crystal	Class	Cut	Direction	Type	Velocity (m/s)	K^2 (%)	$\varepsilon/\varepsilon_0$	$\left\|\frac{1}{\tau}\frac{d\tau}{d\theta}\right\|$ (10^{-6} K^{-1})	References
LiNbO$_3$	$3m$	Y	Z	$\tilde{R}_2$	3488	4.8		94	(Slobodnik 1973)
		$Y+128°$	X	$\tilde{R}_3$	3992	5.3	46	75	(Campbell 1989)
		$Y+131.5°$	X	$\tilde{R}_3$	4000	5.6		72	(Slobodnik 1973)
LiTaO$_3$	$3m$	Y	Z	$\tilde{R}_2$	3230	0.9	47	38	(Feldmann and Hénaff 1977)
Quartz	32	ST	X	$\tilde{R}_3$	3158	0.16	4.5	0	(Slobodnik 1973)
Bi$_{12}$GeO$_{20}$	23	(100)	[011]	$\tilde{R}_2$	1683	1.4	40	140	(Pratt 1972)
AlPO$_4$	32	ST	X	$\tilde{R}_3$	2736	0.56	6.1	0	(Feldmann and Hénaff 1986)
ZnO/glass	$6mm$	Z	$\perp Z$	$\tilde{R}_2$	2576	1.4	10.8	11	(Fujishima 1990)

Table 3.1. Characteristics of Rayleigh waves for different piezoelectric materials

3.2. Interface waves

In this section, we will study the propagation of elastic waves at the interface between an isotropic solid and a perfect fluid, and then at the interface between two isotropic solids. In the first configuration, several modes exist with different characteristics depending on the value of the sound speed in the fluid with respect to the velocity of the transverse wave in the solid. In the second configuration, there is a solution only in a domain for the parameters of the two materials, more or less restricted according to the nature of the contact between the two solids.

3.2.1. *Isotropic solid-perfect fluid interface*

The surface of the isotropic solid, occupying the half-space $x_2 > 0$, is now covered by a fluid that is assumed to be perfect (Figure 3.14). Ox_2 is the normal to the free surface, directed toward the bulk of the solid, and Ox_1 is the propagation direction. A wave, discovered by Scholte, propagates at the interface. This wave transports a fraction of the acoustic energy in the fluid and a fraction in the solid, without attenuation. The distribution of the energy between the two media depends on the velocity V_f of the longitudinal waves in the fluid, compared to the velocity V_T of the transverse waves in the solid. When V_f is smaller than V_T, a Rayleigh wave still propagates on the surface of the isotropic solid, but with an attenuation, due to the radiation of a part of the acoustic energy into the fluid.

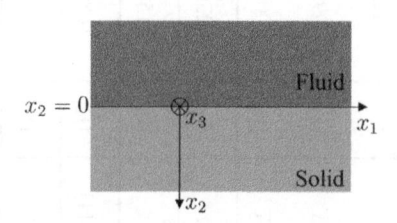

Figure 3.14. *Waves at the solid-fluid interface, disposition of the axes*

3.2.1.1. *Characteristic equation*

The Scholte wave, as well as the leaky Rayleigh wave, are not dispersive: their velocity, independent of frequency, is a solution of a characteristic equation that results from the boundary conditions at the interface $x_2 = 0$. These conditions have been established in section 2.4.2.1 to calculate the coefficients of reflection r_{LL}, conversion r_{LT} and transmission t_{Lf} for a longitudinal incident wave originating from the solid.

The displacement and stress fields in the solid remain the same as in section 3.1.1 and are given, respectively, by expressions [3.7], [3.9] and [3.13] with $k_{2L} = i\alpha_L$ and $k_{2T} = i\alpha_T$. Since the fluid is not viscous, only a longitudinal wave can propagate here, with a mechanical displacement parallel to the wave vector $\underline{k}_f = (k, -k_{2f}, 0)^t$:

$$\underline{u}^f(x_1, x_2, t) = \begin{pmatrix} ik \\ -ik_{2f} \end{pmatrix} a_f e^{-ik_{2f}x_2} e^{i(kx_1 - \omega t)} \quad [3.114]$$

with:

$$k^2 + k_{2f}^2 = k_f^2 = \frac{\omega^2}{V_f^2} \quad [3.115]$$

Either k_{2f} is real and positive and the wave in the fluid is a traveling wave in the direction of the negative x_2, or k_{2f} is imaginary: $k_{2f} = i\alpha_f$ with $\alpha_f > 0$, so that the amplitude, $e^{\alpha_f x_2}$, of the evanescent wave decreases and cancels for $x_2 = -\infty$. The acoustic pressure field, p_f, is given by relation [1.323] with $K = \rho_f V_f^2$:

$$p_f(x_1, x_2, t) = -\rho_f V_f^2 \, \text{div} \, \underline{u}^f = \rho_f \omega^2 a_f e^{-ik_{2f}x_2} e^{i(kx_1 - \omega t)} \quad [3.116]$$

In the case of a solid-perfect fluid interface, the boundary conditions: cancelation of the stress σ_{12}, continuity of the stress σ_{22} and the displacement u_2 normal to the interface lead directly to the equation:

$$\begin{pmatrix} 2kk_{2L} & k_{2T}^2 - k^2 & 0 \\ k^2 - k_{2T}^2 & 2kk_{2T} & \dfrac{\rho_f \omega^2}{\mu} \\ k_{2L} & -k & k_{2f} \end{pmatrix} \begin{pmatrix} a_L \\ a_T \\ a_f \end{pmatrix} = \begin{pmatrix} 0 \\ 0 \\ 0 \end{pmatrix} \quad [3.117]$$

This homogeneous system has a non-trivial solution (a_L, a_T and $a_f \neq 0$) if the determinant of the matrix of coefficients is zero, that is:

$$\left(k^2 - k_{2T}^2\right)^2 + 4k^2 k_{2L} k_{2T} + \frac{\rho_f k_{2L}}{\rho k_{2f}} k_T^4 = 0 \quad [3.118]$$

It must be noted again that this characteristic equation is equivalent to cancel the denominator $d_R + \delta_f$ of the reflection and transmission coefficients for the longitudinal and transverse waves at the isotropic solid-perfect fluid interface (see section 2.4.2). The examination of the different solutions, with the unknown being either the wave number k or the phase velocity V of the interface wave, shows that two types of waves may exist: a Scholte wave associated with a real root of the characteristic equation [3.118] and a leaky Rayleigh wave associated with a complex root.

3.2.1.2. *Scholte wave*

By hypothesis, the velocity V of the interface wave is lower than V_f and V_T, that is, $k = k_1 > k_f, k_T > k_L$. The three normal components of the wave vectors are purely imaginary:

$$\begin{cases} k_{2L} = i\sqrt{k^2 - k_L^2} = i\alpha_L \\ k_{2T} = i\sqrt{k^2 - k_T^2} = i\alpha_T \\ k_{2f} = i\sqrt{k^2 - k_f^2} = i\alpha_f \end{cases} \qquad [3.119]$$

and the three corresponding partial waves are evanescent. The terms of equation [3.118] are all real:

$$\left(2k^2 - k_T^2\right)^2 - 4k^2\sqrt{k^2 - k_L^2}\sqrt{k^2 - k_T^2} + k_T^4 \frac{\rho_f}{\rho}\sqrt{\frac{k^2 - k_L^2}{k^2 - k_f^2}} = 0 \qquad [3.120]$$

Upon dividing by k^4, the equation that gives the phase velocity V of the Scholte wave takes the form:

$$\left(2 - \frac{V^2}{V_T^2}\right)^2 - 4\sqrt{\left(1 - \frac{V^2}{V_L^2}\right)\left(1 - \frac{V^2}{V_T^2}\right)} + \frac{\rho_f V^4}{\rho V_T^4}\sqrt{\frac{1 - (V/V_L)^2}{1 - (V/V_f)^2}} = 0$$

$$[3.121]$$

Considering the system of equations [3.117], the amplitudes a_T and a_f are expressed as functions of the amplitude a_L through the relations:

$$a_T = \frac{2ik\alpha_L}{k^2 + \alpha_T^2} a_L \quad \text{and} \quad a_f = \frac{\alpha_L\left(k^2 - \alpha_T^2\right)}{\alpha_f\left(k^2 + \alpha_T^2\right)} a_L \qquad [3.122]$$

Two situations must be distinguished:

– The velocity V_f of the longitudinal wave in the fluid is lower than that of the transverse wave in the solid ($V_f < V_T < V_L$). Equation [3.121] has a real root V_{Sch} slightly smaller than V_f. In the fluid ($x_2 < 0$), the amplitude of the displacement decreases as $e^{\alpha_f x_2}$, that is, very slowly, compared to the wavelength $\lambda = 2\pi/k_{Sch}$, because $\alpha_f = \sqrt{k_{Sch}^2 - k_f^2}$ is much smaller than the wave number k_{Sch}. According to relation [3.122], the amplitude a_f of the partial wave in the fluid is inversely proportional to α_f and is therefore very large. The Scholte wave transports more energy in the fluid than in the solid (Figure 3.15(a)). Furthermore, according to relation [3.114] and since k_{2f} is very small compared to k, the tangential component u_1^f of the displacement is much larger than the normal component u_2^f, which is comparable to the two components u_1 and u_2 in the solid. The Scholte wave is, therefore, easily measurable, by placing a transducer immersed in the fluid parallel to the interface, in order to probe the tangential displacement u_1^f. Moreover, this wave is more sensitive to the properties of the fluid than those of the solid. Consequently, in terms of application, this type of wave is rarely used in non-destructive evaluation to inspect materials; however, it is worthwhile for characterizing the rheology of fluids (Cegla *et al.* 2005).

– The velocity of the longitudinal wave in the fluid is higher than that of the transverse wave in the solid ($V_T < V_f < V_L$); therefore, the Scholte wave velocity is slightly lower than V_T (Glorieux *et al.* 2001). Thus, the amplitude of the displacements in the fluid and in the solid are comparable (Figure 3.15(b)). It is interesting to note that in this case, the Scholte wave becomes particularly sensitive to the velocity of the transverse waves in the solid and thus to its shear modulus.

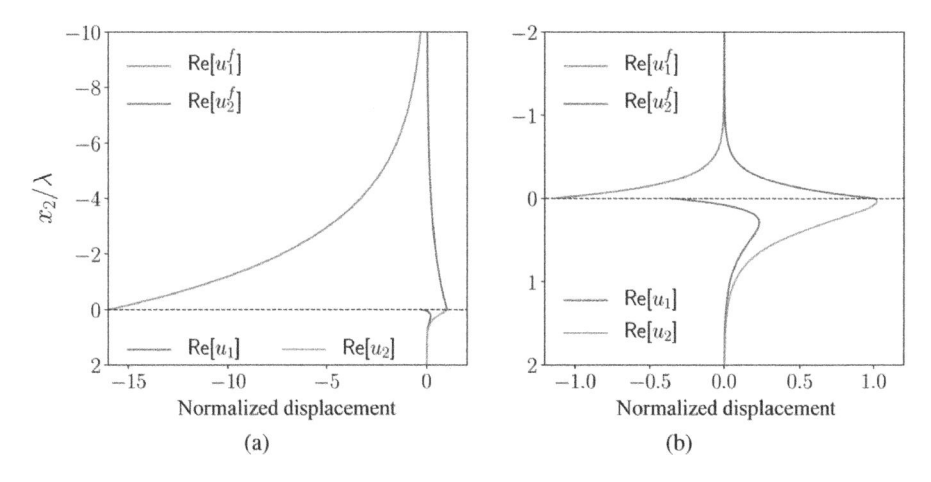

Figure 3.15. *Displacement of the Scholte wave in the liquid (top) and in the solid (bottom) for a) a duralumin–water interface and b) an epoxy–water interface. For a color version of this figure, see www.iste.co.uk/royer/waves1.zip*

In both cases, the three partial waves constituting the Scholte wave are propagative in the x_1 direction and evanescent in the x_2 direction. The energy remains confined at the interface between the two media. Moreover, in agreement with the boundary conditions, the tangential component of the displacement is not continuous at the interface, unlike the normal component (see Figure 3.15).

Since the velocity of the Scholte wave is always lower than the sound speed in the liquid, it cannot be generated by refraction at the interface between the fluid and the solid. This is a major drawback to its use in the field of ultrasonics. On the contrary, the

Scholte wave is widely studied in geophysics, as it propagates at the interface between sediments and water, and similarly in oil wells.

3.2.1.3. Leaky Rayleigh wave

Let us now look for a solution whose phase velocity V lies between V_f and V_T, that is, $k_f > k > k_T > k_L$. It appears that the last term in equation [3.120] is imaginary. The characteristic equation, with complex coefficients, admits a complex root k_{Rf} corresponding to a wave that is attenuated during its propagation. The real part of the wave number, solution of the complete equation, is slightly smaller than k_R: $k_{Rf} \approx k_R + i\alpha_{Rf}$. Its imaginary part, α_{Rf}, expresses an attenuation due to the radiation of a longitudinal wave in the liquid, at an angle $\theta_R = \arcsin(V_f/V_{Rf})$. This angle was introduced during the study of the reflection and transmission at a fluid-solid interface (section 2.4.2.3). It is a *leaky Rayleigh wave*, similar to a pseudo-Rayleigh wave (see section 3.1.2.4), for which the condition of zero energy at an infinite distance from the interface is not satisfied.

The analysis developed in section 4.3.2 to calculate the attenuation coefficient α of the Lamb waves guided in an isotropic plate can be easily transposed to the case of a Rayleigh wave. Indeed, a semi-infinite solid is the limit of a plate whose interface with the fluid is fixed at $x_2 = 0$ and whose other face, located at $x_2 = -h$, moves away to infinity. Since the Rayleigh wave is not dispersive: $V_g = V_R$, the formula [4.131] is written as:

$$\alpha_R \lambda_R \approx \frac{\pi \rho_f V_f}{\rho V_R \cos \theta_R} \frac{|u_2(0)|^2}{\int_{-\infty}^{0} \left[|u_1(X_2)|^2 + |u_2(X_2)|^2\right] dX_2} \quad \text{with} \quad X_2 = kx_2$$

[3.123]

The comparison of expressions [3.28] and [3.35] shows that the second ratio is equal to χ_L/E, so that the attenuation per wavelength, due to the radiation into the fluid, is given by:

$$\alpha_R \lambda_R \approx \frac{\pi \chi_L}{E \cos \theta_R} \frac{\rho_f V_f}{\rho V_R} \quad [3.124]$$

The curves in Figure 3.16(a), plotted for a fused silica substrate ($V_R = 3\,400$ m/s), show that this analysis, which neglects the influence of the fluid on the Rayleigh wave, is a very good approximation of the numerical result $\alpha_{Rf}\lambda_R$, as long as $V_f < V_R/2$ and $\rho_f < \rho/2$, that is $\alpha_R\lambda_R < 0.285$, unlike the ratio of acoustic impedances, $\rho_f V_f/\rho V_R$, which is often used to estimate the Rayleigh wave attenuation (Dransfeld and Salzmann 1970).

ORDER OF MAGNITUDE.– For the fused silica-water interface: $\alpha_R \lambda_R = 0.245$ and for a path of $10\,\lambda_R$, the amplitude of the Rayleigh wave is divided by 11.5, that is, an attenuation of 21 dB. More than 99% of the energy is radiated into the fluid.

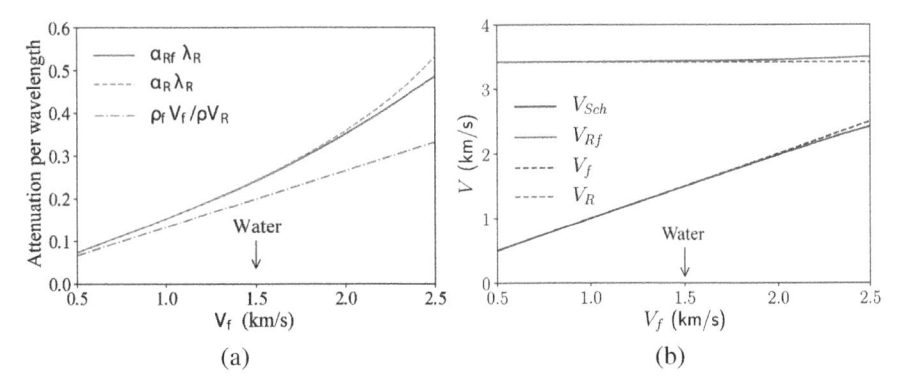

Figure 3.16. *Fused silica-fluid interface. a) Attenuation of the leaky Rayleigh wave versus the speed of sound V_f in the fluid. b) Scholte and Rayleigh wave velocities as functions of V_f*

The velocities of the Scholte and leaky Rayleigh waves are plotted in Figure 3.16(b) as functions of the speed of sound V_f in the liquid. The material considered for these simulations is fused silica, so that the velocity V_f is always lower than the transverse wave velocity V_T in the solid. These velocities must be compared, respectively, with the velocity V_f and with the Rayleigh wave velocity V_R. Thus, as long as V_f is relatively small compared to V_R, the velocity of the Scholte wave is very close to but lower than V_f, and the velocity of the leaky Rayleigh wave is very close to but higher than V_R.

The leaky Rayleigh wave can be easily generated by the interaction of a wave in the liquid, propagating at the angle of incidence θ_R. It is also particularly sensitive to the properties of the solid, which makes it very attractive in the field of ultrasonic non-destructive evaluation. Because of its acoustic radiation into the liquid, it is also used in acousto-fluidics to manipulate small objects or to atomize liquids (Gedge and Hill 2012). It is also the basis of ultrasonic flow meters.

3.2.2. *Interface between two isotropic solids*

The solid occupying the half-space $x_2 \geq 0$ is now loaded by another solid half-space, characterized by its mass density ρ_s and its Lamé constants, λ_s and μ_s (Figure 3.17). The velocities of longitudinal and transverse bulk waves in this solid are $V_L^s = \sqrt{(\lambda_s + 2\mu_s)/\rho_s}$ and $V_T^s = \sqrt{\mu_s/\rho_s}$, respectively.

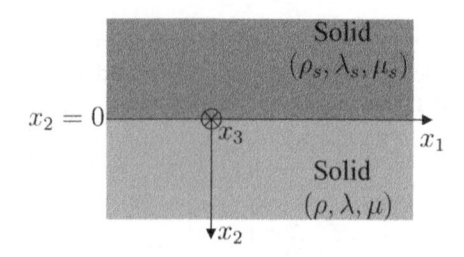

Figure 3.17. *Waves at the interface between two solids, arrangement of the axes*

We are looking for a wave whose mechanical displacement decreases exponentially on both sides of the interface $x_2 = 0$. In the half-space $x_2 \geq 0$, its expression is given by equation [3.7]. In the half-space $x_2 \leq 0$, the mechanical displacement contained in the sagittal plane is of the form:

$$\underline{u}^s(x_1, x_2, t) = \left[\begin{pmatrix} ik \\ \beta_L \end{pmatrix} b_L e^{\beta_L x_2} + \begin{pmatrix} \beta_T \\ -ik \end{pmatrix} b_T e^{\beta_T x_2} \right] e^{i(kx_1 - \omega t)} \qquad [3.125]$$

where the decay factors β_M are defined by:

$$\beta_M = \sqrt{k^2 - (k_M^s)^2} \qquad \text{with} \qquad M = L, T \qquad [3.126]$$

3.2.2.1. *Perfect interface: Stoneley wave*

If the contact between the two half-spaces is assumed to be perfect, the boundary conditions are the continuity of normal and tangential stresses, and the continuity of normal and tangential displacements at the interface.

Taking into account the expressions of the stresses and displacements in each medium, the four equations obtained lead to a matrix system:

$$\begin{pmatrix} -2i\mu k\alpha_L & \mu(k^2 + \alpha_T^2) & -2i\mu_s k\beta_L & -\mu_s(k^2 + \beta_T^2) \\ \mu(k^2 + \alpha_T^2) & 2i\mu k\alpha_T & -\mu_s(k^2 + \beta_T^2) & 2i\mu_s k\beta_T \\ ik & -\alpha_T & -ik & -\beta_T \\ -\alpha_L & -ik & -\beta_L & ik \end{pmatrix} \begin{pmatrix} a_L \\ a_T \\ b_L \\ b_T \end{pmatrix} = \begin{pmatrix} 0 \\ 0 \\ 0 \\ 0 \end{pmatrix}$$

[3.127]

After rather tedious calculations, the Stoneley determinant may be written in the form (Stoneley 1924):

$$d_{Sto} = \mu_s^2 d_R^s \left(\alpha_L \alpha_T - k^2\right) + \mu^2 d_R \left(\beta_L \beta_T - k^2\right)$$

$$+ \mu\mu_s k^2 \left[k^2 \frac{V^4}{(V_T V_T^s)^2}(\alpha_L \beta_T + \alpha_T \beta_L) + 2(k^2 + \alpha_T^2)(k^2 + \beta_T^2)\right] \quad [3.128]$$

$$+ 4\mu\mu_s k^2 \left[2\beta_L \beta_T \alpha_L \alpha_T - \beta_L \beta_T (k^2 + \alpha_T^2) - \alpha_L \alpha_T (k^2 + \beta_T^2)\right]$$

where d_R^s is the determinant of the Rayleigh wave in the material located at $x_2 < 0$.

The displacement components of the Stoneley wave inside the two solids are plotted in Figure 3.18 for a steel-duralumin couple. The decay on either side of the interface is similar to that of a Rayleigh wave.

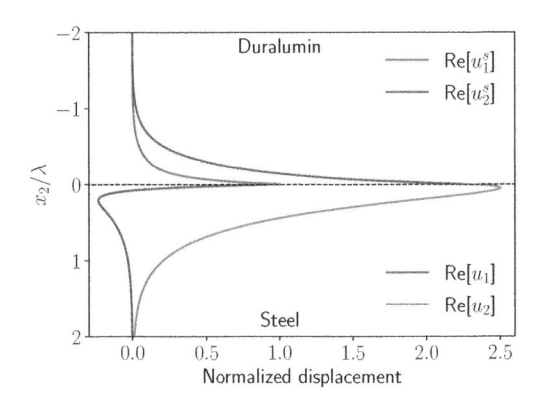

Figure 3.18. *Variation, on both sides of a steel-duralumin interface, of the two displacement components of the Stoneley wave. For a color version of this figure, see www.iste.co.uk/royer/waves1.zip*

3.2.2.2. *Degraded interface*

If the contact is sliding, the boundary conditions are the cancelation of the tangential stresses and the continuity of normal stresses and normal displacements. By applying these conditions, namely $\sigma_{12} = 0$, $\sigma_{12}^s = 0$, $\sigma_{22} = \sigma_{22}^s$ and $u_2 = u_2^s$, we obtain:

$$\begin{pmatrix} -2i\mu k\alpha_L & \mu(k^2 + \alpha_T^2) & 0 & 0 \\ 0 & 0 & -2i\mu_s k\beta_L & -\mu_s(k^2 + \beta_T^2) \\ \mu(k^2 + \alpha_T^2) & 2i\mu k\alpha_T & -\mu_s(k^2 + \beta_T^2) & 2i\mu_s k\beta_T \\ -\alpha_L & -ik & -\beta_L & ik \end{pmatrix} \begin{pmatrix} a_L \\ a_T \\ b_L \\ b_T \end{pmatrix} = \begin{pmatrix} 0 \\ 0 \\ 0 \\ 0 \end{pmatrix}$$

[3.129]

and the expression for the determinant is:

$$d_{Slip} = \tilde{\mu}^2 \alpha_L d_R^s + \tilde{\rho}\beta_L d_R \quad \text{with} \quad \tilde{\mu} = \frac{\mu_s}{\mu} \quad \text{and} \quad \tilde{\rho} = \frac{\rho_s}{\rho} \qquad [3.130]$$

As in the case of a solid half-space loaded by a perfect fluid, there exist two roots with a physical meaning of the characteristic equations resulting from the vanishing of determinants given by equations [3.128] and [3.130]: a real root corresponding to a Stoneley wave and a complex root associated with a leaky Rayleigh wave. The Stoneley wave is not commonly used in the field of ultrasonics as it does not exist for all pairs of materials and it is very difficult to generate with conventional methods. On the other hand, it is widely studied in geophysics to characterize the fracture of rocks. The maps of the Stoneley wave velocity are represented as functions of the ratio of shear moduli $\tilde{\mu}$, on the abscissa, and the ratio of mass densities $\tilde{\rho}$, on the ordinate, in Figure 3.19(a) for a perfect interface, and in Figure 3.19(b) for a sliding interface. These results show that while the range of existence for the Stoneley wave is very restricted when the contact between the two solids is rigid (Chadwick and Borejko 1994), the degradation of this contact favors the existence of this interface wave in a wider domain.

(a) Perfect interface (b) Sliding interface

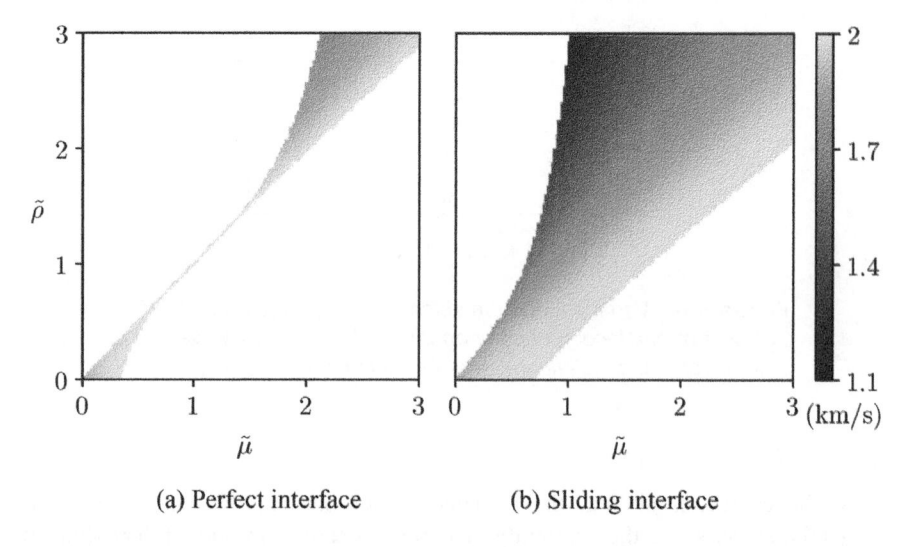

Figure 3.19. *Maps of the Stoneley wave velocity (km/s) as functions of the ratio of shear moduli $\tilde{\mu} = \mu_s/\mu$ on the abscissa and the ratio of mass densities $\tilde{\rho} = \rho_s/\rho$ on the ordinate. For a color version of this figure, see www.iste.co.uk/royer/waves1.zip*

In order to establish the relationship between the characteristic determinants, it is necessary to introduce a new boundary condition corresponding to the continuity of the tangential stresses ($\sigma_{12} = \sigma_{12}^s$) and tangential displacements ($u_1 = u_1^s$), as well as the cancelation of the normal stresses ($\sigma_{22} = \sigma_{22}^s = 0$). The characteristic determinant for this interface of "grip" type is:

$$d_{Grip} = \tilde{\rho}\beta_T d_R + \tilde{\mu}^2 \alpha_T d_R^s \qquad [3.131]$$

The characteristic determinant of the spring model (see equation 2.7), corresponding to the boundary conditions $\sigma_{12} = \sigma_{12}^s = K_t(u_1 - u_1^s)$ and $\sigma_{22} = \sigma_{22}^s = K_l(u_2 - u_2^s)$, is given by (Valier-Brasier et al. 2012):

$$d_{Spring} = \tilde{\mu}^2 d_R d_R^s - \frac{\tilde{\mu}}{\tilde{\rho}}(V/V_T^s)^2 \left(\frac{K_l}{k\mu} d_{Slip} + \frac{K_t}{k\mu} d_{Grip} \right) + \frac{K_l K_t}{(k\mu)^2} d_{Sto}$$

$$[3.132]$$

This determinant shows that the quality of the contact, modeled by the interface stiffnesses K_l and K_t, directly influences the interface waves propagating between the two solids. First of all, this dispersion relation depends on the frequency, indicating that, in this model, the Stoneley wave and the leaky Rayleigh wave are dispersive. The interface stiffnesses being homogeneous to an elasticity divided by a length, a characteristic dimension appears. Moreover, it is interesting to note that when the two interface stiffnesses tend to infinity, the contact becomes perfect and equation [3.132] tends to the Stoneley wave equation. Conversely, when the interface stiffnesses tend to zero, equation [3.132] tends to the Rayleigh wave equation for each half-space.

3.3. Bleustein–Gulyaev wave

According to the decoupling conditions (section 3.1.3.1), a TH wave exists in a semi-infinite piezoelectric solid when the sagittal plane is perpendicular to a binary axis, that is, specifically, in the case of a piezoelectric crystal or ceramic with symmetry $6mm$ (Bleustein 1968; Gulyaev 1969). Therefore, the orientation of the x_1 and x_2 axes is arbitrary in the plane perpendicular to the axis A_6. The expressions for the mechanical displacement u_3 and electric potential $\Phi = u_4$ of a wave whose amplitude decreases with depth ($x_2 > 0$) are:

$$\begin{pmatrix} u_3 \\ u_4 \end{pmatrix} = \begin{pmatrix} p_3 \\ p_4 \end{pmatrix} a e^{imkx_2} e^{i(kx_1 - \omega t)} \quad \text{with} \quad \text{Im}[mk] > 0 \qquad [3.133]$$

Since $C_{45} = 0$ and $e_{14} = e_{25} = 0$, the components of the Christoffel tensor are given by equation [3.70]:

$$\Gamma_{33} = C_{55}^E + m^2 C_{44}^E, \quad \gamma_3 = e_{15} + m^2 e_{24} \quad \text{and} \quad \varepsilon^S = \varepsilon_{11}^S + m^2 \varepsilon_{22}^S$$

$$[3.134]$$

In addition, with $C_{44}^E = C_{55}^E$, $e_{15} = e_{24}$ and $\varepsilon_{11}^S = \varepsilon_{22}^S$, the Christoffel equations [3.76] can be written as:

$$\begin{pmatrix} C_{44}^E(1+m^2) - \rho V^2 & e_{15}(1+m^2) \\ e_{15}(1+m^2) & -\varepsilon_{11}^S(1+m^2) \end{pmatrix} \begin{pmatrix} p_3 \\ p_4 \end{pmatrix} = \begin{pmatrix} 0 \\ 0 \end{pmatrix} \quad [3.135]$$

The compatibility equation for this homogeneous system:

$$(1+m^2)\left[\varepsilon_{11}^S \rho V^2 - (1+m^2)\left(C_{44}^E \varepsilon_{11}^S + e_{15}^2\right)\right] = 0 \quad [3.136]$$

takes the form:

$$(1+m^2)\left(V^2 - V_T^2 - m^2 V_T^2\right) = 0 \quad [3.137]$$

where:

$$V_T = \sqrt{\frac{C_{44}^E + e_{15}^2/\varepsilon_{11}^S}{\rho}} \quad [3.138]$$

is the velocity of the TH wave in the piezoelectric solid. The roots m_1 and m_2 with a positive imaginary part are:

$$\begin{cases} m_1 = i\sqrt{1 - V^2/V_T^2} \;\; \text{si} \;\; V < V_T \;\; \longrightarrow \;\; p_{31} = 1 \;\; \text{and} \;\; p_{41} = \dfrac{e_{15}}{\varepsilon_{11}^S} \\ m_2 = i \;\; \longrightarrow \;\; p_{32} = 0 \;\; \text{and} \;\; p_{42} = \text{any value} \;\; \longrightarrow p_{42} = p_{41} \end{cases} \quad [3.139]$$

The general solution is a linear combination of these two partial waves [3.133] with the same phase velocity $V = \omega/k$:

$$\begin{cases} u_3 = a_1 e^{-\chi k x_2} e^{i(k x_1 - \omega t)} \\ u_4 = \dfrac{e_{15}}{\varepsilon_{11}^S}\left(a_1 e^{-\chi k x_2} + a_2 e^{-k x_2}\right) e^{i(k x_1 - \omega t)} \end{cases} \quad [3.140]$$

by writing:

$$\chi = \sqrt{1 - (V/V_T)^2} \quad [3.141]$$

The coefficients a_1 and a_2 are determined by the mechanical boundary condition $\sigma_{32} = 0$ at $x_2 = 0$. According to equation [1.263], with $u_1 = u_2 = 0$ and $\Phi = u_4$:

$$\sigma_{32} = \frac{\partial}{\partial x_1}\left(C_{45}^E u_3 + e_{14} u_4\right) + \frac{\partial}{\partial x_2}\left(C_{44}^E u_3 + e_{24} u_4\right) \quad [3.142]$$

and since C_{45}^E and e_{14} are zero, we obtain:

$$\sigma_{32}(x_2 = 0) = -k\left[\chi\left(C_{44}^E + \frac{e_{15}^2}{\varepsilon_{11}^S}\right)a_1 + \frac{e_{15}^2}{\varepsilon_{11}^S} a_2\right] = 0 \quad [3.143]$$

By introducing the electromechanical coupling coefficient of the transverse wave (section 1.4.3.2), defined by:

$$K_T^2 = \frac{e_{15}^2}{C_{44}^E \varepsilon_{11}^S + e_{15}^2} \quad \text{we obtain} \quad \chi a_1 + K_T^2 a_2 = 0 \qquad [3.144]$$

and the electric potential $\Phi = u_4$ is given by:

$$\Phi(x_1, x_2, t) = a_1 \frac{e_{15}}{\varepsilon_{11}^S} \left(e^{-\chi k x_2} - \frac{\chi}{K_T^2} e^{-k x_2} \right) e^{i(k x_1 - \omega t)} \qquad [3.145]$$

The phase velocity V is determined by the electrical boundary conditions. According to relation [1.229], with $u_1 = u_2 = 0$, the normal component of the electric induction can be written as:

$$D_2 = \frac{\partial}{\partial x_1} \left(e_{25} u_3 - \varepsilon_{21}^S \Phi \right) + \frac{\partial}{\partial x_2} \left(e_{24} u_3 - \varepsilon_{22}^S \Phi \right) \qquad [3.146]$$

and since ε_{21}^S and e_{25} are zero, and $e_{24} = e_{15}$:

$$D_2(0) = \left. \frac{\partial \left(e_{24} u_3 - \varepsilon_{11}^S \Phi \right)}{\partial x_2} \right|_{x_2 = 0} = -k e_{15} \frac{\chi}{K_T^2} a_1 \qquad [3.147]$$

Considering the value of the potential Φ on the surface $x_2 = 0$, the surface piezoelectric permittivity (equation [3.86]) is expressed by:

$$\tilde{\varepsilon} = \frac{\varepsilon_{11}^S}{1 - K_T^2/\chi} \quad \text{or} \quad \frac{1}{\tilde{\varepsilon}(V)} = \frac{1}{\varepsilon_{11}^S} \left(1 - \frac{K_T^2}{\sqrt{1 - (V/V_T)^2}} \right) \qquad [3.148]$$

Let us examine two typical cases (see section 3.1.3.2):

a) *the surface is covered by a metallic film* whose thickness is very thin compared to the wavelength and brought to zero potential, so that $\tilde{\varepsilon} = \infty$, hence $\chi = K_T^2$. According to equation [3.141], the velocity of the Bleustein–Gulyaev wave:

$$V_\infty = V_T \sqrt{1 - K_T^4} \qquad [3.149]$$

is very close to that of the transverse wave, for example $K_T^2 = 0.2$ imposes $V_\infty = 0.98 V_T$. The decay of the amplitudes of the displacement u_3 and potential Φ becomes more rapid as the coefficient K_T increases. The curves in Figure 3.20 are plotted for a zinc oxide (ZnO) crystal, belonging to the class $6mm$;

b) *the surface is free*: the value of the piezoelectric permittivity $\tilde{\varepsilon} = -\varepsilon_0$ imposes the decay factor χ and the velocity V_B of the Bleustein–Gulyaev wave:

$$\chi = \frac{K_T^2}{1 + \varepsilon_{11}^S/\varepsilon_0} \quad \text{or} \quad V_B = V_T \sqrt{1 - \frac{K_T^4}{\left(1 + \varepsilon_{11}^S/\varepsilon_0\right)^2}} \qquad [3.150]$$

The wave penetrates more deeply into the solid when the surface is not metallized. The penetration depth, $1/\chi$, is a function of the permittivity ε_d of the external medium. Regardless of the electrical boundary conditions at the origin of the decrease in the transverse horizontal displacement, the penetration depth is large compared to the wavelength (10–100 λ), which explains why the velocity V_B differs very little from that of the TH bulk wave. The localization of the Rayleigh wave near the free surface, due to the mechanical boundary conditions, is much stronger.

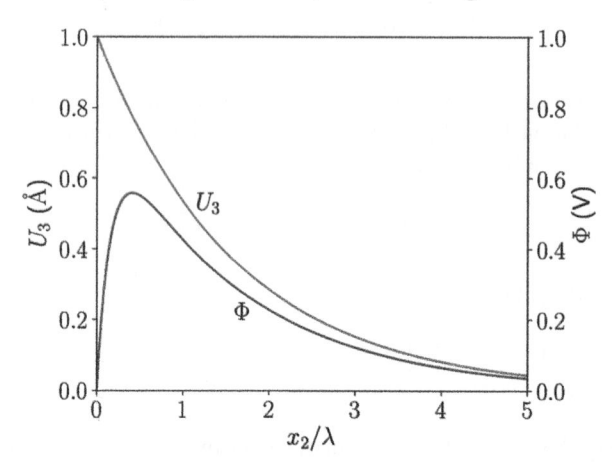

Figure 3.20. *Bleustein–Gulyaev wave in a zinc oxide crystal. Variation of mechanical displacement and electric potential when the surface is metallized*

According to equations [3.140] and [3.145] and by writing $a_1 = a$, the mechanical displacement and electric potential are given by:

$$\begin{cases} u_3 = a e^{-\chi k x_2} e^{i(k x_1 - \omega t)} \\ \Phi = a \dfrac{e_{15}}{\varepsilon_{11}^S} \left(e^{-\chi k x_2} - \dfrac{\varepsilon_d}{\varepsilon_d + \varepsilon_{11}^S} e^{-k x_2} \right) e^{i(k x_1 - \omega t)} \end{cases} \quad [3.151]$$

These general expressions also apply when the surface is metallized ($\varepsilon_d = \infty$).

The absolute amplitude a of the displacement on the surface depends on the power density P_B per unit width w of the beam. Let us use formula [A3.36] with $k = \omega/V_B$ and $X_2 = k x_2$:

$$P_B = \frac{\langle P \rangle}{w} = \frac{1}{2}\rho\omega^2 V_B \int_0^\infty |u_3(x_2)|^2 \, dx_2 = \frac{1}{2}\rho\omega V_B^2 a^2 \int_0^\infty e^{-2\chi X_2} \, dX_2$$

$$[3.152]$$

to calculate, taking into account equation [3.150], the amplitude of the mechanical displacement:

$$a = \sqrt{\frac{4P_B\chi}{\rho\omega V_B^2}} = 2\frac{K_T}{V_B}\sqrt{\frac{\varepsilon_0}{\varepsilon_0 + \varepsilon_{11}^S}}\sqrt{\frac{P_B}{\rho\omega}} \qquad [3.153]$$

and the amplitude of the electric potential at the surface:

$$|\Phi(0)| = a\frac{e_{15}}{\varepsilon_0 + \varepsilon_{11}^S} \qquad [3.154]$$

ORDER OF MAGNITUDE.– For zinc oxide: $K_T = 0.316$, $\rho = 5\,676$ kg.m^{-3}, $V_B = 2\,882$ m/s, with $P_B = 1$ W/cm and $f = 16$ MHz, we obtain $a = 0.95$ nm. Furthermore, as $e_{15} = -0.59$ C.m^{-2} and $\varepsilon_{11} = 7.39 \times 10^{-11}$ F/m: $\Phi(0) = 6.78$ V.

4
Guided Elastic Waves

As shown in the first two chapters, the propagation of elastic waves in an unbounded solid is characterized by a slowness surface. This surface, which is reduced to two spheres for an isotropic solid, is generally composed of three sheets. Each sheet indicates, for a given direction and for each bulk wave, the inverse of the phase velocity, as well as the direction of propagation of energy. The slowness surface represents the plane-wave solutions of the propagation equations. This analysis assumes that the dimensions of the "beam" of waves are small, with respect to the dimensions of the solid.

When the solid is bounded, for instance, by a free plane surface (semi-infinite solid in contact with the air), this boundary imposes conditions on the mechanical and electrical variables. As a result, the waves are reflected, most often with a partial change in their nature, as seen in Chapter 2.

If the crystal is bounded by two plane and parallel free surfaces, that is, in the form of a free plate, the waves progress by reflecting on one surface and then on the other, alternatively: they are guided. These observations, based on reflection and refraction, lead to the notion of *elastic waveguide*, but this approach is difficult to develop in order to find the solutions. To determine the various eigenmodes of a structure, it is preferable to directly solve the propagation equations and the boundary conditions imposed along the frontiers. It has been shown, in Chapter 3, that a single free surface constitutes a guide. The propagation of these surface and interface waves is not dispersive (the phase velocity is independent of the frequency).

One reason for the interest in guided waves is the reduction of losses due to diffraction. Guiding along one of the two lateral dimensions leads to a noticeable improvement since, far from the source, the energy decreases as $1/r$ instead of $1/r^2$. Another advantage, which explains the considerable development of Rayleigh waves, is the accessibility of the mechanical displacement at all points on the free surface. In the field of non-destructive evaluation, elastic waves guided by a structure in the form of a plate or a pipe offer an alternative to the use of bulk waves. They allow a large area to be inspected quickly, either to detect flaws, or to determine the mechanical properties of the material.

In the first section of this chapter, we examine the propagation in an elementary, planar waveguide, in order to introduce the notions of eigenmode, cut-off frequency, dispersion and group velocity. Waves with a transverse component parallel to the surface are studied in the second section. They include the simple wave propagating in a plate, and the Love wave propagating in a layer and its substrate. The third section deals with the propagation of Lamb waves, first in an isotropic plate, where the two components of the mechanical displacement are contained in the sagittal plane. The dispersion equation is established and exploited to specify the behavior of the two families of modes, symmetric and antisymmetric, as well as the displacements of the surfaces of the plate associated with these waves. The decomposition into partial waves, exposed in the case of Rayleigh waves, is used to study the propagation in an anisotropic plate.

Other types of guided waves propagate in a solid of rectangular or circular cross-section. In the last section, we only describe the modes propagating in an isotropic cylinder. These are divided into three families (torsional, compressional and flexural modes).

4.1. Waveguide, group velocity

Figure 4.1 represents the general structure of a one-dimensional waveguide infinite in the x_1 direction and limited laterally by a boundary Σ.

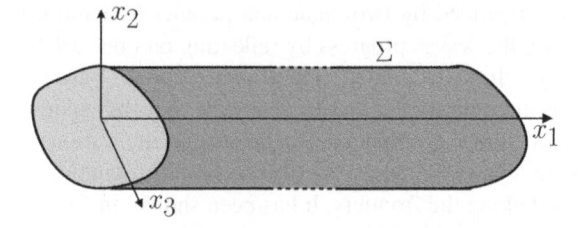

Figure 4.1. *General structure of a guide. The waves propagate along the x_1 axis*

Let $g(x_1, x_2, x_3, t) = g(x_i, t)$ be the space–time function representing the variations of any physical quantity. In a linear regime and for a homogeneous guide, that is, invariant in any translation parallel to Ox_1, the eigenfunctions of the translation operator along x_1 take the form e^{qx_1}:

$$g(x_i, t) = g_1(x_2, x_3, t)e^{qx_1} \quad \text{with} \quad q = -\alpha + i\beta \quad [4.1]$$

If, in addition, the properties of the guide are time-invariant, the eigenfunctions take the form e^{st}:

$$g_1(x_2, x_3, t) = g_0(x_2, x_3)e^{st} \quad \text{with} \quad s = \sigma - i\omega \quad [4.2]$$

In a steady-state harmonic regime ($\sigma = 0$) and for a guide that is non-dissipative both by the material which constitutes it, as well as by the boundaries with the external medium ($\alpha = 0$), the expression for the eigenmodes is:

$$g(x_i, t) = g_0(x_2, x_3)e^{i(\beta x_1 - \omega t)} \quad [4.3]$$

The solution thus defined is a monochromatic wave, propagating in the direction of the positive x_1. In a waveguide, the angular frequency ω is, in general, not proportional to the wave number β. The phase velocity $V = \omega/\beta$ depends on the frequency: the propagation is dispersive. When the wave is no longer monochromatic, it is necessary to introduce another velocity, called the group velocity, which is derived from the dispersion relation $\omega(\beta)$.

In order to illustrate these notions, let us consider the simplest configuration where the guided wave results from the successive reflections of a plane wave on two parallel walls. The hypotheses are:

– an isotropic propagation medium;

– a wave described by a scalar quantity (the acoustic pressure, in the case of a fluid);

– a total reflection on the walls with no change in nature (transverse horizontal [TH] wave in the case of a solid).

4.1.1. *Elementary planar waveguide*

The guide is made up of two infinitely rigid parallel walls, at abscissa $x_2 = 0$ and $x_2 = h$ (Figure 4.2). Without specifying the nature of the vibration u, let us consider an incident wave of wave vector $\underline{k}\,(k_1, k_2, 0)$:

$$u^i = a\cos(k_1 x_1 + k_2 x_2 - \omega t) \quad [4.4]$$

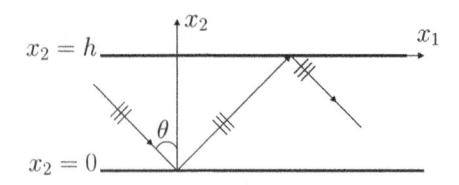

Figure 4.2. *The wave travels in the Ox_1 direction by successive reflections on the two parallel walls*

The wave vector of the wave u^r reflected at $x_2 = 0$ is symmetric with respect to the normal x_2:

$$u^r = -a\cos(k_1 x_1 - k_2 x_2 - \omega t) \qquad [4.5]$$

The sign (–) ensures that the resulting vibration u is zero at all points on the plane $x_2 = 0$:

$$u = u^i + u^r = -2a\sin(k_2 x_2)\sin(k_1 x_1 - \omega t) \qquad [4.6]$$

In order for the vibration u to also be canceled on the second boundary of abscissa $x_2 = h$, it is necessary that:

$$k_2 h = n\pi \quad \text{hence} \quad k_2 = \frac{n\pi}{h} \quad \text{with} \quad n = 1, 2, 3... \qquad [4.7]$$

The final expression:

$$u = -2a\sin\left(\frac{n\pi}{h} x_2\right)\sin(k_1 x_1 - \omega t) \qquad [4.8]$$

is that of a wave propagating along Ox_1, in the corridor delimited by the walls at $x_2 = 0$ and $x_2 = h$, which are nodal planes. This is a *wave guided* by the reflecting walls, with the wave number $\beta = k_1$. Its phase velocity V_φ is higher than that of the free wave V:

$$V_\varphi = \frac{\omega}{\beta} \geq V = \frac{\omega}{k} \quad \text{since} \quad k_1 \leq k = \sqrt{k_1^2 + k_2^2} \qquad [4.9]$$

In the axis of the guide, the wavefronts travel faster, since they move obliquely. Therefore, the fact that the phase velocity of an electromagnetic guided wave is higher than the speed of light does not contradict the principle of relativity, because the quantity that has a physical meaning, as the velocity of energy transport, is the group velocity (section 4.1.2).

Each value of the integer n corresponds to a mode: the nth mode has $n - 1$ nodal planes between the guide walls, equidistant and parallel to them; the mode $n = 2$ represented in Figure 4.3 has one nodal plane at $x_2 = h/2$.

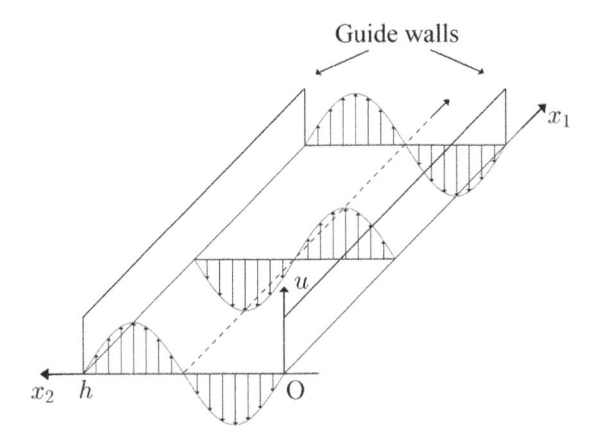

Figure 4.3. *A guided wave presents nodal planes along the walls and also inside the guide, if the order of the mode n is larger than unity. Here $n = 2$ (first antisymmetric mode)*

The wave number $\beta = k_1$ of the guided wave is related to that k of the free wave through:

$$k^2 = \beta^2 + k_2^2 \qquad \text{hence} \qquad \beta^2 = k^2 - \left(\frac{n\pi}{h}\right)^2 \qquad [4.10]$$

The smallest possible real value for k: $k_c = \pi/h$ corresponds to a maximum wavelength: $\lambda_c = 2h$, called the *cut-off wavelength*. Only waves with frequencies higher than the cut-off frequency $f_c = V/2h$ can propagate in the guide. Indeed, for $f < f_c$ the wave number $\beta = i\alpha$ is imaginary and the wave is attenuated as soon as it enters the guide. The cut-off occurs when the angle of incidence θ of the wave vector $\underline{k}$ tends to zero: the phase velocity $V_\varphi = V/\sin\theta$ then tends to infinity. A standing wave regime is established between the two walls of the guide: each cut-off frequency is associated with a transverse resonance along x_2.

A waveguide is an example of a *dispersive structure*, in which the phase velocity of the wave depends on its frequency: the dispersion effect is determined by the geometry of the guide. The angular frequency ω is not proportional to the wave number β. The relation between these two quantities, called the *dispersion equation*:

$$\omega = Vk = V\sqrt{\beta^2 + \left(\frac{n\pi}{h}\right)^2} \qquad [4.11]$$

is represented, in normalized coordinates, in Figure 4.4 for the first modes:

$$\frac{\omega h}{V} = \sqrt{(\beta h)^2 + (n\pi)^2} \qquad \text{or} \qquad \frac{\omega h}{V} = \sqrt{(n\pi)^2 - (\alpha h)^2} \qquad [4.12]$$

The guide is very dispersive near the cut-off frequency and its multiples.

These simple formulae assume that only one wave is reflected and that the nature of this wave is that of an incident wave. They are applicable to an acoustic wave propagating in a fluid between two walls. However, they are not, in general, valid for an elastic wave propagating in a solid, because any incident wave is partially converted, by reflecting on a face of the solid, into several waves of different polarizations, propagating at their own velocities. There is, however, one exception: the TH wave whose mechanical vibration is parallel to the reflector planes. Formulae [4.11] and [4.12], and the curves in Figure 4.4 are applicable to this wave, which conserves its nature upon reflection.

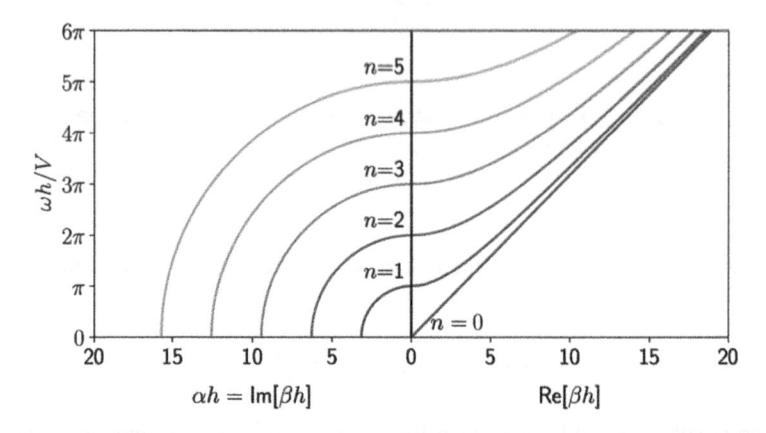

Figure 4.4. *Dispersion curves of an elementary waveguide. At the angular frequency ω, the normalized phase velocity of the nth mode is equal to the slope of the straight line joining the origin to the representative point on the curve of this mode*

4.1.2. *Velocity of a wave packet*

Until now, we have only considered monochromatic waves, that is, sinusoidal vibrations of unlimited duration, with constant amplitude and frequency. In themselves, these pure waves are of little practical interest: indeed, an observer watching a monochromatic wave does not collect more information from its monotonous propagation than from the observation of the regular flow of a homogeneous fluid. The transmission of any information by a wave requires a variation in at least one of its parameters: amplitude or phase. This complex wave, support of information, is no longer monochromatic since its amplitude or frequency is modulated by the signal to be transmitted. However, it results from the superimposition of an infinity of monochromatic waves of different amplitudes and frequencies. Taking the wave number k as the variable, the wave packet or wave group can be written as:

$$u(x,t) = \frac{1}{2\pi} \int_{-\infty}^{\infty} A(k) e^{i(kx-\omega t)} \, dk \qquad [4.13]$$

where $A(k)$ is the amplitude density of the wave packet (Figure 4.5).

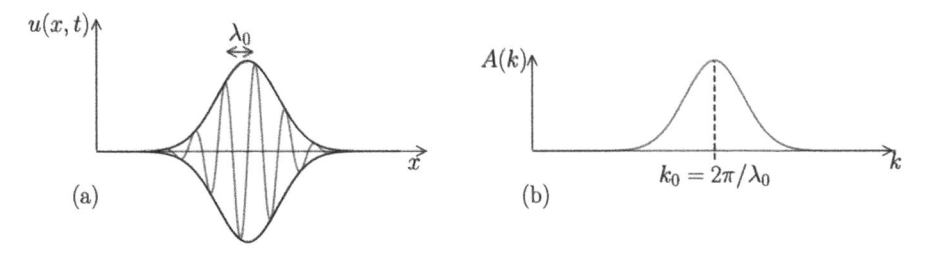

Figure 4.5. *Wave packet: a) Signal at a given time and b) amplitude density as a function of the wave number k*

Two cases need to be distinguished, depending on whether the propagation medium is dispersive or not. For a path ℓ in a non-dispersive medium where the phase velocity V is independent of the frequency, each component of the group is delayed by the same amount $\tau = \ell/V$. After this time τ, the group is unchanged at the exit of the medium. Thus, in a non-dispersive medium, a complex wave propagates without distortion. This property is expressed in several ways:

– the phase velocity is independent of the frequency;

– the phase shift $\varphi = \omega \ell/V$ due to a path ℓ is proportional to the frequency;

– the group delay, defined by $\tau_g = \partial \varphi/\partial \omega = \ell/V$ is independent of frequency.

Let us consider now the propagation of a wave packet in a dispersive medium. In practice, the frequency of the carrier wave is large compared to the frequency of the signal, so that the amplitude density $A(k)$ takes significant values only in a small region around the value $k_0 = \omega_0/V(\omega_0)$. In these conditions, it is sufficient to develop the dispersion equation to the first order around k_0:

$$\omega(k) \approx \omega(k_0) + \left.\frac{d\omega}{dk}\right|_{k_0}(k-k_0) = \omega_0 + V_g(k-k_0) \qquad [4.14]$$

The quantity $V_g = d\omega/dk|_{k_0}$ has the dimension of a velocity. Substituting ω into the expression [4.13] gives:

$$u(x,t) = \frac{1}{2\pi} e^{i(k_0 V_g - \omega_0)t} \int_{-\infty}^{\infty} A(k) e^{ik(x-V_g t)} \, dk \qquad [4.15]$$

or again:

$$u(x,t) = e^{i\Omega_0 t} f(x - V_g t) \quad [4.16]$$

by writing $u(x,0) = f(x)$ and $\Omega_0 = k_0 V_g - \omega_0$. This expression shows that during time t, the group of waves is displaced by $x = V_g t$. Thus, in a dispersive medium, a wave packet centered on the wave number k_0 propagates at the velocity:

$$V_g = \left.\frac{d\omega}{dk}\right|_{k_0} \quad [4.17]$$

called the *group velocity*.

Although the above result does not reveal this, because of the development to the first order of the dispersion equation (equivalent to locally approximate the curve $\omega(k)$ by its tangent), the wave packet can be distorted during its propagation. For example, some elementary waves forming the front of the group, leaving early but traveling slowly, are overtaken by other waves leaving later but traveling faster. The maximum of this group, where the relative position of the waves is always varying, is located at time t at the point x, for which the waves of higher amplitude interfere constructively. For this, the components with wave numbers close to k_0 must be in phase. This condition is expressed by:

$$\varphi(k) = kx - \omega(k)t = \text{Const} \quad \text{for} \quad k \approx k_0 \quad [4.18]$$

or indeed:

$$\left.\frac{d\varphi}{dk}\right|_{k_0} = x - \left.\frac{d\omega}{dk}\right|_{k_0} t = 0 \quad [4.19]$$

This *stationary phase* method, due to Kelvin, shows that the maximum of the wave packet travels at the group velocity V_g given by equation [4.17]. If there is no attenuation, this velocity is also the velocity of the energy transported by the wave packet (Appendix 3).

The following formulae can be deduced from the relation $\omega = kV_\varphi$:

$$\frac{d\omega}{dk} = V_\varphi + k\frac{dV_\varphi}{dk} \quad \text{or} \quad V_g = V_\varphi - \lambda\frac{dV_\varphi}{d\lambda} \quad [4.20]$$

As an example, let us calculate the velocity of a group of waves propagating in the elementary waveguide of section 4.1.1. According to the dispersion equation [4.11]:

$$V_g = \frac{d\omega}{dk} = \frac{V\beta}{\sqrt{\beta^2 + \left(\frac{n\pi}{h}\right)^2}} \quad [4.21]$$

the product:

$$V_g V_\varphi = V^2 \qquad [4.22]$$

is constant and equal to the square of the velocity V of the free wave. The group velocity is always less than V, that is, smaller than the speed of light in the case of an electromagnetic waveguide. It cancels at each cut-off frequency, which corresponds to the physical reality, since the wave can no longer propagate in the guide.

4.1.3. *Propagation of a Gaussian pulse*

Let $s(x,t)$ be a signal carried by a dispersive wave propagating in the x direction with the wave number $k = k(\omega)$. This signal, at the point $x = 0$, is a sinusoidal wave train of central angular frequency ω_0, modulated by a function $h(x = 0, t)$[1]:

$$s(0,t) = e^{i\omega_0 t} h(0,t) \qquad [4.23]$$

The Fourier transform of this signal is:

$$S(0,\omega) = \int_{-\infty}^{\infty} s(0,t) e^{-i\omega t}\, dt = H(0, \omega - \omega_0) \qquad [4.24]$$

$H(x,\omega)$ being the Fourier transform of the function $h(x,t)$. For each spectral component $S(0,\omega)$, the propagation over a distance x introduces a phase shift $-k(\omega)x$, so that the signal $s(x,t)$ is written as:

$$\begin{aligned} s(x,t) &= \frac{1}{2\pi} \int_{-\infty}^{\infty} S(0,\omega) e^{-ik(\omega)x} e^{i\omega t}\, d\omega \\ &= \frac{1}{2\pi} \int_{-\infty}^{\infty} H(0, \omega - \omega_0) e^{i[\omega t - k(\omega)x]}\, d\omega \end{aligned} \qquad [4.25]$$

In order to facilitate calculations, this expression is put in the form:

$$s(x,t) = e^{i[\omega_0 t - k(\omega_0)x]} h(x,t) \qquad [4.26]$$

with:

$$h(x,t) = \frac{1}{2\pi} \int_{-\infty}^{\infty} H(0, \omega - \omega_0) e^{i(\omega - \omega_0)t} e^{-i[k(\omega) - k(\omega_0)]x}\, d\omega \qquad [4.27]$$

---

1 Strictly speaking, the signal $s(0,t)$ should be a real quantity. The function $e^{i\omega_0 t}$ must therefore be replaced by a function of the type $\sin(\omega_0 t)$ or $\cos(\omega_0 t)$, but this unnecessarily lengthens the calculation.

Let $h(0, t)$ be a Gaussian envelope; its Fourier transform is also a Gaussian:

$$h(0,t) = \exp\left(-\frac{\sigma^2 t^2}{2}\right) \longrightarrow H(0,\omega) = \frac{\sqrt{2\pi}}{\sigma} \exp\left(-\frac{\omega^2}{2\sigma^2}\right) \quad [4.28]$$

The function $s(x,t)$ can be written as:

$$s(x,t) = \frac{e^{i[\omega_0 t - k(\omega_0)x]}}{\sqrt{2\pi}\sigma} \int_{-\infty}^{\infty} e^{-\frac{(\omega-\omega_0)^2}{2\sigma^2}} e^{i(\omega-\omega_0)t} e^{-i[k(\omega)-k(\omega_0)]x} \, d\omega \quad [4.29]$$

and the wave number can be approximated by the Taylor expansion around the angular frequency ω_0:

$$k(\omega) = k_0 + (\omega - \omega_0)k_0' + \frac{1}{2}(\omega - \omega_0)^2 k_0'' + \ldots \quad [4.30]$$

with:

$$k_0 = k(\omega_0), \quad k_0' = \left.\frac{dk}{d\omega}\right|_{\omega=\omega_0} = \frac{1}{V_g(\omega_0)} \quad \text{and} \quad k_0'' = \left.\frac{d^2 k}{d\omega^2}\right|_{\omega=\omega_0} \quad [4.31]$$

where V_g is the group velocity defined by relation [4.17]. By denoting $\Omega = \omega - \omega_0$, the signal $s(x,t)$ takes the form:

$$s(x,t) = \frac{e^{i(\omega_0 t - k_0 x)}}{\sigma\sqrt{2\pi}} \int_{-\infty}^{\infty} e^{i\Omega(t-\tau_g)} e^{-\frac{\Omega^2}{2\sigma^2}[1+i k_0'' x]} \, d\Omega \quad [4.32]$$

with τ_g being the group delay associated with the central frequency ω_0 and with the group velocity $V_g(\omega_0)$:

$$\tau_g = k_0' x = \frac{x}{V_g(\omega_0)} \quad [4.33]$$

The calculation of the integral leads to the expression:

$$s(x,t) = e^{i(\omega_0 t - k_0 x)} \frac{\exp\left[-\frac{\sigma^2(t-\tau_g)^2}{2+2i\sigma^2 k_0'' x}\right]}{\sqrt{1+i\sigma^2 k_0'' x}} \quad [4.34]$$

In the case of a non-dispersive wave, the term k_0'' is zero and the signal $s(x,t)$:

$$s(x,t) = e^{i(\omega_0 t - k_0 x)} \exp\left[-\frac{1}{2}\sigma^2(t-\tau_g)^2\right] \quad [4.35]$$

is simply phase-shifted by the quantity $k_0 x$ and delayed by τ_g.

In order to visualize the dispersion effect, we study the same signal carried by a non-dispersive wave and a dispersive wave. A sine function of frequency f_0 windowed by a Gaussian is carried by two Lamb waves: the wave S_0 of constant velocity V_p and the wave A_0, whose velocity V depends on the frequency (section 4.3.1.3.1). Each signal is propagated by multiplying its Fourier transform at $x = 0$ by a phase shift that depends on the wave:

$$S_{S_0}(x,\omega) = S(0,\omega)e^{-i\frac{\omega}{V_p}x} \quad \text{and} \quad S_{A_0}(x,\omega) = S(0,\omega)e^{-i\frac{\omega}{V}x} \quad [4.36]$$

The time signals obtained by calculating the inverse Fourier transform are plotted in Figure 4.6. As expected, the Lamb wave S_0 propagates without dispersion and the signal is unchanged over time. Conversely, the Lamb wave A_0 is highly dispersive, and the signal spreads out over time, because the low-frequency components propagate more slowly than the high-frequency components (section 4.3.1.2).

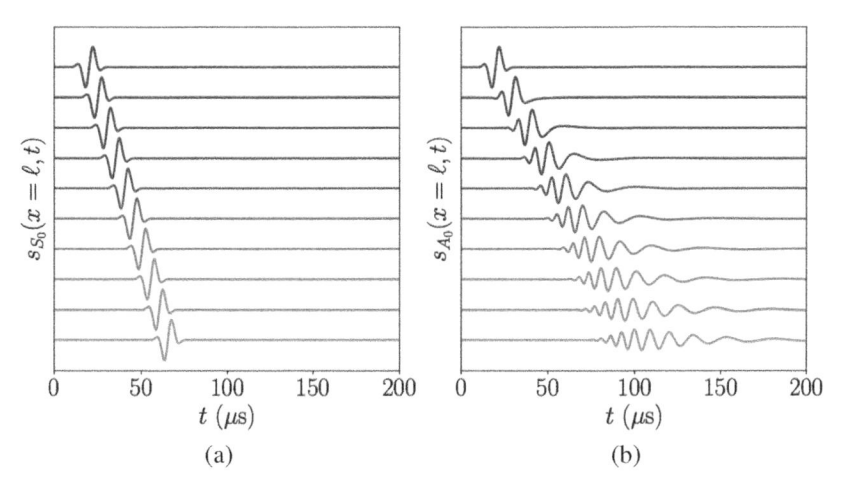

Figure 4.6. *Signals carried by the Lamb waves a) S_0 and b) A_0 for different propagation distances. For a color version of this figure, see www.iste.co.uk/royer/waves1.zip*

4.2. Transverse horizontal waves

According to the results of section 2.2.2, a TH wave, polarized along the x_3 axis, propagates along x_1 independently of the other waves, if the x_1x_2 plane is perpendicular to a direct or inverse binary axis. The case of the Bleustein–Gulyaev surface wave, which propagates in a semi-infinite piezoelectric solid, is discussed at the end of Chapter 3.

In this section, we consider a stratified plane structure made up of different isotropic solids, separated by plane interfaces that are all parallel to the x_1x_3 plane.

We first study the propagation of the TH wave in a plate, then in a layer rigidly bonded to a semi-infinite solid (Love wave) (Love 1911) and finally, in a succession of parallel layers.

4.2.1. Guided TH modes

We will quickly examine the case of the TH wave guided by a simple plate because it is similar to that of the elementary guide given in section 4.1.1. In section 2.3, it was seen that a TH wave, whose displacement is parallel to the interface, is reflected and transmitted without changing its nature. A solid bounded by two free surfaces, that is, a plate, of abscissa $x_2 = 0$ and $x_2 = h$ is thus a guide for a wave polarized along x_3. In the harmonic case:

$$u_3(x_1, x_2, t) = u(x_2) e^{i(kx_1 - \omega t)} \qquad [4.37]$$

the propagation equation along x_1 in the material, assumed to be isotropic, is written as:

$$\frac{d^2 u}{dx_2^2} + k_{2T}^2 u = 0 \quad \text{with} \quad k_{2T}^2 = k_T^2 - k^2 \qquad [4.38]$$

The stress free boundary conditions on the surfaces:

$$\sigma_{23} = \mu \frac{\partial u_3}{\partial x_2} = 0 \quad \text{for} \quad x_2 = 0 \quad \text{and} \quad x_2 = h \qquad [4.39]$$

impose that the displacement will be maximal on these surfaces and thus takes the form:

$$u(x_2) = a \cos[k_{2T}(x_2 - h)] \quad \text{with} \quad k_{2T} = \frac{n\pi}{h} \qquad [4.40]$$

Figure 4.4 shows the dispersion curves for the first six modes. The dispersion equation, which is obtained by squaring k_{2T}, is identical to [4.12]:

$$\frac{\omega h}{V_T} = \sqrt{(kh)^2 + (n\pi)^2} \qquad [4.41]$$

The phase velocity $V = \omega/k$ for the guided TH wave depends on frequency and is always higher than the bulk wave velocity $V_T = \sqrt{\mu/\rho}$, except for the zero-order mode, which propagates at the velocity V_T (Figure 4.7). This mode, without cut-off frequency, which can propagate at very low frequency, has no equivalent in electromagnetism.

4.2.2. Love wave

A guided TH wave can propagate when the plate is rigidly bonded to a semi-infinite solid if the TH wave is evanescent in the substrate (Figure 4.8).

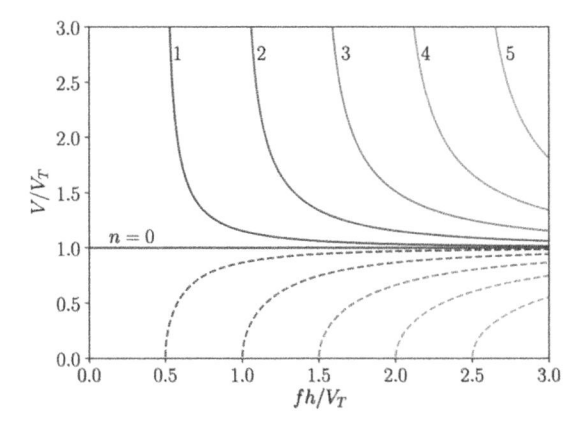

Figure 4.7. *Phase velocities (solid lines) and group velocities (dashed line) of the first six TH modes guided in a plate of thickness h*

The displacement is zero for $x_2 = -\infty$ and, in the substrate, the solution of the propagation equation [4.38] is of the form:

$$u_s(x_2) = a_s e^{\alpha_s x_2} \quad \text{with} \quad \alpha_s = \sqrt{k^2 - (k_T^s)^2} \quad \forall x_2 < 0 \quad [4.42]$$

which implies:

$$k > k_T^s \quad \text{hence} \quad V < V_T^s \quad [4.43]$$

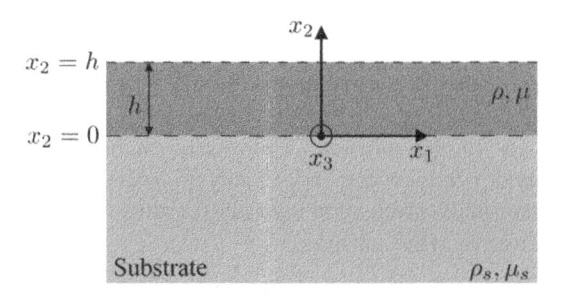

Figure 4.8. *The Love wave propagates in a structure composed of a plate of thickness h, rigidly bonded to a semi-infinite substrate*

The phase velocity V is necessarily lower than the velocity $V_T^s = \sqrt{\mu_s/\rho_s}$ of the bulk TH wave, propagating in the substrate of mass density ρ_s and shear modulus μ_s. In the plate (or layer), the mechanical stress σ_{32} is zero on the free surface $x_2 = h$, and the displacement is maximum here. The solution is, therefore, identical to [4.40] with:

$$k_{2T} = \sqrt{k_T^2 - k^2} \quad \text{which implies} \quad k < k_T \quad \text{hence} \quad V > V_T \quad [4.44]$$

The inequalities [4.43] and [4.44] require that the velocity of the TH wave in the substrate V_T^s has to be higher than that of the TH wave in the layer, V_T. Thus, the phase velocity V of the guided wave, called the *Love wave* (named after a British geophysicist), lies between these two values: $V_T < V < V_T^s$.

At the interface $x_2 = 0$, the continuity of the displacement and of the stress σ_{32} imposes the relations:

$$a_s = a\cos(k_{2T}h) \quad \text{and} \quad \mu_s \alpha_s a_s = \mu k_{2T} a \sin(k_{2T}h) \quad [4.45]$$

They lead to the *dispersion equation*, relating ω and k through the parameters α_s and k_{2T}:

$$\tan(k_{2T}h) = \frac{\alpha_s \mu_s}{k_{2T}\mu} \quad [4.46]$$

For a particular phase velocity between V_T and V_T^s, that is, for α_s and k_{2T} given by equations [4.42] and [4.44], the dispersion equation admits an infinite number of solutions:

$$(k_{2T}h)_n = \tan^{-1}\left(\frac{\mu_s}{\mu}\sqrt{\frac{1-(V/V_T^s)^2}{(V/V_T)^2-1}}\right) + n\pi = k_n h\sqrt{(V/V_T)^2 - 1} \quad n \in \mathbb{N}$$

[4.47]

In Figure 4.9, the evolution of the phase and group velocities of the first Love modes is plotted as functions of the frequency, for a plexiglass plate of thickness $h = 2$ mm, rigidly bonded to a duralumin substrate. When the frequency is equal to the cut-off frequency of a mode, the phase velocity tends toward the velocity V_T^s of the transverse waves in the substrate. Conversely, at high frequency, the phase velocity of each mode tends toward the velocity V_T of the transverse waves in the plate. Physically, the small wavelengths are only sensitive to the layer, while the long wavelengths are sensitive to the substrate.

The displacements and stresses of the first two Love modes for $f = 0.5$ MHz are plotted in Figures 4.10(a) and (b), respectively. Let us note that the decay in the depth is much faster for the fundamental mode.

4.2.3. *Love wave in an inhomogeneous medium*

Let us study the propagation of Love waves in an inhomogeneous or stratified layer, modeling, for example, the earth's crust, whose mass density ρ and shear modulus μ depend on the depth x_2. The displacement is polarized along the direction $\underline{e}_3$ and does not depend on x_3. We will denote u the displacement component along x_3 and τ the tangential stress σ_{23}. Since the expressions for the stresses σ_{13} and σ_{23} are:

$$\sigma_{13} = \mu(x_2)\frac{\partial u_3(x_1, x_2, t)}{\partial x_1} \quad \text{and} \quad \sigma_{23} = \mu(x_2)\frac{\partial u_3(x_1, x_2, t)}{\partial x_2} = \tau \quad [4.48]$$

the propagation equation [1.73] is written as:

$$\rho(x_2)\frac{\partial^2 u}{\partial t^2} = \frac{\partial \sigma_{13}}{\partial x_1} + \frac{\partial \sigma_{23}}{\partial x_2} = \mu(x_2)\frac{\partial^2 u}{\partial x_1^2} + \frac{\partial \tau}{\partial x_2} \quad [4.49]$$

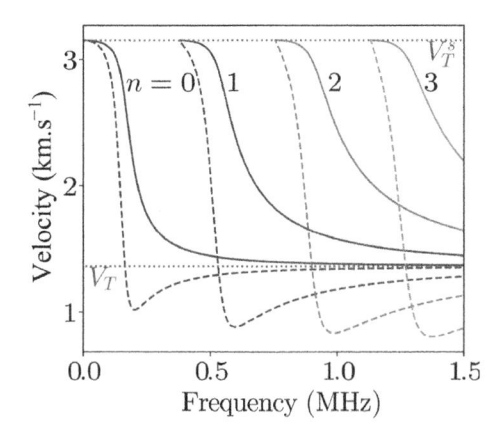

Figure 4.9. *Phase velocity (solid lines) and group velocity (dashed lines) for the first Love modes, depending on the frequency for a plexiglass plate of thickness h = 2 mm rigidly bonded to a duralumin block. For a color version of this figure, see www.iste.co.uk/royer/waves1.zip*

The solution is sought in the form of a time-harmonic guided wave, propagating in the x_1 direction. By introducing the *state vector* $\underline{\zeta}(x_2)$, whose two components $u(x_2)$ and $\tau(x_2)$ are continuous at an interface of normal x_2:

$$\begin{pmatrix} u_3(x_1, x_2, t) \\ \sigma_{23}(x_1, x_2, t) \end{pmatrix} = \begin{pmatrix} u(x_2) \\ \tau(x_2) \end{pmatrix} e^{i(kx_1-\omega t)} = \underline{\zeta}(x_2)e^{i(kx_1-\omega t)} \quad [4.50]$$

equations [4.48] and [4.49] lead to the matrix system:

$$\frac{d\underline{\zeta}(x_2)}{dx_2} = \boldsymbol{A}\underline{\zeta}(x_2) \quad \text{with} \quad \boldsymbol{A}(x_2) = \begin{pmatrix} 0 & \frac{1}{\mu(x_2)} \\ \mu(x_2)k^2 - \rho(x_2)\omega^2 & 0 \end{pmatrix} \quad [4.51]$$

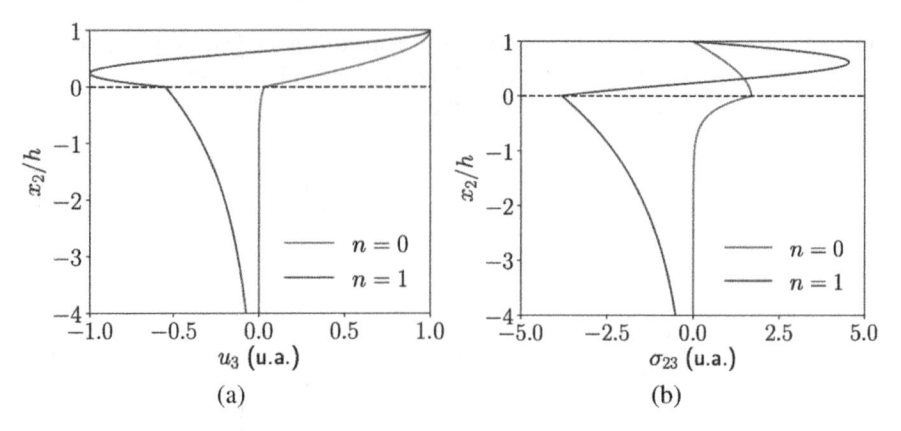

Figure 4.10. Love waves: a) tangential displacement u_3 and b) stress σ_{23} in the depth for the first two modes

When the state matrix $\boldsymbol{A}$ is constant, the solution to the differential system [4.51] is expressed using the exponential of the matrix $\boldsymbol{A}$; by writing $y = x_2$, we obtain:

$$\underline{\zeta}(y) = e^{\boldsymbol{A}(y-y_0)}\underline{\zeta}(y_0) \quad \text{with} \quad e^{\boldsymbol{A}y} = \sum_{k=0}^{\infty}\frac{1}{k!}(\boldsymbol{A}y)^k \qquad [4.52]$$

When the matrix $\boldsymbol{A}$ is a function of y, the resolution is more complex; it needs a convolution integral. To bypass this difficulty, the medium is divided into N slices, of thickness $h_n = y_n - y_{n-1}$ small compared to the wavelength, so that the parameters of the medium ρ_n and μ_n, and therefore the matrix $\boldsymbol{A} = \boldsymbol{A}_n$, can be considered as constant in the thickness of the slice n (Figure 4.11). The solution is given by:

$$\underline{\zeta}(y_n) = e^{\boldsymbol{A}_n h_n}\underline{\zeta}(y_{n-1}) \qquad [4.53]$$

The matrix $\boldsymbol{A}$ is of the form:

$$\boldsymbol{A} = \begin{pmatrix} 0 & \frac{1}{\mu} \\ -\mu k_2^2 & 0 \end{pmatrix} \quad \text{with} \quad k_2^2 = \frac{\rho\omega^2}{\mu} - k^2 \qquad [4.54]$$

Noting that:

$$\boldsymbol{A}^2 = -k_2^2 \boldsymbol{1} \quad \text{hence} \quad \boldsymbol{A}^{2n} = k_2^{2n}\boldsymbol{1} \quad \text{and} \quad \boldsymbol{A}^{2n+1} = (-1)^n k_2^{2n}\boldsymbol{A} \qquad [4.55]$$

the matrix exponential [4.52] takes the form:

$$e^{\boldsymbol{A}h} = \left[1 - \frac{(k_2 h)^2}{2!} + \frac{(k_2 h)^4}{4!} + \ldots\right]\boldsymbol{1} + \left[k_2 h - \frac{(k_2 h)^3}{3!} + \ldots\right]\frac{1}{k_2}\boldsymbol{A} \qquad [4.56]$$

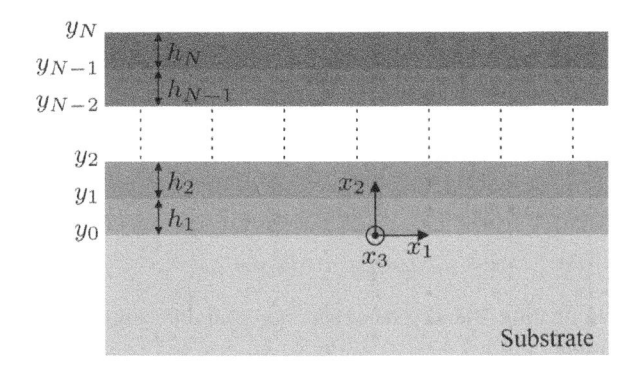

Figure 4.11. *The propagation medium, inhomogeneous along x_2, is divided into N slices whose thickness is small compared to the wavelength*

For the nth layer of thickness h_n, we get:

$$e^{A_n h_n} = \cos(k_{2n} h_n)\mathbf{1} + \frac{\sin(k_{2n} h_n)}{k_{2n}} A_n \quad \text{with} \quad k_{2n} = \sqrt{\frac{\rho_n}{\mu_n}\omega^2 - k^2} \quad [4.57]$$

The vector $\underline{\zeta}_n$ between two positions y_{n+1} and y_n is expressed by:

$$\underline{\zeta}_n(y_{n+1}) = T_n \underline{\zeta}_n(y_n) \quad \text{with} \quad T_n = \begin{pmatrix} \cos(k_{2n} h_n) & \dfrac{\sin(k_{2n} h_n)}{\mu_n k_{2n}} \\ -\mu_n k_{2n} \sin(k_{2n} h_n) & \cos(k_{2n} h_n) \end{pmatrix}$$

[4.58]

hence, taking into account the continuity $\underline{\zeta}_{n-1}(y_n) = \underline{\zeta}_n(y_n)$ at each interface:

$$\underline{\zeta}_1(y_0) = \prod_{n=1}^{N} T_n \underline{\zeta}_N(y_N) \quad [4.59]$$

In the substrate, the mechanical displacement must be cancel at infinity:

$$u_s(x_2) = a e^{\alpha_s x_2} \quad \text{with} \quad \alpha_s > 0 \quad [4.60]$$

enforcing the stress:

$$T_s(x_2) = \mu_s \alpha_s a e^{\alpha_s x_2} \quad [4.61]$$

and the vector $\underline{\zeta}_s$ at $x_2 = y_0 = 0$:

$$\underline{\zeta}_s = a \begin{pmatrix} 1 \\ \mu_s \alpha_s \end{pmatrix} \quad [4.62]$$

The boundary condition at the free surface $x_2 = y_N$, where $y_N = \sum_n h_n$ is the total thickness of the layer, is $\tau_N = 0$. It follows that equation [4.59] takes the form:

$$a \begin{pmatrix} 1 \\ \mu_s \alpha_s \end{pmatrix} = M \begin{pmatrix} u_N \\ 0 \end{pmatrix} \quad \text{with} \quad M = \prod_{n=1}^{N} T_n \qquad [4.63]$$

This matrix equation yields two relations:

$$a = M_{11} u_N \quad \text{and} \quad \mu_s \alpha_s a = M_{21} u_N \qquad [4.64]$$

which, after eliminating the displacement u_N and the amplitude a, lead to the dispersion equation:

$$M_{21} - \mu_s \alpha_s M_{11} = 0 \qquad [4.65]$$

between ω and k, through the intermediary of the components k_{2n} [4.57].

4.3. Lamb waves

The TH wave studied in the previous section propagates alone in a stratified plane structure, because its polarization is conserved during a reflection or a transmission. This is not the same for the longitudinal (L) or transverse vertical (TV) wave: they are coupled at an interface. Therefore, the analysis of the propagation in a free plate of a wave polarized in the sagittal plane is more complex than that of the TH wave. We first treat the isotropic case, which is simpler, using the potential method, and then the case of an anisotropic plate, using the partial wave method already used to study the Rayleigh wave propagation. In many circumstances, the plate is immersed or covered by a fluid. The interaction with a fluid has several consequences on the guided modes: a modification (generally small) in their velocity, an attenuation by radiation (under certain conditions), as well as the appearance of an interface mode of the Scholte wave type.

4.3.1. *Free isotropic plate*

In an isotropic solid, the Lamb wave (named after an English geophysicist) is composed of a longitudinal displacement and of a transverse vertical displacement contained in the sagittal plane. It propagates in a plate of thickness d, whose two surfaces are free. The presence of the two components is easily explained, since a longitudinal wave that encounters a surface is partially transformed into a transverse wave, and vice versa. Another explanation arises from the possible propagation of a Rayleigh wave on each free surface. These waves travel independently as long as the distance between the free surfaces is large compared to the Rayleigh wavelength. When the thickness of the plate is no longer very large, the waves on the two surfaces couple and give rise to Lamb waves.

We will begin by establishing the *Rayleigh–Lamb equations* or dispersion equations (Rayleigh 1888) for the two families of modes (symmetric and antisymmetric). These are illustrated by the dispersion curves $\omega(k)$ for usual materials. Two fundamental modes, which propagate at any frequency, are distinguished from the other modes of a higher order, which do not propagate below a certain cut-off frequency (Lamb 1917). It is, therefore, necessary to study the behavior of Lamb waves at low frequency and at high frequency, that is, when the plate thickness d is small or large compared to the wavelength λ of the guided wave. The group velocity of some Lamb modes vanishes at particular points. Once locally excited, the energy of these ZGV (*zero group velocity*) modes does not propagate, resulting in a phenomenon of local resonance. Using dimensionless quantities along the ω and k axes, it appears that the dispersion curves of two isotropic solids with the same Poisson's ratio are identical and pass through regularly spaced points, corresponding to the so-called Lamé modes.

4.3.1.1. *Rayleigh–Lamb equations*

The orthogonal reference frame is given in Figure 4.12. The waves guided by the plate of thickness $d = 2h$ propagate along Ox_1 with a wave number k and the free surfaces are normal to Ox_2. By hypothesis, the mechanical quantities do not depend on x_3.

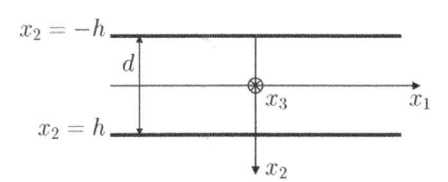

Figure 4.12. *Arrangement of the axes. The plate thickness is $d = 2h$*

In an isotropic solid, the mechanical displacement derives from a scalar potential ϕ_L and a vector potential $\underline{\psi}$. In the harmonic case, these potentials satisfy the Helmholtz equations [3.1]:

$$\Delta \phi_L + k_L^2 \phi_L = 0 \quad \text{and} \quad \Delta \underline{\psi} + k_T^2 \underline{\psi} = 0 \qquad [4.66]$$

It has been shown in section 3.1.1.1 that the components u_1 and u_2 of a wave polarized in the sagittal plane are expressed only as functions of ϕ_L and $\psi_3 = \phi_T$ (equations [3.2]):

$$\begin{cases} u_1 = ik\phi_L + \dfrac{\partial \phi_T}{\partial x_2} \\ u_2 = \dfrac{\partial \phi_L}{\partial x_2} - ik\phi_T \end{cases} \qquad [4.67]$$

The functions:

$$\phi_M(x_1, x_2, t) = f_M(x_2) e^{i(kx_1 - \omega t)} \quad \text{with} \quad M = L, T \quad [4.68]$$

satisfy the equations [3.4], which are written as:

$$\frac{d^2 f_M(x_2)}{dx_2^2} + k_{2M}^2 f_M(x_2) = 0 \quad [4.69]$$

where k_{2M} is the normal component of the longitudinal wave vector (M = L) and the transverse wave vector (M = T) in the isotropic solid:

$$k_{2M}^2 = k_M^2 - k^2 \quad \text{with} \quad k_M = \frac{\omega}{V_M} \quad \text{and} \quad M = L, T \quad [4.70]$$

The solutions f_L and f_T must satisfy the boundary conditions $\sigma_{12} = \sigma_{22} = 0$ on the two free planes $x_2 = \pm h$. The tangential stress $\sigma_{12} = 2\mu \varepsilon_{12}$ is expressed by:

$$\sigma_{12} = \mu \left(\frac{\partial u_1}{\partial x_2} + \frac{\partial u_2}{\partial x_1} \right) = \mu \left(\frac{d^2 f_T}{dx_2^2} + k^2 f_T + 2ik \frac{df_L}{dx_2} \right) e^{i(kx_1 - \omega t)} \quad [4.71]$$

or, since $d^2 f_T / dx_2^2 = -k_{2T}^2 f_T$:

$$\sigma_{12} = \mu \left[\left(k^2 - k_{2T}^2 \right) f_T + 2ik \frac{df_L}{dx_2} \right] e^{i(kx_1 - \omega t)} \quad [4.72]$$

The normal stress σ_{22} is given by the relation:

$$\sigma_{22} = \lambda \Delta \phi_L + 2\mu \frac{\partial u_2}{\partial x_2} = \left[-\lambda k_L^2 f_L + 2\mu \left(\frac{d^2 f_L}{dx_2^2} - ik \frac{df_T}{dx_2} \right) \right] e^{i(kx_1 - \omega t)} \quad [4.73]$$

As it was shown (equation [2.77]) that the multiplicative factor of $f_L : -(\lambda k_L^2 + 2\mu k_{2L}^2)$ is equal to $\mu(k^2 - k_{2T}^2)$, the stress σ_{22} is also proportional to the shear modulus μ:

$$\sigma_{22} = \mu \left[\left(k^2 - k_{2T}^2 \right) f_L - 2ik \frac{df_T}{dx_2} \right] e^{i(kx_1 - \omega t)} \quad [4.74]$$

To satisfy the boundary conditions simultaneously at $x_2 = \pm h$, the stresses σ_{12} and σ_{22} must be even or odd functions of x_2. Since the derivative of an even function is an odd function and vice versa, the solutions to equations [4.69] must be of different parity:

$$f_L(x_2) = a_L \cos(k_{2L} x_2 + \alpha) \quad \text{and} \quad f_T(x_2) = a_T \sin(k_{2T} x_2 + \alpha) \quad [4.75]$$

with $\alpha = 0$ (σ_{22} even, σ_{12} odd) or $\alpha = \pi/2$ (σ_{22} odd, σ_{12} even). The displacement components are deduced from relations [4.67]:

$$\begin{cases} u_1 = ika_L \cos(k_{2L}x_2 + \alpha) + k_{2T}a_T \cos(k_{2T}x_2 + \alpha) \\ u_2 = -k_{2L}a_L \sin(k_{2L}x_2 + \alpha) - ika_T \sin(k_{2T}x_2 + \alpha) \end{cases} \quad [4.76]$$

These expressions reveal two families of modes (Figure 4.13):

– the symmetric modes defined by $\alpha = 0$. Their longitudinal component is an even function of x_2, and their transverse component is an odd function;

– the antisymmetric modes defined by $\alpha = \pi/2$. Their longitudinal component is an odd function of x_2, and their transverse component is an even function.

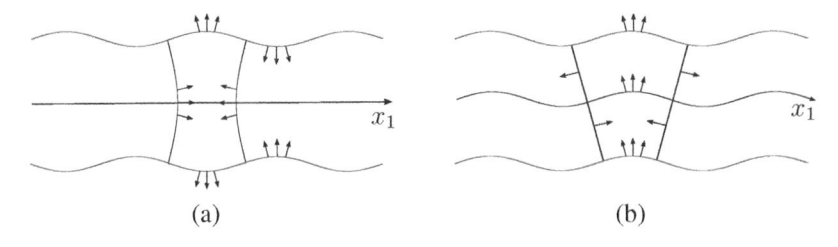

Figure 4.13. Lamb wave displacement. a) Symmetric: on either side of the median plane, the longitudinal components are equal and the transverse components are opposite. b) Antisymmetric: on either side of the median plane, the transverse components are equal and the longitudinal components are opposite

The stresses σ_{12} and σ_{22} are obtained by substituting the functions $f_L(x_2)$ and $f_T(x_2)$ in equations [4.72] and [4.74]:

$$\begin{cases} \sigma_{12} = \mu \left[-2ikk_{2L}a_L \sin(k_{2L}x_2 + \alpha) + \left(k^2 - k_{2T}^2\right) a_T \sin(k_{2T}x_2 + \alpha) \right] \\ \sigma_{22} = \mu \left[\left(k^2 - k_{2T}^2\right) a_L \cos(k_{2L}x_2 + \alpha) - 2ikk_{2T}a_T \cos(k_{2T}x_2 + \alpha) \right] \end{cases}$$
[4.77]

The *dispersion equation* for Lamb waves is derived from the boundary conditions $\sigma_{12}(h) = \sigma_{22}(h) = 0$, that is, in matrix form:

$$\begin{pmatrix} -2ikk_{2L}\sin(k_{2L}h + \alpha) & \left(k^2 - k_{2T}^2\right)\sin(k_{2T}h + \alpha) \\ \left(k^2 - k_{2T}^2\right)\cos(k_{2L}h + \alpha) & -2ikk_{2T}\cos(k_{2T}h + \alpha) \end{pmatrix} \begin{pmatrix} a_L \\ a_T \end{pmatrix} = \begin{pmatrix} 0 \\ 0 \end{pmatrix} \quad [4.78]$$

The compatibility condition for this linear and homogeneous system:

$$\left(k^2 - k_{2T}^2\right)^2 \cos(k_{2L}h + \alpha)\sin(k_{2T}h + \alpha) \\ + 4k^2 k_{2L}k_{2T}\sin(k_{2L}h + \alpha)\cos(k_{2T}h + \alpha) = 0 \quad [4.79]$$

provides, in implicit form, the dispersion equation $\omega(k)$ for the symmetric modes ($\alpha = 0$) and the antisymmetric modes ($\alpha = \pi/2$). An equivalent form, called the *Rayleigh–Lamb equations*, is obtained by revealing the tangents:

$$\frac{\tan(k_{2T}h)}{\tan(k_{2L}h)} = -\frac{4k^2 k_{2T} k_{2L}}{(k_{2T}^2 - k^2)^2} \quad \text{symmetric modes} \quad [4.80]$$

and:

$$\frac{\tan(k_{2T}h)}{\tan(k_{2L}h)} = -\frac{(k_{2T}^2 - k^2)^2}{4k^2 k_{2L} k_{2T}} \quad \text{antisymmetric modes} \quad [4.81]$$

From these relations between ω and k, it is not possible to obtain analytic expressions for the different branches of the dispersion curve. In the (ω, k) plane, three sectors can be distinguished, depending on the value of the phase velocity $V = \omega/k$, with respect to the velocity V_L of the longitudinal waves and to that V_T of the transverse waves (Figure 4.14). Indeed, according to relations [4.70]:

– $V_L < V$, hence $k < k_L < k_T$: k_{2L} and k_{2T} are real;

– $V_T < V < V_L$, hence $k_L < k < k_T$: k_{2L} is imaginary and k_{2T} is real;

– $V < V_T$, hence $k_L < k_T < k$: k_{2L} and k_{2T} are imaginary.

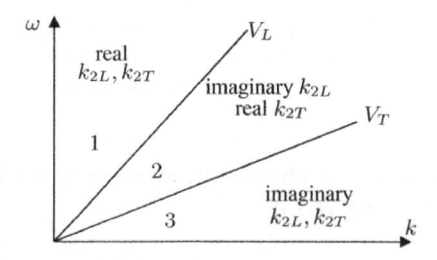

Figure 4.14. *The (ω, k) plane is divided into three sectors, depending on the value of the phase velocity V with respect to V_L and V_T*

The components of mechanical displacement are obtained by replacing, for instance, a_L taken from the second equation [4.78] in equations [4.76]:

$$\begin{cases} u_1 = k_{2T} a_T \left[\cos(k_{2T} x_2 + \alpha) - \frac{2k^2}{k^2 - k_{2T}^2} \frac{\cos(k_{2T} h + \alpha)}{\cos(k_{2L} h + \alpha)} \cos(k_{2L} x_2 + \alpha) \right] \\ u_2 = -ika_T \left[\sin(k_{2T} x_2 + \alpha) + \frac{2k_{2L} k_{2T}}{k^2 - k_{2T}^2} \frac{\cos(k_{2T} h + \alpha)}{\cos(k_{2L} h + \alpha)} \sin(k_{2L} x_2 + \alpha) \right] \end{cases}$$

[4.82]

It is easy to show that in all cases the two components are in phase quadrature. As for the Rayleigh waves, the particles describe an ellipse that is deformed, since the amplitudes vary differently with the position x_2 in the thickness of the plate.

At a given frequency, the displacement field and the stress field constitute the eigenmodes of a homogeneous, isotropic, elastic plate, whose faces are free. In the case of a local modification of the characteristics of the solid (presence of inhomogeneities), of the plate (thickness variation) or in the boundary conditions (presence of mechanical sources), the solutions can be sought in the form of an expansion on all Lamb modes. The existence of an orthogonality relation between these eigenmodes simplifies the determination of the coefficients of the development by individualizing their calculation. Fraser (1976) has shown that two Lamb modes, n and m, with the same frequency and of wave numbers k_n and k_m, satisfy the orthogonality relation:

$$\int_{-h}^{h} \left[u_2^{(n)} \sigma_{12}^{(m)} - u_1^{(m)} \sigma_{11}^{(n)} \right] \, dx_2 = \delta_{mn} C_n \qquad [4.83]$$

where the stresses $\sigma_{11}^{(n)}$ and $\sigma_{12}^{(m)}$ are obtained from equation [4.77] and where C_n is a normalization coefficient.

4.3.1.2. *Dispersion curves*

The numerical resolution of the Rayleigh–Lamb equations [4.80] and [4.81] provides the pairs (ω, k) for which the symmetric S_n and antisymmetric A_n Lamb modes can propagate. By introducing the "frequency × thickness" product and the "thickness/wavelength" ratio:

$$\Omega = \frac{\omega h}{\pi} = fd \quad \text{and} \quad K = \frac{kh}{\pi} = \frac{d}{\lambda} \qquad [4.84]$$

the terms $k_{2T} h$ and $k_{2L} h$ take the form:

$$k_{2M} h = \sqrt{\left(\frac{\omega h}{V_M}\right)^2 - (kh)^2} = \pi \sqrt{\frac{\Omega^2}{V_M^2} - K^2} \quad \text{with} \quad M = L, T$$

$$[4.85]$$

For a material characterized by the bulk wave velocities V_L and V_T, and for a positive real value of K, equations [4.80] and [4.81] are satisfied by a countable infinity of values Ω_n. The dispersion curves $\Omega_n(K)$ represented in the (Ω, K) plane are independent of the thickness d of the plate. Since $\Omega/K = \omega/k$, the slope of the line joining a point to the origin is equal to the phase velocity V of the mode and the slope of the tangent at a point on a curve represents the group velocity $V_g = d\omega/dk$. The "frequency × thickness" product is commonly expressed in MHz.mm, which is equivalent to a velocity in km/s. The abscissa K is a dimensionless variable.

The dispersion curves of the first Lamb modes for a duralumin plate of thickness $d = 2h$ are plotted in Figure 4.15. The curves in solid lines represent the symmetric Lamb modes and those in dashed lines represent the antisymmetric Lamb modes. It is interesting to note that the dispersion curves of the same family of modes do not intersect. In contrast, the curve of a symmetric (antisymmetric) mode can intersect that of an antisymmetric (symmetric) mode.

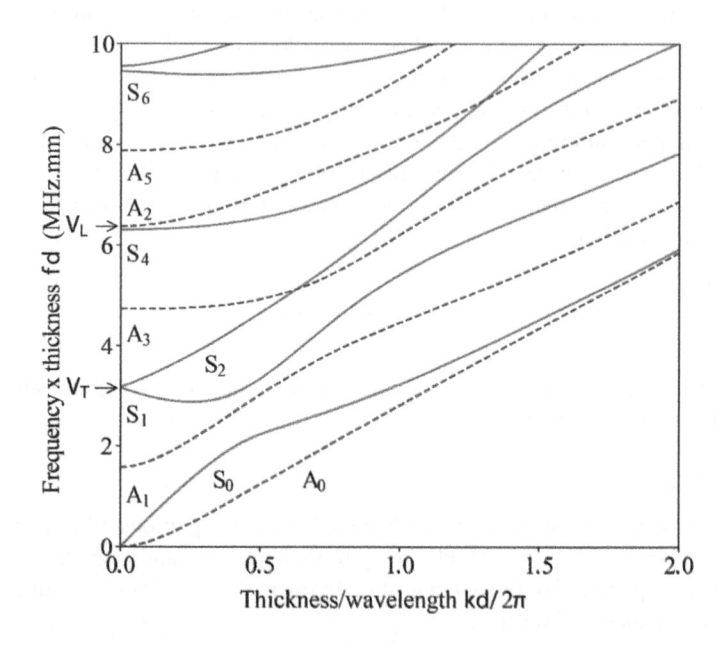

Figure 4.15. *Dispersion curves of the first Lamb modes in a duralumin plate. Symmetric modes (solid lines) and antisymmetric modes (dashed lines)*

The index of each mode is equal to the number of nodes in the thickness of the plate, at the cut-off frequency, of the longitudinal or transverse vibration (see section 4.3.1.3). Therefore, the order of the curves of the same family depends on the ratio V_T/V_L. For example, the curve of the mode S_2 is above or below that of the mode S_1 depending on whether V_T is larger or smaller than $V_L/2$.

The curves are generally increasing, that is, the product fd increases when kd increases. There are exceptions, for example, the frequency of the mode S_1 decreases for a product kd less than 2 before passing through a minimum. In this case, the group velocity changes sign when canceled at this particular frequency, corresponding to a zero group velocity (ZGV) mode. A negative group velocity means that it is opposite to the phase velocity, arbitrarily chosen to be positive (ω and $k > 0$): the acoustic energy propagates in the opposite direction to the wave vector; the mode is said to be contra-propagative. A zero group velocity corresponds to a local resonance, that is, to a stationary vibration (section 4.3.1.4).

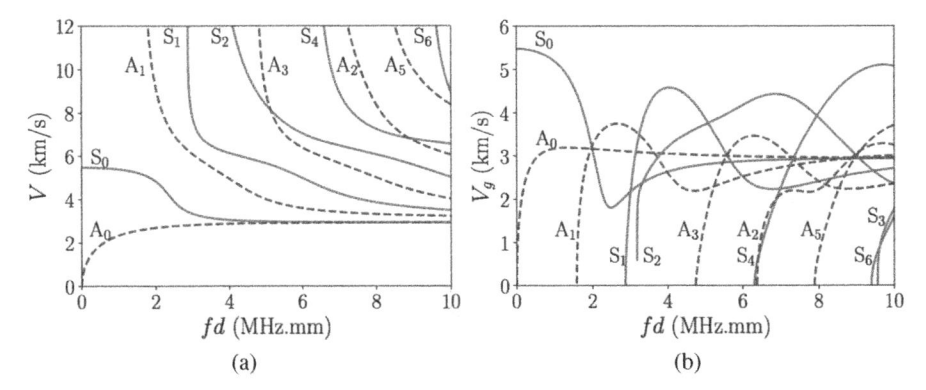

Figure 4.16. a) Phase velocities and b) group velocities of the first Lamb modes in a duralumin plate depending on the product fd. Symmetric (solid lines) and antisymmetric (dashed lines) modes. For a color version of this figure, see www.iste.co.uk/royer/waves1.zip

The evolution of the phase and group velocities of the first Lamb modes is plotted in Figure 4.16 as a function of the product fd. The behavior of the two *fundamental modes*, A_0 and S_0, is very different from that of the *higher order modes*, A_n and S_n, with $n \geq 1$.

– fundamental modes exist at all frequencies, unlike higher order modes, which only exist above a cut-off frequency;

– when kd tends to zero, the phase velocity of the A_0 mode tends to zero, and that of the S_0 mode tends to a finite value V_P, while the phase velocity of the higher order modes tends to infinity; their group velocity is canceled out;

– at high frequencies, the phase velocity of the A_0 and S_0 modes tends toward the Rayleigh wave velocity V_R, while the velocity of the higher order modes tends to V_T.

The displacement of matter for the fundamental Lamb modes S_0 and A_0 are represented in Figure 4.17 for a duralumin plate of thickness $d = 2$ mm, at the frequency $f = 1$ MHz. The displacement of the S_0 mode is symmetric with respect to the median plane of the plate, while that of the A_0 mode is antisymmetric.

4.3.1.3. *Behavior at low and high frequencies*

An important distinction between plate modes arises from their behavior when k tends to zero (in practice, $kh \ll 1$), that is, when the wavelength λ is very large compared to the plate thickness $d = 2h$. Two cases must be distinguished, depending on whether their frequency approaches zero or a cut-off frequency of finite value f_c.

(a) S_0 Mode   (b) A_0 Mode

Figure 4.17. *Particle displacements for a duralumin plate of thickness $d = 2$ mm, at the frequency $f = 1$ MHz, associated with a) the symmetric (S_0) and b) the antisymmetric (A_0) Lamb mode*

4.3.1.3.1. Modes without a cut-off frequency

When k and ω tend to zero, then according to the relation [4.70], the components k_{2L} and k_{2T} also tend to zero.

– *Symmetric modes*: $\alpha = 0$. In the dispersion equation [4.80], the development of the tangents, restricted to the first order:

$$\left(2k^2 - k_T^2\right)^2 = 4k^2\left(k^2 - k_L^2\right) \quad \text{hence} \quad k_T^4 = 4k^2\left(k_T^2 - k_L^2\right) \quad [4.86]$$

shows that the phase velocity $V = \omega/k$ tends to a finite limit, called the *plate velocity*:

$$V_P = 2V_T\sqrt{1 - \kappa^2} = V_T\sqrt{\frac{2}{1-\nu}} = V_L\frac{\sqrt{1-2\nu}}{1-\nu} \quad \text{with} \quad \kappa = \frac{V_T}{V_L} \quad [4.87]$$

For example, for some typical values of the Poisson's ratio, the plate velocity is:

$$\begin{cases} \nu = 0 & \text{(rigid solid)} & V_P = \sqrt{2}V_T = V_L \\ \nu = 1/3 & \text{(usual metals)} & V_P = \sqrt{3}V_T = V_L\sqrt{3}/2 \\ \nu = 1/2 & \text{(soft material)} & V_P = 2V_T \ll V_L \end{cases} \quad [4.88]$$

The plate velocity is always between $V_T\sqrt{2}$ and V_L. Consequently, k_{2T} is real and k_{2L} is imaginary (sector 2 in Figure 4.14):

$$k_{2T}^2 = k^2\left(\frac{V_P^2}{V_T^2} - 1\right) = k^2\frac{1+\nu}{1-\nu} \quad \text{and} \quad k_{2L}^2 = k^2\left(\frac{V_P^2}{V_L^2} - 1\right) = -k^2\left(\frac{\nu}{1-\nu}\right)^2$$

[4.89]

The mechanical displacement is deduced from equations [4.82] with $\alpha = 0$ and all the cosines are equal to unity ($|k_{2L}h|$ and $k_{2T}h << 1$):

$$\begin{cases} u_1 = k_{2T}a_T \left(1 + \dfrac{2k^2}{k_{2T}^2 - k^2}\right) = \dfrac{k_{2T}a_T}{\nu} \\ u_2 = -ik_{2T}a_T \left(1 + \dfrac{2k_{2L}^2}{k^2 - k_{2T}^2}\right)kx_2 = -i\dfrac{k_{2T}a_T}{1-\nu}kx_2 \end{cases} \quad [4.90]$$

The longitudinal component u_1 is uniform, the transverse component ($\pi/2$ phase-shifted) is maximum on the surfaces $x_2 = \pm h$. Since $kh << 1$, it is very small with respect to u_1, the symmetric mode without cut-off frequency S_0 is practically longitudinal. Its velocity V_P is lower than the velocity V_L of the bulk wave, because the solid is less rigid due to the absence of matter on both sides of the plate. A "surprising" result: according to the last line in [4.88], the velocity V_P of the longitudinal wave in a rubber strip, equal to $2V_T$, is very small compared to the velocity V_L of the bulk wave.

– *Antisymmetric mode*: $\alpha = \pi/2$. In the domain of long wavelengths, $k_{2L}h$, $k_{2T}h$ and kh are much smaller than unity. The first-order development of equation [4.81] leads to the relation $k_{2T}^2 = -k^2$. In order to determine the velocity V, we must develop the tangents to the third-order, which leads to:

$$k_T^2 = \frac{\omega^2}{V_T^2} = \frac{4}{3}(1-\kappa^2)k^4h^2 \quad [4.91]$$

The phase velocity tends to zero like the product kh and considering equation [4.87] for V_P, it comes:

$$V \approx \frac{2V_T}{\sqrt{3}}\sqrt{1-\kappa^2}kh = \frac{V_P}{\sqrt{3}}kh, \quad \text{hence} \quad \omega = \frac{V_P}{\sqrt{3}}k^2h \quad [4.92]$$

Near the origin, the dispersion curve $\omega(k)$ is a parabola. The mechanical displacement is given by equations [4.82] with $\alpha = \pi/2$ and $k_{2L}h$, $k_{2T}h$, kh tending to zero:

$$\begin{cases} u_1 = -k_{2T}^2 a_T A x_2 = a_T A k^2 x_2 \\ u_2 = ika_T A \end{cases} \quad \text{with} \quad A = \frac{k_{2T}^2 + k^2}{k_{2T}^2 - k^2} \quad [4.93]$$

The transverse component u_2 is uniform in the thickness of the plate, the longitudinal component u_1, which is maximum on the two free surfaces, is very small with respect to u_2. In the low-frequency domain, the antisymmetric mode without cut-off frequency A_0 is a flexural mode. The dispersion equation [4.92] can be found from the Love–Kirchhoff equation, which governs the transverse motion of thin plates:

$$\rho d \frac{\partial^2 u_2}{\partial t^2} + D \frac{\partial^4 u_2}{\partial x_1^4} = 0 \quad \text{with} \quad D = \frac{Ed^3}{12(1-\nu^2)} \quad \text{and} \quad V_P = \sqrt{\frac{E}{\rho(1-\nu^2)}} \quad [4.94]$$

D is the flexural rigidity of the plate and V_P is another expression for the plate velocity [4.87].

4.3.1.3.2. Modes with a cut-off frequency

When k tends to zero, ω tends to ω_c and the components k_{2L} and k_{2T} tend to $k_L = \omega_c/V_L$ and $k_T = \omega_c/V_T$. Equation [4.78] reduces to:

$$a_T \sin(k_T h + \alpha) = 0 \quad \text{and} \quad a_L \cos(k_L h + \alpha) = 0 \quad [4.95]$$

and the displacement components [4.76] become:

$$u_1 = k_T a_T \cos(k_T h + \alpha) \quad \text{and} \quad u_2 = -k_L a_L \cos(k_L h + \alpha) \quad [4.96]$$

For each value of α (0 or $\pi/2$), equations [4.95] admit two types of solutions, such that one of the two coefficients a_L and a_T is non-zero. The results are grouped in Table 4.1.

– *Symmetric modes*: $\alpha = 0$. Let us start with the even solutions S_{2n}, such that $k_T h = n\pi$, that is, $a_T \neq 0$ and $a_L = 0$. The mechanical displacement is longitudinal. For the odd solutions S_{2m+1}, such that $k_L h = (2m+1)(\pi/2)$, that is, $a_L \neq 0$ and $a_T = 0$, the displacement is transverse.

– *Antisymmetric modes*: $\alpha = \pi/2$. Let us start with the even solutions A_{2n}, such that $k_L h = n\pi$, that is, $a_L \neq 0$ and $a_T = 0$: the displacement is transverse. For the odd solutions A_{2m+1}, such that $k_T h = (2m+1)(\pi/2)$, that is, $a_T \neq 0$ and $a_L = 0$, the displacement is longitudinal.

Mode	Frequency $f_c d$	Mechanical displacement	
		Longitudinal u_1	Transverse u_2
S_{2n}	nV_T	$a_T \cos\left(n\pi \dfrac{x_2}{h}\right)$	0
S_{2m+1}	$\left(m+\dfrac{1}{2}\right) V_L$	0	$a_L \sin\left[\left(m+\dfrac{1}{2}\right)\pi \dfrac{x_2}{h}\right]$
A_{2n}	nV_L	0	$a_L \cos\left(n\pi \dfrac{x_2}{h}\right)$
A_{2m+1}	$\left(m+\dfrac{1}{2}\right) V_T$	$a_T \cos\left[\left(m+\dfrac{1}{2}\right)\pi \dfrac{x_2}{h}\right]$	0

Table 4.1. *Cut-off frequency and mechanical displacement of higher order Lamb modes*

Figure 4.18 illustrates the results for the first four symmetric and antisymmetric Lamb modes. The cut-off frequencies correspond to a stretch or a shear resonance of the plate, for which its thickness d is equal to an even or an odd multiple of the half-wavelength λ_L or λ_T:

– $d = n\lambda$ for the modes S_{2n} ($\lambda = \lambda_T = V_T/f_c$) or A_{2n} ($\lambda = \lambda_L = V_L/f_c$);

– $d = (2m+1)(\lambda/2)$ for the modes S_{2m+1} ($\lambda = \lambda_L$) or A_{2m+1} ($\lambda = \lambda_T$).

Even	Odd	Even	Odd
S_0	S_1	A_0	A_1
$fd \approx V_p \dfrac{kh}{\pi}$	$f_c d = \dfrac{V_L}{2}$	$fd \approx V_p \dfrac{(kh)^2}{\pi\sqrt{3}}$	$f_c d = \dfrac{V_T}{2}$
S_2	S_3	A_2	A_3
$f_c d = V_T$	$f_c d = \dfrac{3}{2} V_L$	$f_c d = V_L$	$f_c d = \dfrac{3}{2} V_T$

a) Symmetric modes   b) Antisymmetric modes

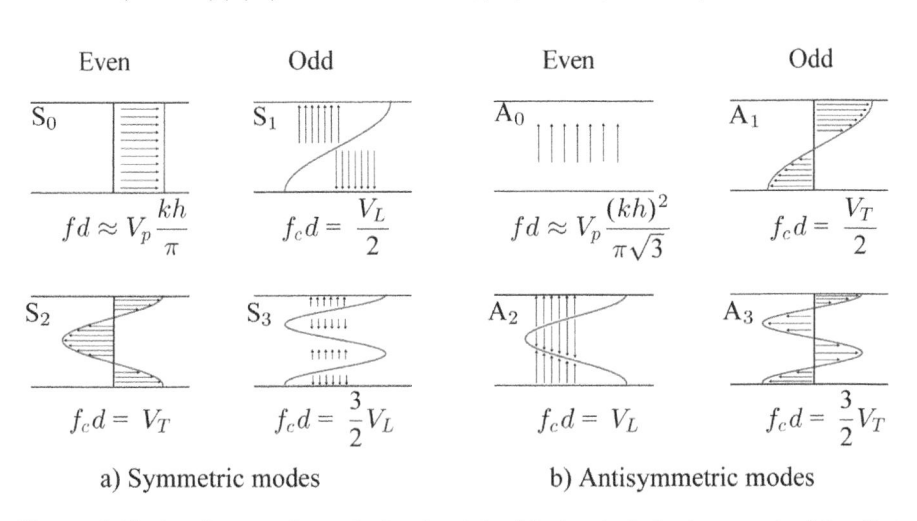

Figure 4.18. *Lamb wave in an isotropic plate. Mechanical displacement of the first modes for $kh \ll 1$. a) Symmetric modes. b) Antisymmetric modes. The cut-off frequency of modes S_{2m+1} and A_{2n} corresponds to a thickness-stretch resonance. The cut-off frequency of modes S_{2n} and A_{2m+1} corresponds to a thickness-shear resonance*

The index of each mode is equal to the number of nodes in the plate thickness, both for the longitudinal displacement (modes S_{2n} and A_{2m+1}) as well as for the transverse displacement (modes S_{2m+1} and A_{2n}). This numbering, different from that of other authors, is essential to understand the conditions of existence of the ZGV modes. The lowest cut-off frequency is always that of the A_1 mode: it corresponds to a product fd equal to half the transverse wave velocity. For the other modes of the same family, the order of the dispersion curves depends on the velocity ratio $\kappa = V_T/V_L$, that is, on the Poisson's ratio ν. For example, the modes S_1 and S_2 interchange their relative positions for the critical value $\kappa = 1/2$, corresponding to $\nu = 1/3$.

The distribution of the mechanical displacements inside the plate varies substantially with frequency, that is, with the product kh. Nonetheless, the distinction between the fundamental modes S_0 and A_0, on the one hand, and the higher order modes, S_n and A_n ($n \geq 1$), on the other hand, remains at high frequency.

4.3.1.3.3. High-frequency behavior

The plate thickness is assumed to be very large compared to the wavelength ($kh = \pi d/\lambda \to \infty$). If the phase velocity V is less than V_T (sector 3 in Figure 4.14), the roots k_{2L} and k_{2T} are imaginary: $k_{2L} = i\alpha_L$ and $k_{2T} = i\alpha_T$. In the development of trigonometric functions, increasing and decreasing exponentials appear, for example:

$$\cos(k_{2M}h) = \frac{1}{2}\left(e^{-\alpha_M h} + e^{\alpha_M h}\right) \quad \text{with} \quad M = L, T \quad \text{and} \quad \alpha_M > 0 \qquad [4.97]$$

When $kh \to \infty$, the terms $e^{-\alpha_M h}$ tend to zero and the dispersion equation [4.79] becomes:

$$\left[(k^2 + \alpha_T^2)^2 - 4k^2\alpha_L\alpha_T\right]e^{(\alpha_L + \alpha_T)h} = 0 \qquad [4.98]$$

Since the exponential does not get canceled, it leads to the Rayleigh equation [3.15] established in section 3.1.1.1. This is the case for the A_0 and S_0 modes, whose phase velocity tends to the Rayleigh wave velocity V_R when $kh \to \infty$. The displacement of matter of the fundamental mode S_0 (A_0) is localized near the free surfaces of the plate, in the form of two Rayleigh waves arranged symmetrically (antisymmetrically) with respect to the median plane.

If only k_{2L} is imaginary (sector 2 in Figure 4.14): $k_{2L} = i\alpha_L$ and $\tan(k_{2L}h) = i\tanh(\alpha_L h)$ tends to the imaginary, when $kh \to \infty$. For the symmetric modes, the dispersion equation [4.80] becomes:

$$\left(1 - \frac{k_{2T}^2}{k^2}\right)^2 \frac{\tan(k_{2T}h)}{k_{2T}h} = \frac{4\alpha_L}{k^2 h} \qquad [4.99]$$

It is satisfied when $kh \to \infty$, with a component $k_{2T} = n\pi/h$, which cancels the tangent. For the antisymmetric modes, the dispersion equation [4.81] becomes:

$$\left(1 - \frac{k_{2T}^2}{k^2}\right)^2 k^2 = -4\alpha_L k_{2T} \tan(k_{2T}h) \qquad [4.100]$$

It is satisfied when $kh \to \infty$, with a component $k_{2T} = (n + 1/2)\pi/h$ which makes the tangent infinite. In both cases, according to equation [4.70], the velocity of the higher order modes ($n \geq 1$) slowly approaches V_T through higher values:

$$\frac{\omega^2}{V_T^2} = k^2 + \frac{m^2\pi^2}{h^2} \quad \text{hence} \quad V = V_T\sqrt{1 + \frac{m^2\pi^2}{k^2 h^2}} \quad \text{with} \quad m = n \quad \text{or} \quad n + \frac{1}{2}$$

$$[4.101]$$

4.3.1.4. Zero group velocity modes

By examining the dispersion curves in Figure 4.15, we have observed that the frequency of the S_1 mode passes through a minimum at a particular point of coordinates (k_0, f_0), where the group velocity $d\omega/dk$ vanishes ($k_0 d = 1.58$ and $f_0 d = 2.866$ MHz.mm for a duralumin plate).

Let us note that in a narrow frequency range between f_0 and the cut-off frequency of the S_1 mode, $f_c d = V_L/2 = 3.17$ MHz.mm and for a given direction of propagation of energy (from the source to the observation point), two waves can propagate, with opposite wave numbers. For many years, these modes, called ZGV, remained a curiosity. In the following sections, we show that a ZGV mode results from a strong repulsion, in the vicinity of the cut-off frequencies, between a pair of modes, and we examine the conditions of existence of these modes.

4.3.1.4.1. Origin of ZGV modes

The multiple reflections of longitudinal or transverse waves between the top and bottom faces of the plate give rise to thickness-shear resonances (modes S_{2n} or A_{2m+1}) or to thickness-stretch resonances (modes S_{2m+1} or A_{2n}), whose frequencies, given in Table 4.1, are the cut-off frequencies f_c of Lamb modes. For example, for the mode S_1: $f_c = V_L/2d$ and for the mode S_2: $f_c = V_T/d$; these modes interchange their relative positions for the critical value $\nu = 1/3$, such that $V_L = 2V_T$. Then, the S_1 and S_2 branches intersect the vertical axis at the same point with equal and opposite slopes, and the frequency of the lower branch undergoes a minimum at a non-zero wave number k_0. The coincidence of two cut-off frequencies plays an essential role in the existence of ZGV modes (Prada *et al.* 2008).

Let us consider two symmetric modes of different parity, like S_5 and S_8. The difference between their cut-off frequencies, respectively, $2.5V_L/d$ and $4V_T/d$, depends on ν. For $\nu = 0.13$, this difference is relatively large and the modes are weakly interacting (Figure 4.19(a)). For $\nu = 0.155$, a stronger interaction leads to a flat lower branch (Figure 4.19(b)). When the velocity ratio is equal to the critical value $V_L/V_T = 8/5 = 1.6$ for $\nu = 0.179$, the modes are degenerated and a very strong repulsion creates a ZGV mode in the lower branch for $k = k_0$ (Figure 4.19(c)). For $\nu = 0.20$, the mode S_5 passes beyond the mode S_8 and the weaker coupling gives rise to a less pronounced trough (Figure 4.19(d)).

A dispersion curve of a mode cannot intersect the curve of another mode belonging to the same family. However, in Figure 4.19(c), it can be observed that the symmetric modes S_5 and S_8 are uncoupled for $k = 0$. This property, valid for any waveguide uniform along the x_1 axis, results from the symmetry with respect to the $x_2 x_3$ plane. For $k \neq 0$, the propagation along one of the $+x_1$ or $-x_1$ direction breaks this symmetry: the dispersion curves of the two modes repel each other. The lower branch thus presents a zone with a negative slope and a point where the group velocity vanishes.

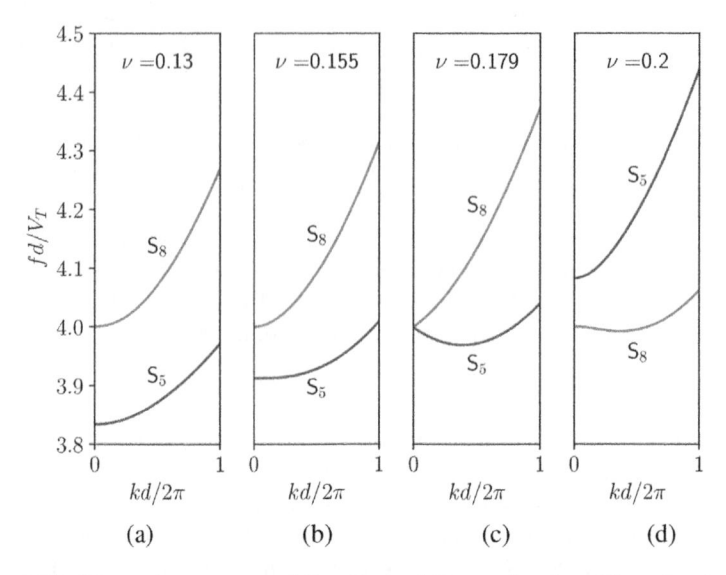

Figure 4.19. Dispersion curves of the S_5 and S_8 modes for different values of the Poisson's ratio. a) Weakly interacting modes. b) Stronger interaction leading to a nearly flat lower branch. c) Coincidence: the strong repulsion creates a ZGV mode. d) S_5 and S_8 modes are reversed: the repulsion is weaker and the trough is less pronounced. For a color version of this figure, see www.iste.co.uk/royer/waves1.zip

4.3.1.4.2. Selection rules

Using the classification in section 4.3.1.3, where, at the cut-off frequency, the index of each mode is equal to the number of nodes in the plate thickness, it is possible to state the following rules for selecting ZGV modes:

1) symmetric and antisymmetric modes are uncoupled and since there is no coincidence between the cut-off frequencies of modes with the same parity, a ZGV mode results from the repulsion between symmetric modes S_{2m+1} and S_{2n}, or between antisymmetric modes A_{2n} and A_{2m+1}, of different parity;

2) for symmetric modes and considering the cut-off frequencies of even modes, S_{2n} and odd modes S_{2m+1}: $f_c d = nV_T$ and $(2m+1)V_L/2$, respectively; the repulsion, i.e., the difference $f_c - f_0$, is maximum for a critical value ν_c of the Poisson's ratio, such that:

$$\left(\frac{V_L}{V_T}\right)_S = \frac{2n}{2m+1} \quad \text{with} \quad n \geq 1 \quad \text{and} \quad m \geq 0 \quad [4.102]$$

3) for antisymmetric modes A_{2n} and A_{2m+1}: $f_c d = nV_L$ and $(2m+1)V_T/2$, respectively, the repulsion is maximum for a value of the Poisson's ratio, such that:

$$\left(\frac{V_L}{V_T}\right)_A = \frac{2m+1}{2n} \quad \text{with} \quad n \geq 1 \quad \text{and} \quad m \geq 0 \quad [4.103]$$

ZGV modes appear when the velocity ratio V_L/V_T is equal to a rational number. For the first symmetric ZGV mode ($m = 0, n = 1 \to V_L/V_T = 2$), the repulsion between Lamb modes S_1 and S_2 is maximum for $\nu_1 = 1/3$. For the first antisymmetric ZGV mode ($m = 1, n = 1 \to V_L/V_T = 3/2$), the repulsion between modes A_2 and A_3 is maximum for $\nu_2 = 1/10$. The domain of existence of ZGV modes is limited on either side of the critical value ν_c.

4.3.1.4.3. Range of existence

For a given value of the Poisson's ratio and of the velocity V_T, it is possible, from the Rayleigh–Lamb equations, to calculate the group velocity by numerical differentiation. The frequencies f_0, corresponding to the ZGV points, are then determined by the zero crossings of the group velocity. The results are presented in Figure 4.20, as a plot of the dimensionless frequency $F = fd/V_T$ (for $f = f_0$ or f_c) in the range $0 \leq F \leq 4$, *versus* the Poisson's ratio ($0 \leq \nu < 0.5$).

The ZGV S_1 and A_2 modes exist between $\nu = 0$ and $\nu = 0.45$ and 0.32, respectively. The ranges of existence of the other ZGV modes are narrower. It should also be noted that the symmetric ZGV modes (S_3/S_6 or S_5/S_{10}) and antisymmetric ZGV modes (A_6/A_9) correspond to the same critical values ν_1 and ν_2. Since the indices of these pairs of modes are odd multiples (3 or 5) of the indices of ZGV modes (S_1/S_2) and (A_2/A_3), the bulk wave velocity ratio is unchanged: $V_L/V_T = 2$ or 3/2. These higher order modes are harmonics of the fundamental ZGV modes S_1/S_2 and A_2/A_3.

The critical values of the velocity ratio V_L/V_T, of the Poisson's ratio ν and the range of existence $[\nu_{min}, \nu_{max}]$ of the first four symmetric and antisymmetric ZGV modes are given in Table 4.2.

In the absence of dissipation, the energy transport velocity is equal to the group velocity, so that any mechanical energy, deposited locally at the frequency of a ZGV mode, does not propagate: it is trapped under the source zone, without transfer to the adjacent matter. This local resonance effect, predicted by Tolstoy and Usdin (1957), can be easily observed using laser ultrasonic techniques based on optical generation and detection, without mechanical contact with the sample. An experimental result is shown after the sections devoted to these methods (Volume 2, Figure 3.41).

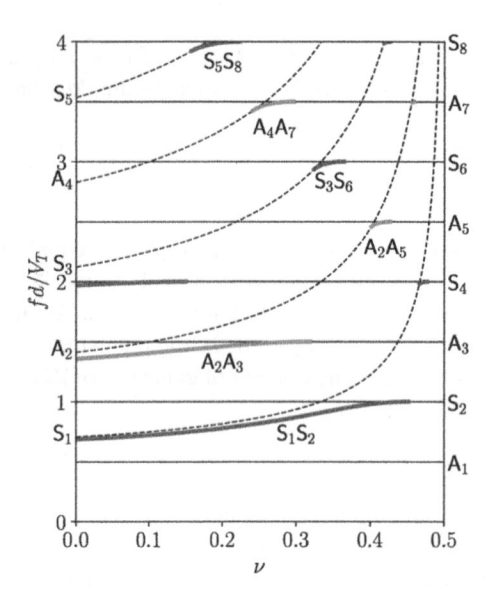

Figure 4.20. *Normalized cut-off frequencies $f_c d/V_T$ (solid lines) and $f_c d/V_L$ (dashed curves) and minimum frequency $f_0 d/V_T$, versus the Poisson's ratio ν. The ZGV branches appear around the crossing points of these curves for modes belonging to the same family: symmetric (in blue) or antisymmetric (in orange). For a color version of this figure, see www.iste.co.uk/royer/waves1.zip*

ZGV mode	S_3/S_4	S_7/S_{10}	S_5/S_8	S_1/S_2	A_2/A_3	A_6/A_9	A_4/A_7	A_4/A_9
V_L/V_T	4/3	10/7	8/5	2	3/2	3/2	7/4	9/4
ν_c	−0.143	0.020	0.179	0.333	0.100	0.100	0.258	0.377
ν_{min}	−0.973	−0.012	0.156	−0.540	−0.740	0.073	0.241	0.375
ν_{max}	0.149	0.072	0.223	0.451	0.319	0.148	0.296	0.387

Table 4.2. *Critical values of the velocity ratio V_L/V_T, of Poisson's ratio ν and range of existence $[\nu_{min}, \nu_{max}]$ of the first four ZGV symmetric and antisymmetric Lamb modes*

4.3.1.5. *Influence of Poisson's ratio*

The elastic behavior of an isotropic material is characterized by two constants λ and μ, related to the velocities of the longitudinal and transverse waves V_L and V_T, respectively. However, the use of dimensionless frequency $F = fd/V_T$ and wave number $K = kd/2\pi$ allows us to express Lamb wave propagation in terms of only one material parameter, the ratio of these velocities $\kappa = V_T/V_L$ or the Poisson's ratio ν:

$$\nu = \frac{1 - 2\kappa^2}{2(1 - \kappa^2)} \quad \text{with} \quad \kappa = \frac{V_T}{V_L} = \sqrt{\frac{1 - 2\nu}{2(1 - \nu)}} \quad [4.104]$$

The study is limited to usual isotropic materials for which $\nu \geq 0$ (Royer *et al.* 2009). The normalized frequency F was numerically determined *versus* the normalized wave number K. The dispersion curves are plotted for $0 \leq K \leq 5$ and $0 \leq F \leq 6$, for six values of Poisson's ratio ν, from 0 to 0.49 (the limiting case of a fluid, with $V_T = 0$ and $\nu = 0.5$ is excluded) The first five values are spaced apart by 0.1. Results are presented in Figure 4.21 for the first symmetric and antisymmetric modes. Since $F/K = V/V_T$, a straight line of unit slope corresponds to a phase velocity $V = \omega/k$ equal to V_T. For a given mode, the lower curve in a bundle corresponds to $\nu = 0$ and the upper one to $\nu = 0.49$. The dashed line corresponds to $V = V_T\sqrt{2}$. It should be noted that by plotting the normalized frequency fd/V_T on the ordinate instead of the product fd, the order of modes is unchanged, whatever the value of Poisson's ratio. This justifies the convention chosen to index the Lamb modes (section 4.3.1.3). These two networks of normalized curves report on the propagation of Lamb waves in any isotropic plate with thickness d, characterized by its transverse velocity V_T and its Poisson's ratio ν: it is *the complete spectrum of plate modes*.

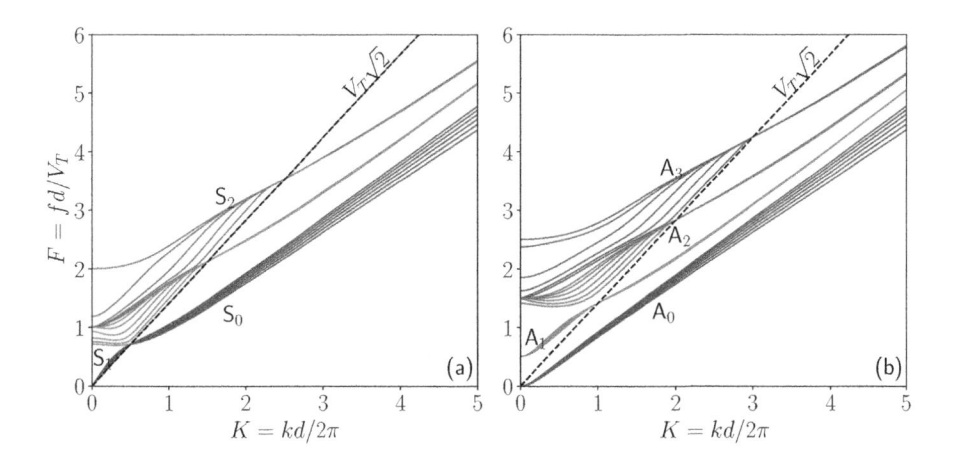

Figure 4.21. *Variation with Poisson's ratio $(0 \leq \nu \leq 0.49)$ of Lamb waves in a free isotropic plate of thickness d. Bundle of dispersion curves for a) symmetric S_0, S_1 and S_2 modes and b) antisymmetric A_0, A_1, A_2 and A_3 modes. For a color version of this figure, see www.iste.co.uk/royer/waves1.zip*

Except for the A_0 mode, the (ω, k) plane can be divided into two angular sectors. In the upper sector $(V > V_T\sqrt{2})$, dispersion curves are very sensitive to the material parameters. In the lower sector $(V < V_T\sqrt{2})$, all the curves, regardless of the Poisson's ratio value, are gathered into a thin pencil. The velocity $V_T\sqrt{2}$ represents the lowest possible value for the longitudinal wave velocity V_L, reached when $\nu = 0$. In this sector, the phase velocity is always less than V_L, which drastically reduced the influence of the velocity ratio and, therefore, of the Poisson's ratio, on the dispersion curves of Lamb waves. For high values of kd, that is, $d \gg \lambda$, the branches of the symmetric and antisymmetric modes of order $n \geq 1$ tend asymptotically to a line of unit slope corresponding to a phase velocity equal to V_T. Lamb waves are often used to characterize the material properties in plate-like structures; for modes of order greater than or equal to unity, the sensitivity to the parameters ν or κ is much larger in the upper sector. Curves of a given Lamb mode intersect the line with slope $V_T\sqrt{2}$ at a fixed point independent of the Poisson's ratio. These points are regularly spaced on this line and their abscissa kd are equal to $(2n+1)\pi$ for the symmetric mode S_n (or $K = n + 1/2$) and $2n\pi$ for the antisymmetric mode A_n (or $K = n$). Curves of a bundle do not cross each other, they intersect the line $F = \sqrt{2}K$ at the coincidence point with the same slope and their order in the bundle does not change. This property is valid for all symmetric and antisymmetric modes.

Figure 4.22(a) shows that symmetric Lamb waves exhibit a particular behavior in the case of a rigid solid ($\nu = 0$): the segments of the line $F = \sqrt{2}K$ belong successively to these modes. The change from mode S_n to mode S_{n+1} occurs at the coincidence point of abscissa $kd = (2n+1)\pi$, giving rise to a discontinuity in the slope and, therefore, in the group velocity $V_g = d\omega/dk$. These remarks do not apply to antisymmetric modes (figure 4.22(b)).

Most of these observations regarding the Lamb mode spectrum can be explained from the specific propagation conditions in an isotropic plate of a purely transverse wave (Lamé modes) or a purely longitudinal wave (degenerate Lamé mode).

4.3.1.6. *Lamé modes*

The Rayleigh–Lamb equations result from the boundary conditions on the free surfaces $x_2 = \pm h$ of the plate. The *Lamé modes* are particular solutions of these equations, for which $k_T^2 = 2k^2$, that is, for normal components of the wave vectors, such that:

$$k_{2T}^2 = k_T^2 - k^2 = k^2 \quad \text{and} \quad k_{2L}^2 = k_L^2 - k^2 = \kappa^2 k_T^2 - k^2 = -\frac{\nu}{1-\nu}k^2 \quad [4.105]$$

Since the term $k_{2T}^2 - k^2$ is zero, the boundary conditions (equation [4.78]) are satisfied for the symmetric modes ($\alpha = 0$) when:

$$a_L = 0 \quad \text{and} \quad k_{2T}h = kh = \left(n + \frac{1}{2}\right)\pi \quad [4.106]$$

and for the antisymmetric modes ($\alpha = \pi/2$) when:

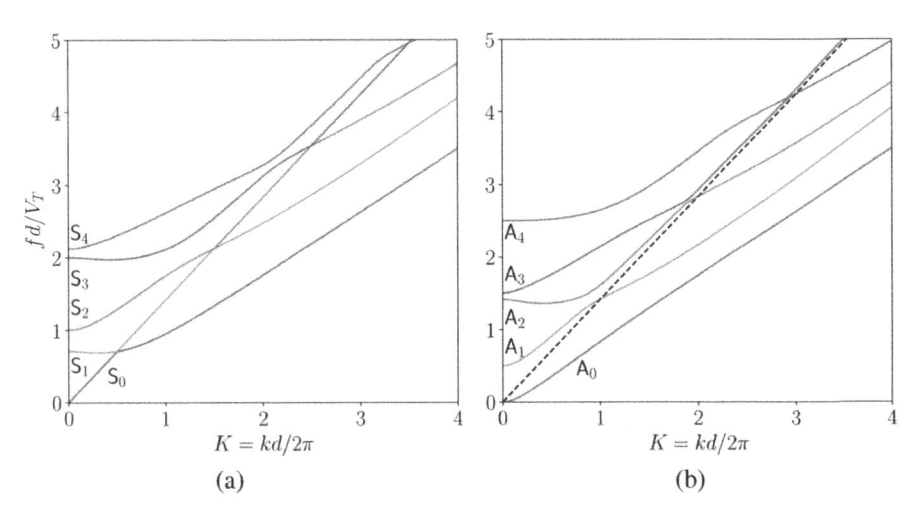

$$a_L = 0 \quad \text{and} \quad k_{2T}h = kh = n\pi \qquad [4.107]$$

Figure 4.22. *Specific behavior for $\nu = 0$ (rigid solid). a) Segments of the line $F = \sqrt{2}K$ belong successively to different symmetric modes. The change from mode S_n to mode S_{n+1} occurs at the coincidence point of abscissa $K = n + 1/2$ and gives rise to a discontinuity in the slope. b) This discontinuous behavior is not observed on curves for antisymmetric modes. For a color version of this figure, see www.iste.co.uk/royer/waves1.zip*

In both cases, the amplitude a_L of the scalar potential is zero. As $k = k_T/\sqrt{2}$, the phase velocity $V = V_T\sqrt{2}$ is that of a purely transverse wave that propagates, as shown in Figure 4.23(a), along a direction at 45° from the plate axis, and which is reflected on the faces without conversion into longitudinal waves. The absence of mode conversion at this angle of incidence was highlighted in section 2.4.1.2. According to equation [4.87], the plate velocity V_P is always greater than $V_T\sqrt{2}$. The equality $V = V_T\sqrt{2}$ is possible for the mode S_0, and for all modes with a cut-off frequency. The only exception is the antisymmetric mode A_0, whose phase velocity is always less than V_R, and thus less than V_T.

With $a_L = 0$ and $k_{2T} = k$, the components of the mechanical displacement [4.76] take the form:

$$u_1 = ka_T \cos(kx_2 + \alpha) \quad \text{and} \quad u_2 = -ika_T \sin(kx_2 + \alpha) \qquad [4.108]$$

Since $\cos(kh + \alpha) = 0$, the tangential component u_1 is zero on both faces: at an incidence of 45°, the transverse vertical wave does not create any displacement in the plane of the plate. Figure 4.23(b) shows the variations in the plate thickness of the displacement components u_1 and u_2 of the S_0 mode.

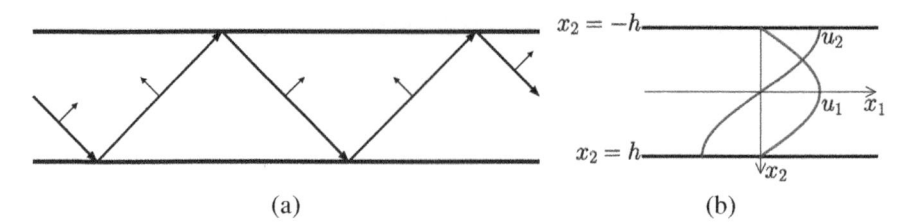

Figure 4.23. *Lamé modes. a) The guided wave is a transverse wave propagating at 45° from the plate axis. b) Displacement components of the Lamé mode S_0*

The Lamé modes exist for any positive value of ν; they appear at equally spaced wave numbers and frequency × thickness products:

$$kh = \frac{m\pi}{2} \quad \text{and} \quad fd = m\frac{V_T}{\sqrt{2}}, \quad m = 1, 2, 3... \quad [4.109]$$

or, for normalized coordinates:

$$K = \frac{kd}{2\pi} = \frac{m}{2} \quad \text{and} \quad F = \frac{fd}{V_T} = \frac{m}{\sqrt{2}} \quad [4.110]$$

with m being odd for symmetric modes and even for antisymmetric modes. The acoustic energy is transported at a velocity equal to the projection of the transverse wave velocity on the plane of the plate: therefore, the group velocity of the Lamé modes is equal to $V_T/\sqrt{2}$, that is, half the phase velocity.

Since the positions and slopes of the Lamé modes in the (F, K) plane are independent of the Poisson's ratio, each curve in a given bundle crosses the Lamé line $F = \sqrt{2}K$ at the same point, with the same slope. This behavior can be observed in Figures 4.21(a) and 4.21(b) for all the Lamb modes, except for the A_0 mode, whose phase velocity is always less than V_T.

In the specific case $\nu = 0$, the velocity of the longitudinal wave is equal to that of the Lamé modes: $V_L = V_T\sqrt{2}$. Since k_{2L} and $k_{2T}^2 - k^2$ are zero and if $a_T = 0$, the boundary conditions (equation [4.78]) are satisfied with a non-zero scalar potential ($a_L \neq 0$). The mechanical displacement (equation [4.76]):

$$u_1 = ika_L \cos\alpha \quad \text{and} \quad u_2 = 0 \quad [4.111]$$

is longitudinal and uniform for the symmetric modes ($\alpha = 0$) and identically zero for the antisymmetric modes ($\alpha = \pi/2$). In a rigid solid, this *degenerate Lamé mode* satisfies the boundary conditions for any value of k: the Lamé line is the locus of the roots of the Rayleigh–Lamb equation for the symmetric modes when $\nu = 0$ (Figure 4.22(a)). Each mode describes a segment of this line; the transition from the mode S_n to the mode S_{n+1} occurs at the coincidence point $K = n + 1/2$. At this point, the slope of the branch S_n changes, which leads to a discontinuity in the group velocity from $V_T\sqrt{2}$ to $V_T/\sqrt{2}$. The inverse change occurs for the mode S_{n+1}. As expected, this behavior is specific to symmetric modes (Figure 4.22(b)).

4.3.1.7. *Plate with finite dimensions*

When the lateral dimensions of the plate are finite, particular effects appear around the edges. In this section, we will describe two of them: the vibration at a free end, called edge resonance, and the propagation of waves guided by a ribbon, in particular, a crest wave of the Rayleigh wave type.

4.3.1.7.1. *Edge resonance*

A resonance phenomenon appears on the contour of a free plate (or disk). This vibration, called *edge mode*, is symmetrical; it exists in a narrow band around the resonance frequency and the mechanical displacement is confined in a small region near the free edge. In the case of a plate, the edge resonance frequency is lower than the minimum frequency corresponding to the ZGV-S_1S_2 mode (section 4.3.1.4.3). In this frequency range, only the symmetric Lamb mode S_0 can propagate. On the free edge ($x_1 = 0$) of a semi-infinite plate ($-\infty < x_1 \leq 0$), the stresses:

$$\sigma_{11} = ik(\lambda + 2\mu)u_1 + \lambda \frac{\partial u_2}{\partial x_2} \quad \text{and} \quad \sigma_{12} = ik\mu u_2 + \mu \frac{\partial u_1}{\partial x_2} \quad [4.112]$$

must cancel. According to equations [4.82] for the mechanical displacement (with $\alpha = 0$), σ_{11} is an odd function of the wave number k, and σ_{12} is an even function. The change in the propagation direction at the end of the plate changes the sign only for the stress σ_{11}. Consequently, the boundary conditions at the free edge cannot be fulfilled by superimposing an incident mode S_0 and a reflected mode S_{-0} alone. Complex modes must be introduced and the Rayleigh–Lamb equations [4.80] are therefore solved for complex wave numbers. Two branches appear, starting on either side of the ZGV point located on the real branch of the Lamb mode S_1 (Figure 4.24(a)). The points on the three lower branches indicate the edge resonance frequency. Near the edge of the plate, the extensional mode S_0 is coupled with the inhomogeneous Lamb modes corresponding to the complex roots of the dispersion equation. Taking into account these two complex modes explains the significant increase in the displacement amplitude at the end of the plate, but only gives an approximate value for the resonance frequency (Mindlin and Medick 1959). More recent analyses, involving a large number of real and complex modes, lead to a much better agreement with experimental data. Indeed, the resonance frequency is complex, that is, the edge resonance is damped, except for two values, $\nu_1 = 0$ and $\nu_2 = 0.2248$, of the Poisson's ratio, which cancel the imaginary part of the frequency. A good approximation of the real part of the normalized angular frequency $\Omega = \omega h/V_T$ is given by the formula (Pagneux 2006):

$$\text{Re}[\Omega_R] = 0.652\nu^2 + 0.898\nu + 1.9866 \quad [4.113]$$

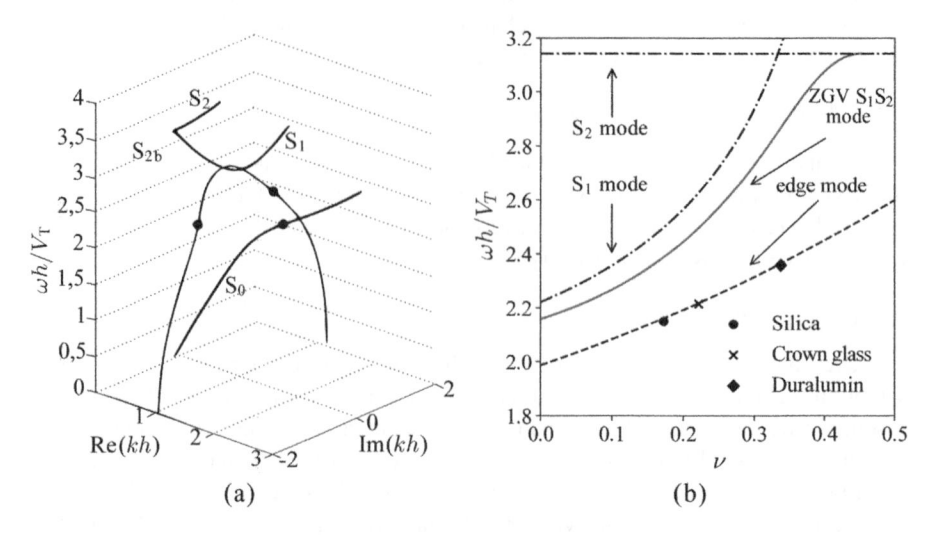

Figure 4.24. a) *Dispersion curves of Lamb modes S_0, S_1, S_2, S_{2b} and of the first two complex modes for a duralumin plate. The points correspond to the edge resonance frequency.* b) *Real part of the normalized frequency $\omega h/V_T$ as a function of the Poisson's ratio for the edge mode, the ZGV-$S_1 S_2$ mode and the thickness modes S_1 and S_2.*

The variations of the normalized resonance frequencies of the edge mode (Ω_R), the ZGV-$S_1 S_2$ mode (Ω_0) and the thickness modes S_1 and S_2 (Ω_c) are plotted as functions of the Poisson's ratio in Figure 4.24(b). For usual materials ($\nu > 0$), the angular frequency Ω_R is lower than that of the ZGV mode. For example, in the case of a duralumin plate ($V_T = 3\,134$ m/s, $\nu = 0.340$) of thickness $d = 1.51$ mm, the value of the edge resonance frequency, 1.566 MHz, measured using laser ultrasonic techniques (Cès *et al.* 2011) is very close to that predicted by equation [4.113]:

$$\mathrm{Re}[\Omega_R] = 2.367 \quad \text{hence} \quad f_R = \frac{\Omega_R V_T}{\pi d} = 1.564 \text{ MHz} \qquad [4.114]$$

The frequency of the ZGV mode is equal to 1.901 MHz. The quality factor Q of the edge resonance is deduced from the imaginary part of Ω_R: $Q = |\mathrm{Re}(\Omega_R)/2\mathrm{Im}(\Omega_R)|$. With $\mathrm{Im}[\Omega_R] = -3.35 \times 10^{-3}$, the theoretical value ($Q = 350$) indicates that the mechanical displacement at the end of the plate is much larger than in its center. With $\mathrm{Im}[kh]$ of the order of unity, the complex modes in Figure 4.24(a) decrease very rapidly; at a distance equal to twice the plate thickness, the amplitude of the vibration is divided by a factor of about 50. From this distance from the plate edge the ZGV resonance dominates. The transition between edge resonance at 1.564 MHz and ZGV resonance at 1.901 MHz is abrupt.

4.3.1.7.2. Waves guided by a ribbon

The waveguide has now the shape of a thin rectangular ribbon infinite in the x_1 direction; its width b (along x_3) is large compared to its thickness d, along x_2. In an isotropic plate and at low frequency, three modes can propagate along x_1: the flexural wave of A_0-type, the transverse horizontal mode TH_0 and the compressional mode S_0. Unlike the A_0 mode, the symmetric modes TH_0 and S_0 are polarized in the plane $(x_1 x_3)$ of the plate. They are non-dispersive: the velocity of the TH_0 mode is rigorously equal to the transverse wave velocity V_T (section 4.2.1), and the velocity of the S_0 mode is approximately equal to the plate velocity V_P, as long as the frequency is much smaller than the first cut-off frequency (section 4.3.1.3.1):

$$V_{S_0} \approx V_p = V_T \sqrt{\frac{2}{1-\nu}} \quad \text{if} \quad fd << \frac{V_T}{2} \quad [4.115]$$

In a ribbon of finite width b, the two in-plane motions (S_0 and TH_0) are coupled through successive reflections on the edges $x_3 = \pm b/2$ of the ribbon. This coupling is similar to that of bulk waves in a plate. Thus, the dispersion curves for the low-frequency waves guided by the ribbon are close to those of a plate of thickness b and bulk wave velocities V_P (instead of V_L) and V_T. Everything happens as if the Poisson's ratio ν of the material was modified. Considering relation [4.104], the equivalent Poisson's ratio $\tilde{\nu}$, defined by:

$$\tilde{\nu} = \frac{1 - 2\tilde{\kappa}^2}{2(1 - \tilde{\kappa}^2)}, \quad \text{with} \quad \tilde{\kappa}^2 = \frac{V_T^2}{V_P^2} = \frac{1-\nu}{2} \quad [4.116]$$

is smaller than ν:

$$\tilde{\nu} = \frac{\nu}{1+\nu}, \quad \text{that is, for instance } \tilde{\nu} = \frac{1}{4} \quad \text{if} \quad \nu = \frac{1}{3} \quad [4.117]$$

At a very low frequency, the velocity of the longitudinal mode $\tilde{S}_0$ in the ribbon is constant and given by an expression similar to equation [4.115]:

$$V_{\tilde{S}_0} \approx \tilde{V}_P = V_T \sqrt{\frac{2}{1-\tilde{\nu}}} \quad \text{if} \quad fb << \frac{V_T}{2} \quad [4.118]$$

Substituting $\tilde{\nu}$ by its expression [4.117], the "ribbon" velocity $\tilde{V}_P$:

$$\tilde{V}_P = V_T \sqrt{2(1+\nu)} = \sqrt{\frac{2\mu(1+\nu)}{\rho}} = \sqrt{\frac{E}{\rho}} \quad [4.119]$$

is none other than the bar velocity V_b, that is, the velocity of a longitudinal wave propagating in an elongated structure (bar, ribbon, cylinder, etc.) whose lateral dimensions are very small compared to the wavelength. Since the stiffness of the bar is equal to the Young's modulus E of the isotropic solid, the bar velocity is

$V_b = \sqrt{E/\rho}$, whatever the shape of its cross-section. In the specific case ($\nu = 1/3$) corresponding to a duralumin ribbon ($\nu = 0.334$), the velocity $\tilde{V}_P = (8/9)V_P$ is slightly lower than V_P.

At high frequencies ($b > 2\lambda_T$), the longitudinal mode $\tilde{S}_0$ splits into two Rayleigh waves, propagating independently on each free surfaces $x_3 = \pm b/2$ of the ribbon. These crest waves result from the coupling between the longitudinal mode S_0 and the transverse mode TH_0. If the condition [4.115], written as $d \ll \lambda_T/2$, is still valid, the velocity $\tilde{V}_R$ of these Rayleigh waves propagating along the edges of the ribbon is obtained by replacing V_L by V_P. The two conditions cannot be realized unless the ribbon width b is very large compared to its thickness d, for example $b > 10d$. Since the Rayleigh wave velocity depends little on the longitudinal wave velocity: $\tilde{V}_R \approx V_R$.

> ORDER OF MAGNITUDE.– In the case of duralumin ($V_L = 6\,340$ m/s and $V_T = 3\,140$ m/s), we obtain for a plate: $V_P = 5\,460$ m/s and $V_R = 2\,910$ m/s, and for a thin ribbon: $\tilde{V}_P = V_b = 5\,150$ m/s and $\tilde{V}_R = 2\,850$ m/s.

The dispersion curves of the planar modes in a ribbon, of form factor $b/d > 10$, plotted in the dimensionless coordinate system fb/V_T and kb, are thus similar to the curves for Lamb waves in a material of Poisson's ratio $\tilde{\nu}$. The comparison with the exact curves shows that the approximation is valid as long as $fb < 3V_T$.

It should be noted that the analysis of elastic wave propagation in an isotropic solid has revealed five characteristic velocities: V_L, V_P, V_b, V_T and V_R. These velocities are listed in decreasing order in Table 4.3.

Type of wave	Velocity	Conditions of existence
Longitudinal	V_L	Infinite solid with dimensions $\gg \lambda$
Longitudinal	$V_T\sqrt{2} < V_P < V_L$	Thin plate of thickness $d \ll \lambda$
Longitudinal	$V_b < V_P$	Bar with lateral dimensions b and $d \ll \lambda$
Transverse	$V_T < V_L/\sqrt{2}$	Infinite solid with dimensions $\gg \lambda$
Rayleigh	$V_R < V_T$	Semi-infinite solid

Table 4.3. *Characteristic velocities of elastic waves propagating in an isotropic solid*

Each time, the decrease in velocity is due to a reduction in the stiffness of the solid by the removal of matter:

– V_P is smaller than V_L due to the lack of matter on both sides of the plate;

– V_b is smaller than V_P due to the lack of matter on all lateral faces of the bar;

– V_R is lower than V_T due to the absence of matter above the free surface.

4.3.2. *Isotropic plate immersed in a fluid*

The plate in Figure 4.25 is now totally immersed in a fluid of mass density ρ_f and sound speed V_f, and therefore acoustic impedance $Z_f = \rho_f V_f$.

On either side of the plate, the expression for the mechanical displacement contained in the $x_1 x_2$ plane is:

$$\underline{u}^{(f)}(x_1, x_2, t) = \frac{V_f}{\omega} b_f \begin{pmatrix} k_1 \\ -k_{2f} \end{pmatrix} e^{-ik_{2f}(x_2+h)} e^{i(k_1 x_1 - \omega t)} \quad \text{at} \quad x_2 \leq -h \quad [4.120]$$

and:

$$\underline{u}^{(f)}(x_1, x_2, t) = \frac{V_f}{\omega} a_f \begin{pmatrix} k_1 \\ k_{2f} \end{pmatrix} e^{ik_{2f}(x_2-h)} e^{i(k_1 x_1 - \omega t)} \quad \text{at} \quad x_2 \geq h \quad [4.121]$$

Considering equation [1.323], the acoustic pressure $p_a^{(f)}$ is given by:

$$p_a^{(f)}(x_1, x_2, t) = \begin{cases} -i\omega Z_f b_f e^{-ik_{2f}(x_2+h)} e^{i(k_1 x_1 - \omega t)} & \text{at} \quad x_2 \leq -h \\ -i\omega Z_f a_f e^{ik_{2f}(x_2-h)} e^{i(k_1 x_1 - \omega t)} & \text{at} \quad x_2 \geq h \end{cases} \quad [4.122]$$

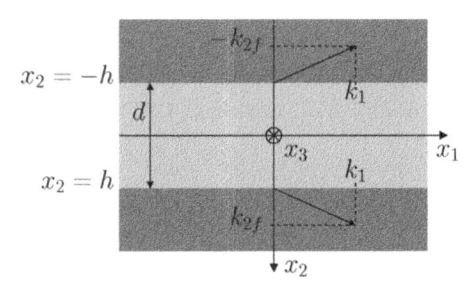

Figure 4.25. *Immersed plate. Disposition of axes and components of the wave vector in the fluid*

Given the symmetry of the fluid/plate/fluid structure, the decomposition into symmetric and antisymmetric Lamb modes is conserved. The mechanical displacements and stresses in the solid are given by equations [4.76] and [4.77], respectively. The dispersion equations result from the boundary conditions on the tangential stress σ_{12}, which cancels at $x_2 = \pm h$ (non-viscous fluid), on the normal stress σ_{22}, which is the opposite of the acoustic pressure, and on the component u_2 of the displacement, which is continuous at $x_2 = \pm h$.

For the *symmetric modes* ($\alpha = 0$), σ_{22} is an even function of x_2, while u_2 is an odd function. From equations [4.120] to [4.122], these conditions impose that in the fluid $b_f = a_f$. The boundary conditions are, therefore, simultaneously fulfilled on both interfaces; by writing them in the order $\sigma_{12}(h) = 0$, $\sigma_{22}(h) = -p_a^{(f)}(h)$ and $u_2(h) = u_2^{(f)}(h)$, they lead to the system:

$$\begin{pmatrix} -2ikk_{2L}\sin(k_{2L}h) & (k^2 - k_{2T}^2)\sin(k_{2T}h) & 0 \\ (k^2 - k_{2T}^2)\cos(k_{2L}h) & -2ikk_{2T}\cos(k_{2T}h) & -\dfrac{iZ_f\omega}{\mu} \\ k_{2L}\sin(k_{2L}h) & ik\sin(k_{2T}h) & \dfrac{V_f}{\omega}k_{2f} \end{pmatrix} \begin{pmatrix} a_L \\ a_T \\ a_f \end{pmatrix} = \begin{pmatrix} 0 \\ 0 \\ 0 \end{pmatrix}$$

[4.123]

For the *antisymmetric modes* ($\alpha = \pi/2$), σ_{22} is an odd function of x_2, while u_2 is an even function. It follows that $b_f = -a_f$. The boundary conditions are thus simultaneously satisfied at $x_2 = \pm h$:

$$\begin{pmatrix} -2ikk_{2L}\cos(k_{2L}h) & (k^2 - k_{2T}^2)\cos(k_{2T}h) & 0 \\ (k^2 - k_{2T}^2)\sin(k_{2L}h) & -2ikk_{2T}\sin(k_{2T}h) & \dfrac{iZ_f\omega}{\mu} \\ k_{2L}\cos(k_{2L}h) & ik\cos(k_{2T}h) & \dfrac{V_f}{\omega}k_{2f} \end{pmatrix} \begin{pmatrix} b_L \\ b_T \\ a_f \end{pmatrix} = \begin{pmatrix} 0 \\ 0 \\ 0 \end{pmatrix}$$

[4.124]

The *dispersion equations* for the symmetric and antisymmetric modes result from the cancelation of the determinant of equations [4.123] and [4.124]:

$$d_S - i\delta_f \tan(k_{2L}h)\tan(k_{2T}h) = 0 \quad \text{and} \quad d_A + i\delta_f = 0 \qquad [4.125]$$

where the determinants d_S and d_A of a free plate are deduced from the Rayleigh–Lamb equations [4.80] and [4.81]:

$$\begin{cases} d_S = 4k^2 k_{2L} k_{2T} \tan(k_{2L}h) + (k^2 - k_{2T}^2)^2 \tan(k_{2T}h) \\ d_A = 4k^2 k_{2L} k_{2T} \tan(k_{2T}h) + (k^2 - k_{2T}^2)^2 \tan(k_{2L}h) \end{cases} \qquad [4.126]$$

It also appears the factor δ_f, given by equation [2.100], is characteristic of the coupling of acoustic waves at a fluid–solid interface.

The phase velocities of the fundamental Lamb modes for a duralumin plate immersed in water are plotted in Figure 4.26(a). The comparison of these curves (solid lines) with those for the free plate (dashed lines) shows that the liquid has very little effect on the velocity of Lamb waves. The immersion in the fluid leads to the appearance of an additional mode: the quasi-Scholte mode (QS). This mode, which is very close to the Scholte wave propagating at the interface between a solid half-space and a fluid half-space (section 3.2.1.2), does not radiate energy into the liquid. Its energy is confined inside the fluid. At low frequencies, its phase velocity is very dispersive and at higher frequency, it tends toward the Scholte wave velocity. The immersion of the plate also leads to an attenuation of the Lamb waves (Figure 4.26(b)). Thus, at low frequency, the attenuation of the A_0 mode is very high, expressing an acoustic radiation into the fluid on both sides of the plate, like the leaky Rayleigh wave studied in section 3.2.1.3.

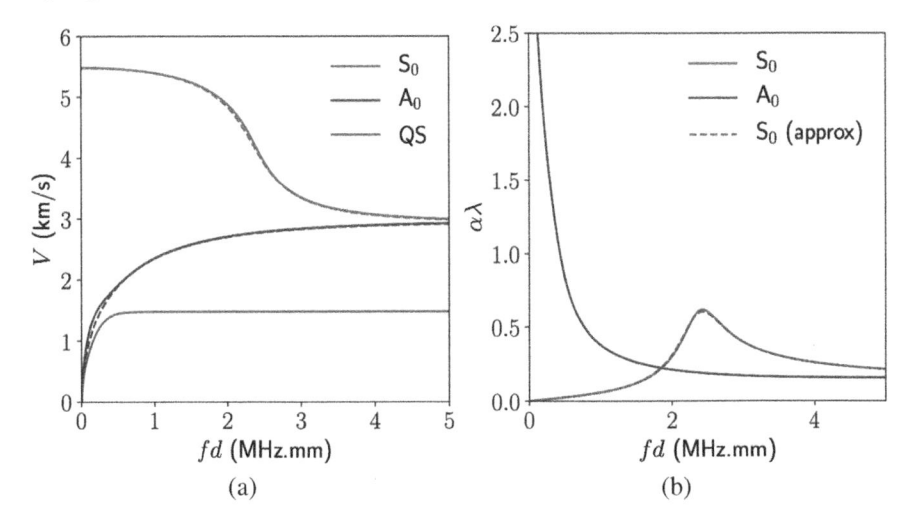

Figure 4.26. *Duralumin plate immersed in water. a) Phase velocity of the fundamental Lamb modes S_0 and A_0 and of the quasi-Scholte (QS) mode. b) Attenuation per wavelength. For a color version of this figure, see www.iste.co.uk/royer/waves1.zip*

The attenuation of the S_0 mode passes through a maximum for a frequency × thickness product of 2.4 MHz.mm corresponding to the minimum of its group velocity (Figure 4.16(b)). This result can be explained by an approximate model, valid for any wave propagating in the plate with a phase velocity $V > V_f$ and which radiates energy into the fluid, in the form of a plane wave, along the angle θ, such that:

$$\theta = \arcsin\left(\frac{V_f}{V}\right) \qquad [4.127]$$

After a path x_1, the wave amplitude is multiplied by $e^{-\alpha x_1}$; therefore, the average power P transported in the plate varies as $e^{-2\alpha x_1}$. With $P_f = -\mathrm{d}P/\mathrm{d}x_1$ denoting the average power radiated into the fluid per unit length, the attenuation coefficient per wavelength $\alpha\lambda$ is given by:

$$\alpha = \frac{P_f}{2P} \quad \text{hence} \quad \alpha\lambda = \frac{\pi P_f}{kP} \qquad [4.128]$$

The acoustic power P_f is equal to the flux of the Poynting vector ($J_2 = -\sigma_{2i}\dot{u}_i$) across the interface $x_2 = h$. In the fluid, the only non-zero stress is

$\sigma_{22} = -p_a = -Z_f \dot{u}_f$. Since $u_f = u_2(h)/\cos\theta$, for a time-harmonic wave and a beam of width w, we obtain:

$$P_f = \frac{w}{2} \operatorname{Re}\left[p_a \dot{u}_2^*(h)\right] = \frac{w}{2} \frac{Z_f \omega^2}{\cos\theta} |u_2(h)|^2 \qquad [4.129]$$

The average power P transported by a guided wave is expressed by equation [A3.36] as a function of the components of mechanical displacement and of the group velocity V_g:

$$P = \frac{w}{2} \rho_s V_g \omega^2 \int_{-h}^{+h} \left[|u_1(x_2)|^2 + |u_2(x_2)|^2\right] dx_2 \qquad [4.130]$$

The attenuation per wavelength, deduced from equation [4.128]:

$$\alpha\lambda = \frac{\pi \rho_f V_f}{\rho_s V_g \cos\theta} \frac{|u_2(h)|^2}{k \int_{-h}^{+h} \left[|u_1(x_2)|^2 + |u_2(x_2)|^2\right] dx_2} \qquad [4.131]$$

is proportional to the square of the displacement normal to the surface and inversely proportional to the group velocity V_g. The attenuation of the mode S_0 passes through a minimum in the vicinity of the inflection point of its phase velocity, at the origin of the peak observed in Figure 4.26(b). If the plate is completely immersed, $\alpha\lambda$ must be multiplied by two to take into account the radiation from both faces.

The *generation of Lamb modes* in an immersed plate is generally achieved through the interaction of an ultrasonic beam with the plate. The angle of incidence of the beam makes it possible to select one or more modes *via* the Snell–Descartes law. This phenomenon is studied here for an incident plane wave interacting with a plate bounded by the planes $x_2 = -h$ and $x_2 = h$ (Figure 4.25). The acoustic displacement [4.120] in the fluid at $x_2 \leq -h$ is modified to take into account the incident wave:

$$\underline{u}^{(f)}(x_1, x_2, t) = \frac{V_f}{\omega} A e^{i(k_1 x_1 - \omega t)} \left[\begin{pmatrix} k_1 \\ k_{2f} \end{pmatrix} e^{ik_{2f} x_2} + r \begin{pmatrix} k_1 \\ -k_{2f} \end{pmatrix} e^{-ik_{2f} x_2} \right]$$
$$[4.132]$$

where r is the amplitude reflection coefficient. By returning to equations [4.121] and [4.122] and, considering equations [4.76] and [4.77], the application of the boundary conditions on the two surfaces of the plate leads to a system of six equations with six unknowns. This system is solved numerically and the map for the intensity reflection coefficient $R = |r|^2$ is represented in Figure 4.27(a) as a function of the angle of incidence, on the abscissa, and of the frequency × thickness product, on the ordinate. The dispersion curves for the free plate are plotted in Figure 4.27(b) as a function of the angle of incidence, given by relation [4.127].

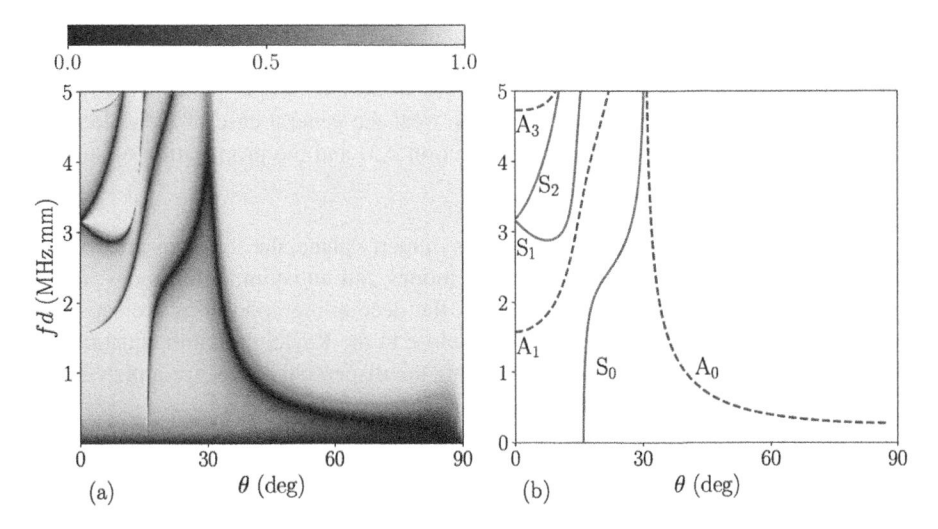

Figure 4.27. *a) Plate immersed in water, map of the intensity reflection coefficient $R = |r|^2$. b) Dispersion curves for the same duralumin plate in vacuum, as a function of the angle of incidence on the abscissa and of the frequency × thickness product on the ordinate. For a color version of this figure, see www.iste.co.uk/royer/waves1.zip*

The angle θ is fixed by the Snell–Descartes law: $\sin\theta = V_f/V$, which relates the velocity in the fluid to that of the Lamb waves. The following values of the angle θ correspond, for:

– $\theta = 0$, to an infinite phase velocity, i.e., to the cut-off frequencies ($k = 0$) in Table 4.1;

– $\theta = 90°$, to $V = V_f$ (1 480 m/s for water): only the A_0 mode has such a low phase velocity;

– $\theta \approx 15°$ such that $\sin\theta = V_f/V_P$, to the plate velocity of the S_0 mode at low frequency ($V_P = 5\,720$ m/s for duralumin).

If the plate is loaded on only one side by a fluid, the structure is asymmetrical, the modes are coupled and are determined from the dispersion equation (Aubert *et al.* 2016):

$$2d_S d_A + i\delta_f \left[\tan(k_{2L}h)\tan(k_{2T}h)d_A - d_S\right] = 0 \qquad [4.133]$$

4.3.3. *Free anisotropic plate*

To study the propagation of guided waves in an anisotropic plate, we will use the partial wave method, already exposed to treat the general case of the reflection-transmission phenomena at an interface (section 2.5) and the propagation of surface waves (section 3.1.2).

When the median plane of the plate is a symmetry plane, the solutions are divided into two independent families: symmetric modes and antisymmetric modes. If the sagittal plane is also a plane of symmetry, the decoupling conditions are satisfied. Formally, the dispersion equations are very close to the Rayleigh–Lamb equations of an isotropic plate. We will conclude by giving the dispersion curves for a plate made up of a carbon-fiber composite material.

4.3.3.1. *Partial wave method*

The reference frame is shown in Figure 4.12. The guided waves propagate along Ox_1 and the free surfaces are normal to Ox_2. By hypothesis, the mechanical quantities do not depend on x_3. Let us search for a solution similar to [3.38]:

$$u_l(x_1, x_2, t) = a p_l e^{ikmx_2} e^{i(kx_1 - \omega t)} \quad \text{with} \quad l = 1, 2, 3 \quad [4.134]$$

corresponding to a time-harmonic plane wave, of amplitude a and of polarization vector $\underline{p}$, propagating in the direction $(1, m, 0)$ contained in the sagittal plane $x_1 x_2$. As in section 2.5.1, substituting this solution in the propagation equations leads to the Christoffel equation:

$$\left(\Gamma_{il} - \rho V^2 \delta_{il}\right) p_l = 0 \quad \text{with} \quad \Gamma_{il} = c_{i11l} + (c_{i12l} + c_{i21l}) m + c_{i22l} m^2 \quad [4.135]$$

The characteristic equation, which expresses the compatibility condition of the homogeneous system [4.135], is identical to equation [2.127]. The coefficients α_k of the polynomial of the sixth-degree in m depend on the stiffnesses of the material and on the phase velocity $V = \omega/k$. Since these coefficients are real, the roots are real (homogeneous wave) or complex conjugates (inhomogeneous or evanescent wave). A root $m = m' + im''$ with a positive (negative) imaginary part m'' represents a wave whose amplitude $a \exp(-m'' k x_2)$ decreases (increases) exponentially when x_2 increases. In a semi-infinite medium, increasing solutions are not physically acceptable. In the case of a plate of finite thickness, all solutions must be retained.

Each root m_K ($K = 1... 6$) corresponds to a partial wave of polarization p_{lK} and amplitude a_K. The *mechanical displacement* is the sum of the displacement of these six partial waves, having the same phase velocity $V = \omega/k$, that is, for each component:

$$u_l(x_1, x_2, t) = \left[\sum_{K=1}^{6} a_K p_{lK} e^{im_K k x_2}\right] e^{i(kx_1 - \omega t)}, \quad \text{with} \quad l = 1, 2, 3$$

[4.136]

As in section 2.5.2, the *mechanical stress* σ_{i2} is of the form:

$$\sigma_{i2} = ikA_i(m)ae^{imkx_2} e^{i(kx_1 - \omega t)}$$

[4.137]

that is, for the total wave:

$$\sigma_{i2} = ik\left[\sum_{K=1}^{6} A_{iK} a_K e^{im_K k x_2}\right] e^{i(kx_1 - \omega t)}$$

[4.138]

The coefficients $A_{iK} = A_i(m_K)$ are arranged in a (3×6) matrix.

The boundary conditions: $\sigma_{i2} = 0$ with $i = 1, 2, 3$, on the two free surfaces $x_2 = \pm h$, lead to a homogeneous linear system of dimension 6×6, connecting the amplitudes a_K:

$$\begin{pmatrix} A_{11}E_1 & A_{12}E_2 & A_{13}E_3 & A_{14}E_4 & A_{15}E_5 & A_{16}E_6 \\ A_{21}E_1 & A_{22}E_2 & A_{23}E_3 & A_{24}E_4 & A_{25}E_5 & A_{26}E_6 \\ A_{31}E_1 & A_{32}E_2 & A_{33}E_3 & A_{34}E_4 & A_{35}E_5 & A_{36}E_6 \\ A_{11}\overline{E}_1 & A_{12}\overline{E}_2 & A_{13}\overline{E}_3 & A_{14}\overline{E}_4 & A_{15}\overline{E}_5 & A_{16}\overline{E}_6 \\ A_{21}\overline{E}_1 & A_{22}\overline{E}_2 & A_{23}\overline{E}_3 & A_{24}\overline{E}_4 & A_{25}\overline{E}_5 & A_{26}\overline{E}_6 \\ A_{31}\overline{E}_1 & A_{32}\overline{E}_2 & A_{33}\overline{E}_3 & A_{34}\overline{E}_4 & A_{35}\overline{E}_5 & A_{36}\overline{E}_6 \end{pmatrix} \begin{pmatrix} a_1 \\ a_2 \\ a_3 \\ a_4 \\ a_5 \\ a_6 \end{pmatrix} = \begin{pmatrix} 0 \\ 0 \\ 0 \\ 0 \\ 0 \\ 0 \end{pmatrix}$$

[4.139]

by writing:

$$\begin{cases} E_K = e^{im_K kh} \\ \overline{E}_K = e^{-im_K kh} = E_K^{-1}, \end{cases} \quad \text{with} \quad K = 1\ldots 6$$

[4.140]

The first three lines come from the boundary conditions at $x_2 = h$; the last three lines come from these conditions at $x_2 = -h$. The phase velocity $V = \omega/k$, which results from the cancelation of the matrix determinant, depends on the normalized wave number kh. According to the symmetry of the material constituting the plate, two cases can be distinguished:

– the median plane is a symmetry plane: the solutions separate into two independent families, corresponding to a symmetric or antisymmetric motion of the plate;

– the sagittal plane x_1x_2 is also a symmetry plane: the propagation of the waves polarized in the x_1x_2 plane is decoupled from that of the TH waves polarized along x_3.

4.3.3.2. *Solid with a monoclinic symmetry*

The normal Ox_2 to the plate is parallel to a direct or inverse binary axis of symmetry. The median plane x_1x_3 is a symmetry plane and the characteristic equation [2.143] is a third-degree equation in $z = m^2$. The six roots are pairwise opposite because the propagation directions of reflected and incident waves of the same nature are symmetric with respect to the interfaces.

Five possibilities may be listed, depending on whether:

– the three roots z are real and positive; the six partial waves are homogeneous;

– two roots are real and positive, the others negative; four waves are homogeneous, two are evanescent;

– two roots are real and negative, the others positive; four waves are evanescent, two are homogeneous;

– all the roots z are real and negative; the six partial waves are evanescent;

– two roots are complex conjugate, resulting in four inhomogeneous partial waves; the other two waves are homogeneous or evanescent, depending on whether the other root, necessarily real, is positive or negative.

Given the direction of the x_2 axis and the expression [4.134] of the mechanical displacement:

$$u_l(x_1, x_2, t) = a p_l e^{-m'' k x_2} e^{i(kx_1 + m' k x_2 - \omega t)} \qquad [4.141]$$

a negative (positive) value of m or m' corresponds to a rising (descending) wave. A negative (positive) value of m'' corresponds to an increasing (decreasing) wave with x_2. Three cases have to be distinguished:

– to a positive real root z correspond a rising homogeneous wave (root $m_K < 0$) and a descending homogeneous wave (root $m_J = -m_K > 0$);

– to a negative real root z correspond a decreasing wave (root $m_K = i\chi$ with $\chi > 0$) and an increasing wave (root $m_J = -i\chi$);

– to a pair of complex conjugate roots z and z^* are associated four roots (m, m^*, $-m$ and $-m^*$) corresponding to inhomogeneous waves. The roots m_K and m_K^* with a negative real part correspond to rising waves (increasing or decreasing). The other two roots $m_J = -m_K$ and $m_J^* = -m_K^*$ with a positive real part correspond to descending waves (decreasing or increasing).

In all cases, there exist three partial waves, labeled by the index K, of amplitude a_K and polarization vector $\underline{p}_K$, and three partial waves labeled by the index J, of amplitude b_J and polarization vector $\underline{q}_J$, such that $m_J = -m_K$. The expression [4.136] of the mechanical displacement becomes:

$$u_l(x_1, x_2, t) = \left[\sum_{K=1}^{3} a_K p_{lK} e^{im_K k x_2} + \sum_{J=1}^{3} b_J q_{lJ} e^{im_J k x_2}\right] e^{i(kx_1 - \omega t)} \quad [4.142]$$

As in section 2.5.2, the stresses σ_{i2} are in the form:

$$\sigma_{i2}(x_1, x_2, t) = ik A_i(m) a e^{imkx_2} e^{i(kx_1 - \omega t)} \text{ or } ik B_i(m) b e^{imkx_2} e^{i(kx_1 - \omega t)} \quad [4.143]$$

that is, by omitting the propagation factor $e^{i(kx_1 - \omega t)}$, common to all terms:

$$\sigma_{i2}(x_2) = ik \left[\sum_{K=1}^{3} a_K A_{iK} e^{im_K k x_2} + \sum_{J=1}^{3} b_J B_{iJ} e^{im_J k x_2}\right] \quad [4.144]$$

The coefficients $A_{iK} = A_i(m_K)$ and $B_{iJ} = B_i(m_J)$ are arranged in (3×3) matrices and the relations [2.147] between these coefficients are written (without summation over the indices i and J) as:

$$A_{iJ} = \varepsilon_{iJ} B_{iJ} \quad \text{with} \quad \varepsilon_{iJ} = \begin{pmatrix} 1 & -1 & -1 \\ -1 & 1 & 1 \\ 1 & -1 & -1 \end{pmatrix} \quad [4.145]$$

By combining the equal and opposite roots m_J and $m_K = -m_J$, the stresses σ_{i2} are divided into an even function and an odd function of x_2: $\sigma_{i2} = \sigma_{i2}^{(s)} + \sigma_{i2}^{(a)}$, with:

$$\begin{cases} \sigma_{i2}^{(s)} = ik \sum_{J=1}^{3} B_{iJ} (b_J + \varepsilon_{iJ} a_J) \cos(m_J k x_2) \\ \sigma_{i2}^{(a)} = -k \sum_{J=1}^{3} B_{iJ} (b_J - \varepsilon_{iJ} a_J) \sin(m_J k x_2) \end{cases} \quad [4.146]$$

The three boundary conditions on the two free surfaces $x_2 = \pm h$:

$$\begin{cases} \sigma_{i2}(h) = \sigma_{i2}^{(s)}(h) + \sigma_{i2}^{(a)}(h) = 0 \\ \sigma_{i2}(-h) = \sigma_{i2}^{(s)}(h) - \sigma_{i2}^{(a)}(h) = 0 \end{cases} \quad [4.147]$$

are equivalent to the cancelation of the even components $\sigma_{i2}^{(s)}$ and odd components $\sigma_{i2}^{(a)}$ on the single face $x_2 = +h$. By denoting $C_J = \cos(m_J kh)$ and $S_J = \sin(m_J kh)$, we obtain:

$$\begin{cases} \sigma_{12}^{(s)} = ik\left[B_{11}\left(b_1+a_1\right)C_1 + B_{12}\left(b_2-a_2\right)C_2 + B_{13}\left(b_3-a_3\right)C_3\right] \\ \sigma_{22}^{(s)} = ik\left[B_{21}\left(b_1-a_1\right)C_1 + B_{22}\left(b_2+a_2\right)C_2 + B_{23}\left(b_3+a_3\right)C_3\right] \\ \sigma_{32}^{(s)} = ik\left[B_{31}\left(b_1+a_1\right)C_1 + B_{32}\left(b_2-a_2\right)C_2 + B_{33}\left(b_3-a_3\right)C_3\right] \\ \sigma_{12}^{(a)} = -k\left[B_{11}\left(b_1-a_1\right)S_1 + B_{12}\left(b_2+a_2\right)S_2 + B_{13}\left(b_3+a_3\right)S_3\right] \\ \sigma_{22}^{(s)} = -k\left[B_{21}\left(b_1+a_1\right)S_1 + B_{22}\left(b_2-a_2\right)S_2 + B_{23}\left(b_3-a_3\right)S_3\right] \\ \sigma_{32}^{(s)} = -k\left[B_{31}\left(b_1-a_1\right)S_1 + B_{32}\left(b_2+a_2\right)S_2 + B_{33}\left(b_3+a_3\right)S_3\right] \end{cases} \quad [4.148]$$

By sorting the lines in the following order: $\sigma_{12}^{(a)}, \sigma_{22}^{(s)}, \sigma_{32}^{(a)}, \sigma_{12}^{(s)}, \sigma_{22}^{(a)}, \sigma_{32}^{(s)}$, the boundary conditions can be written in matrix form as:

$$\begin{pmatrix} B_{11}S_1 & B_{12}S_2 & B_{13}S_3 & 0 & 0 & 0 \\ B_{21}C_1 & B_{22}C_2 & B_{23}C_3 & 0 & 0 & 0 \\ B_{31}S_1 & B_{32}S_2 & B_{33}S_3 & 0 & 0 & 0 \\ 0 & 0 & 0 & B_{11}C_1 & B_{12}C_2 & B_{13}C_3 \\ 0 & 0 & 0 & B_{21}S_1 & B_{22}S_2 & B_{23}S_3 \\ 0 & 0 & 0 & B_{31}C_1 & B_{32}C_2 & B_{33}C_3 \end{pmatrix} \begin{pmatrix} b_1-a_1 \\ b_2+a_2 \\ b_3+a_3 \\ b_1+a_1 \\ b_2-a_2 \\ b_3-a_3 \end{pmatrix} = \begin{pmatrix} 0 \\ 0 \\ 0 \\ 0 \\ 0 \\ 0 \end{pmatrix}$$

[4.149]

They are divided into two independent groups involving distinct combinations $(b_J \pm a_J)$ of the partial wave amplitudes.

Given relations [2.145] and [2.146] between the polarizations p_{lJ} and q_{lJ}: $p_{lJ} = -\varepsilon_{lJ} q_{lJ}$ (without summation over the indices l and J), the mechanical displacement also splits into an even and an odd function of x_2:

$$u_l = \sum_{J=1}^{3} q_{lJ}\left(b_J - \varepsilon_{lJ} a_J\right)\cos(m_J k x_2) + i \sum_{J=1}^{3} q_{lJ}\left(b_J + \varepsilon_{lJ} a_J\right)\sin(m_J k x_2)$$

[4.150]

For the *symmetric modes*, the components $u_1^{(s)}$, $u_2^{(a)}$ and $u_3^{(s)}$ are linear combinations of the amplitudes: $b_1 - a_1$, $b_2 + a_2$ and $b_3 + a_3$:

$$\begin{pmatrix} u_1^{(s)} \\ u_2^{(a)} \\ u_3^{(s)} \end{pmatrix} = \begin{pmatrix} q_{11}\cos(m_1 k x_2) & q_{12}\cos(m_2 k x_2) & q_{13}\cos(m_3 k x_2) \\ iq_{21}\sin(m_1 k x_2) & iq_{22}\sin(m_2 k x_2) & iq_{23}\sin(m_3 k x_2) \\ q_{31}\cos(m_1 k x_2) & q_{32}\cos(m_2 k x_2) & q_{33}\cos(m_3 k x_2) \end{pmatrix} \begin{pmatrix} b_1-a_1 \\ b_2+a_2 \\ b_3+a_3 \end{pmatrix}$$

[4.151]

The motion of the plate is symmetric with respect to the median plane $x_1 x_3$: the components u_1 and u_3, contained in the plane of the plate, are even functions of x_2; the perpendicular component u_2 is an odd function. The dispersion equation of these symmetric plate modes results from the cancelation of the determinant:

$$D_S(\omega, k) = \det \begin{vmatrix} B_{11} \sin(m_1 k h) & B_{12} \sin(m_2 k h) & B_{13} \sin(m_3 k h) \\ B_{21} \cos(m_1 k h) & B_{22} \cos(m_2 k h) & B_{23} \cos(m_3 k h) \\ B_{31} \sin(m_1 k h) & B_{32} \sin(m_2 k h) & B_{33} \sin(m_3 k h) \end{vmatrix} = 0 \quad [4.152]$$

For the *antisymmetric modes*, the components $u_1^{(a)}$, $u_2^{(s)}$ and $u_3^{(a)}$ are linear combinations of the amplitudes: $b_1 + a_1$, $b_2 - a_2$ and $b_3 - a_3$:

$$\begin{pmatrix} u_1^{(a)} \\ u_2^{(s)} \\ u_3^{(a)} \end{pmatrix} = \begin{pmatrix} i q_{11} \sin(m_1 k x_2) & i q_{12} \sin(m_2 k x_2) & i q_{13} \sin(m_3 k x_2) \\ q_{21} \cos(m_1 k x_2) & q_{22} \cos(m_2 k x_2) & q_{23} \cos(m_3 k x_2) \\ i q_{31} \sin(m_1 k x_2) & i q_{32} \sin(m_2 k x_2) & i q_{33} \sin(m_3 k x_2) \end{pmatrix} \begin{pmatrix} b_1 + a_1 \\ b_2 - a_2 \\ b_3 - a_3 \end{pmatrix}$$

[4.153]

They correspond to an antisymmetric or flexural motion of the plate: the components u_1 and u_3, contained in the plane of the plate, are odd functions of x_2, the perpendicular component u_2 is an even function. The dispersion equation $D_A(\omega, k) = 0$ of these antisymmetric modes is deduced from equation [4.152] by interchanging the sine and cosine. As in the isotropic case, the plate modes are divided into two independent families. The first involves the stresses $\sigma_{12}^{(a)}$, $\sigma_{22}^{(s)}$, $\sigma_{32}^{(a)}$; the second involves the stresses $\sigma_{12}^{(s)}$, $\sigma_{22}^{(a)}$, $\sigma_{32}^{(s)}$.

4.3.3.3. *Solid with an orthorhombic symmetry*

If, in addition, the Ox_3 axis is parallel to a direct or inverse binary axis, the propagation of the waves polarized in the sagittal plane is decoupled from that of the TH waves polarized along x_3. For the waves polarized in the sagittal plane, the characteristic equation is a second-degree equation in $z = m^2$. Four possibilities can be listed, depending on whether:

– the two roots z are real positive; the four partial waves are homogeneous;

– one root is real positive, the other is negative; two waves are homogeneous, the two others are evanescent;

– the two roots are real negative; the four partial waves are evanescent;

– the roots z are complex conjugate; the four partial waves are inhomogeneous.

The boundary conditions relate only to σ_{12} and σ_{22}. Since the symmetric and antisymmetric modes are independent, it is not necessary to distinguish the amplitudes

$b_1 + a_1$ and $b_1 - a_1$, on the one hand, $b_2 + a_2$ and $b_2 - a_2$, on the other hand. By denoting them by c_1 and c_2, respectively, we obtain:

$$\begin{pmatrix} B_{11}\sin(m_1kh+\alpha) & B_{12}\sin(m_2kh+\alpha) \\ B_{21}\cos(m_1kh+\alpha) & B_{22}\cos(m_2kh+\alpha) \end{pmatrix} \begin{pmatrix} c_1 \\ c_2 \end{pmatrix} = \begin{pmatrix} 0 \\ 0 \end{pmatrix} \quad [4.154]$$

with $\alpha = 0$ for the symmetric modes and $\alpha = \pi/2$ for the antisymmetric modes. The compatibility condition for this linear and homogeneous system requires the determinant of the matrix to be zero:

$$B_{11}B_{22}\sin(m_1kh+\alpha)\cos(m_2kh+\alpha) - B_{12}B_{21}\cos(m_1kh+\alpha)\sin(m_2kh+\alpha) = 0$$
$$[4.155]$$

The coefficients B_{iJ} and the roots m_1 and m_2 of the characteristic equation depend only on the phase velocity $V = \omega/k$ and on the elastic constants of the material. Equation [4.155] provides V as a function of the dimensionless variable kh: it is the dispersion equation for the Lamb waves. Since $kh = \pi f d/V$, the velocity of the Lamb waves is a function of the frequency $\times$ plate thickness product.

Once the velocity V is determined, the displacement components:

$$\begin{cases} u_1 = q_{11}c_1\cos(m_1kx_2+\alpha) + q_{12}c_2\cos(m_2kx_2+\alpha) \\ u_2 = iq_{21}c_1\sin(m_1kx_2+\alpha) + iq_{22}c_2\sin(m_2kx_2+\alpha) \end{cases} \quad [4.156]$$

are obtained by substituting the coefficient c_1, taken from equation [4.154]:

$$\begin{cases} u_1 = c_2\left[q_{12}\cos(m_2kx_2+\alpha) - q_{11}\dfrac{B_{22}\cos(m_2kh+\alpha)}{B_{21}\cos(m_1kh+\alpha)}\cos(m_1kx_2+\alpha)\right] \\ u_2 = ic_2\left[q_{22}\sin(m_2kx_2+\alpha) - q_{21}\dfrac{B_{22}\cos(m_2kh+\alpha)}{B_{21}\cos(m_1kh+\alpha)}\sin(m_1kx_2+\alpha)\right] \end{cases}$$
$$[4.157]$$

Given the correspondence: $m_1k = k_{2L}$ and $m_2k = k_{2T}$, these expressions are similar to those in [4.82], established for an isotropic solid. The two components are still in phase quadrature; in the sagittal plane, the particles describe an ellipse that is deformed because the amplitudes vary differently with the position x_2 in the thickness of the plate.

4.3.3.4. *Composite material*

Figure 4.28 shows the phase velocity of Lamb waves propagating in a plate of thickness $2h = 4$ mm and made of a composite material used in aeronautics (T300/914). It is a unidirectional assembly of 32 plies, of thickness 125 µm, parallel to the x_1x_3 plane. The carbon fibers, aligned along the x_1 axis, are embedded in an epoxy resin. At low frequencies (up to a few MHz), this strongly anisotropic material may be considered as homogeneous.

The elastic constants used in the simulations for this composite, of mass density $\rho = 1\,560$ kg/m^3, are those of a transversely isotropic material, whose axis of isotropy is parallel to x_1 (tensor [1.157]):

$$C_{11} = 143.8 \text{ GPa} \quad C_{12} = C_{13} = 6.2 \text{ GPa} \quad C_{22} = C_{33} = 13.3 \text{ GPa}$$
$$C_{23} = 6.5 \text{ GPa} \quad C_{44} = 3.6 \text{ GPa} \quad C_{55} = C_{66} = 5.7 \text{ GPa}$$

When the propagation direction is parallel (or perpendicular) to the fiber axis x_1, transverse modes TH$_n$, polarized along x_3 (or x_1), are decoupled from Lamb modes S$_n$ and A$_n$, polarized in the sagittal plane. For example, in Figure 4.28(a), the curves of modes S$_1$ and TH$_2$ cross each other without any lifting of the degeneracy, and it is similar for A$_1$ and TH$_3$.

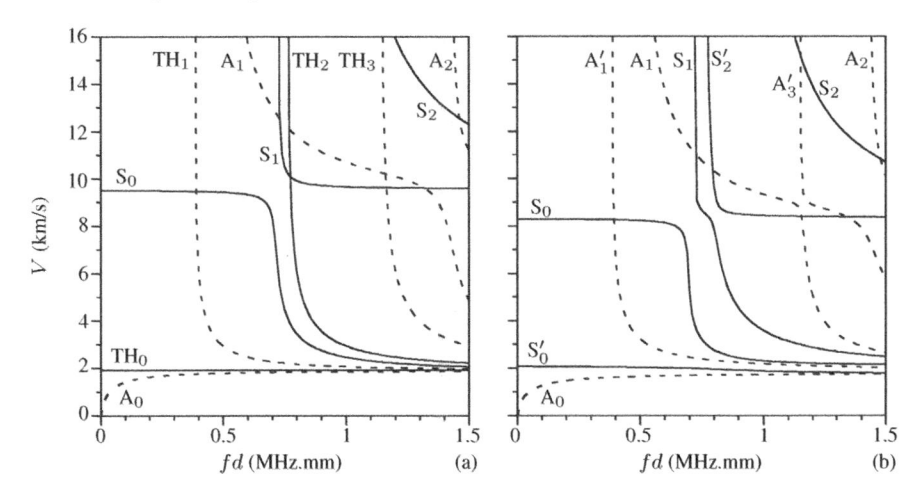

Figure 4.28. *Unidirectional composite material made of carbon fibers embedded in an epoxy resin. Phase velocity versus the frequency × thickness product for Lamb modes and transverse waves TH$_n$ propagating along the fiber axis a) and along a direction $\theta = 30°$ off this axis b). Symmetric modes (solid lines) and antisymmetric modes (dashed lines)*

When the propagation direction is tilted by an angle θ with respect to x_1, the stiffness tensor is that of a monoclinic material [A2.42]. There are no more TH waves: modes TH$_{2n}$ become symmetric modes S$'_{2n}$ and modes TH$_{2n+1}$ become antisymmetric modes A$'_{2n+1}$. All the waves are coupled (Figure 4.28(b)), for example, the curves of modes S$_1$ and S$'_2$ (ex mode TH$_2$) no longer intersect; they repel each other.

4.3.3.5. *Group velocity and slowness curves*

In an unbounded anisotropic solid, the propagation of elastic waves is illustrated by the slowness surface (section 1.3.5). This surface, which gives the inverse of the phase velocity k/ω as a function of the propagation direction, consists of three sheets: one for the quasi-longitudinal wave and two for quasi-transverse waves. The energy velocity of these three waves is normal to the corresponding sheet. Let us examine these notions for an anisotropic plate, which presents both a frequency and a spatial dispersion. Let us return to the definition (equation [1.189]) of the group velocity vector $\underline{V}^g$:

$$V_i^g = \left(\frac{\partial \omega}{\partial k_i}\right)_{k_j} \quad \text{with} \quad i,j = 1,3 \quad \text{and} \quad j \neq i \quad [4.158]$$

where k_i are the components of the wave vector $\underline{k}$, in the $x_1 x_3$ plane of the plate. The implicit differentiation of the dispersion equation $D(\omega, k_i) = 0$, for any variation of the components dk_i, leads to:

$$\left(\frac{\partial D}{\partial \omega}\right)_{k_i} d\omega + \left(\frac{\partial D}{\partial k_i}\right)_{\omega, k_j} dk_i = 0 \quad \text{hence} \quad \left(\frac{\partial D}{\partial k_i}\right)_{\omega, k_j} = -V_i^g \left(\frac{\partial D}{\partial \omega}\right)_{k_i}$$

[4.159]

For a variation dk of the wave number in a given direction defined by the unit vector n_i: $dk_i = n_i\, dk$, the previous relation is written as:

$$\left(\frac{\partial D}{\partial \omega}\right)_{k_i} d\omega = -n_i \left(\frac{\partial D}{\partial k_i}\right)_{\omega, k_j} dk = n_i V_i^g \left(\frac{\partial D}{\partial \omega}\right)_{k_i} dk \quad [4.160]$$

hence:

$$d\omega = n_i V_i^g\, dk \quad [4.161]$$

With ψ denoting the angle between the group velocity vector $\underline{V}^g$ and the wave vector (also called the beam steering angle), it comes:

$$\left(\frac{\partial \omega}{\partial k}\right)_{n_i} = \underline{n}\cdot\underline{V}^g = V^g \cos\psi \quad [4.162]$$

where the subscript n_i indicates that the derivative is calculated for a fixed propagation direction. The group velocity V^g is given by a relation:

$$V^g = \frac{1}{\cos\psi}\left(\frac{\partial \omega}{\partial k}\right)_{n_i} \quad [4.163]$$

that generalizes both the relation $V^g = d\omega/dk$, valid in a dispersive isotropic medium ($\psi = 0$), and the relation $V = V^g \cos \psi$, established (section 1.3.4) in the case of a non-dispersive anisotropic solid ($d\omega/dk = \omega/k = V$). If the boundary conditions and the material are non-dissipative, the energy velocity and the group velocity are identical (section A3.3 in Appendix 3) and equation [4.163] provides the energy velocity, that is, the power flux. In a dispersive medium, the deviation angle ψ of the energy depends on the angular frequency ω.

In an anisotropic plate, each mode is represented by a slowness curve, locus at a given frequency of the end of the slowness vector $\underline{s} = \underline{k}/\omega$, when the direction of propagation varies in the plane of the plate. Due to the dispersion effect, the phase slowness $\underline{s}$ changes with frequency. The number of curves, equal to three at low frequency (modes S_0, A_0 and TH_0), increases with the appearance of higher order modes at high frequency. With $d\omega = 0$ and $dk_i = \omega\, ds_i$, relation [4.159] leads to:

$$\left(\frac{\partial D}{\partial k_i}\right)_{\omega,k_j} \omega\, ds_i = 0 \quad \text{hence} \quad \left(\frac{\partial D}{\partial \omega}\right)_{k_i} V_i^g\, ds_i = 0 \qquad [4.164]$$

that is, since the partial derivative of D with respect to ω does not cancel (the velocity V^g cannot become infinite):

$$V_i^g\, ds_i = 0, \quad \text{when} \quad d\omega = 0 \qquad [4.165]$$

The group velocity vector is still normal to the slowness curve plotted at a given frequency. This property makes it possible to find the deviation angle ψ of the acoustic beam, i.e., the direction of the acoustic power flux, without performing the complete calculation of the Poynting vector.

The slowness curves for the first Lamb modes of the unidirectional composite of thickness 4 mm studied in the previous section are plotted in Figure 4.29 for two frequencies. As expected, the number of curves increases with frequency. Only modes with a longitudinal (S_0), transverse vertical (A_0) or transverse horizontal (TH_0) polarization can propagate up to zero frequency.

4.4. Cylindrical guides

The waves studied so far propagate near a plane surface (Rayleigh, Bleustein–Gulyaev, Scholte, or Stoneley waves) or between parallel plane surfaces (TH, Love, or Lamb waves). However, waveguides with a finite cross-section are used, especially if low-frequency waves must travel long distances. The classic example is the guide with a circular cross-section. The analysis of the propagation of elastic waves is more complex than in a plate of infinite width, since an additional dimension is involved.

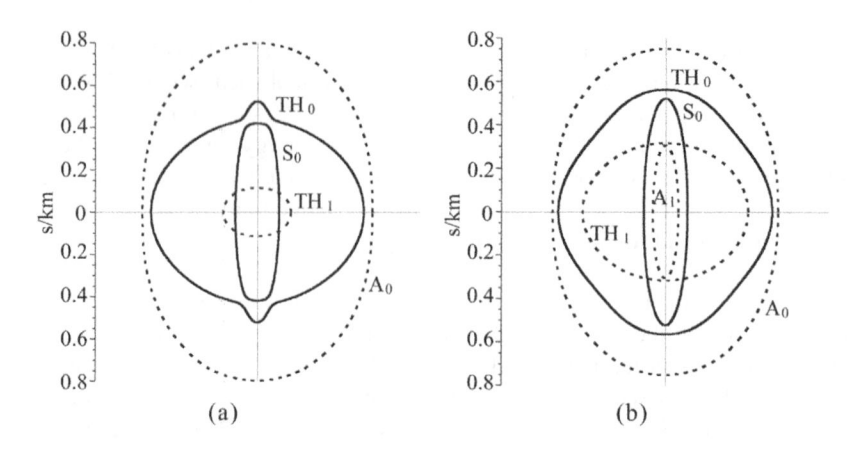

Figure 4.29. *Lamb wave slowness curves for a unidirectional composite plate of thickness 4 mm, at frequencies of a) 400 kHz and b) 600 kHz*

Since the material is assumed to be isotropic, the displacement vector $\underline{u}$ derives from a scalar potential ϕ and a vector potential $\underline{\psi}$ (equation [1.77]) and it is convenient to use cylindrical coordinates r, θ and z. In the harmonic case, these potentials satisfy the Helmholtz equations [1.87]. For a guided wave propagating along the axis z of the cylinder (Figure 4.30) and using the separation of variable method, the scalar potential is written as:

$$\phi(r,\theta,z,t) = R(r)\Theta(\theta)e^{i(kz-\omega t)} \qquad [4.166]$$

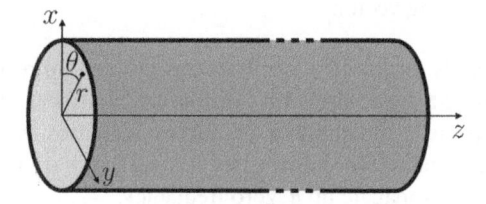

Figure 4.30. *Cylindrical waveguide of radius a and coordinates r, θ, z*

Given the expression [A1.7] for the Laplacian in cylindrical coordinates, the functions $\Theta(\theta)$ and $R(r)$ satisfy the equations:

$$\frac{d^2\Theta}{d\theta^2} + n^2\Theta = 0 \qquad [4.167]$$

and:

$$\frac{d^2 R}{dr^2} + \frac{1}{r}\frac{dR}{dr} + \left(p^2 - \frac{n^2}{r^2}\right)R = 0 \quad \text{with} \quad p^2 = k_L^2 - k^2 \qquad [4.168]$$

For $\Theta(\theta)$, the solutions are $\sin(n\theta)$ or $\cos(n\theta)$ where n must be an integer to ensure the continuity of the potential ϕ and its derivatives. Since the different quantities have a finite value at $r = 0$, the solutions for $R(r)$ are Bessel functions of the first kind, that is $R_n(r) = J_n(pr)$. The complete solution is, for example:

$$\phi = AJ_n(pr)\cos(n\theta)e^{i(kz-\omega t)} \qquad [4.169]$$

The development of the vector Laplacian $\underline{\Delta}\,\underline{\psi}$ (equation [A1.8]) leads to three equations:

$$\begin{cases} \Delta\psi_z + k_T^2\psi_z = 0 \\ \Delta\psi_r - \dfrac{\psi_r}{r^2} - \dfrac{2}{r^2}\dfrac{\partial\psi_\theta}{\partial\theta} + k_T^2\psi_r = 0 \\ \Delta\psi_\theta - \dfrac{\psi_\theta}{r^2} + \dfrac{2}{r^2}\dfrac{\partial\psi_r}{\partial\theta} + k_T^2\psi_\theta = 0 \end{cases} \qquad [4.170]$$

The first equation, which contains only the component ψ_z, is similar to the Helmholtz equation for the scalar potential; the solution is thus obtained by replacing p by q in [4.169]:

$$\psi_z = BJ_n(qr)\cos(n\theta)e^{i(kz-\omega t)} \quad \text{with} \quad q^2 = k_T^2 - k^2 \qquad [4.171]$$

The two other components ψ_r and ψ_θ are coupled: their expressions are, *a priori*, of the form:

$$\psi_r = \Psi_r(r)\sin(n\theta)e^{i(kz-\omega t)} \quad \text{and} \quad \psi_\theta = \Psi_\theta(r)\cos(n\theta)e^{i(kz-\omega t)} \qquad [4.172]$$

The presence of $\sin(n\theta)$ in the first expression leads to that of $\cos(n\theta)$ in the second one, because the coupling terms in the last two equations [4.170] contain derivatives, with respect to θ, with opposite signs. The substitution gives:

$$\frac{\partial^2\Psi_r}{\partial r^2} + \frac{1}{r}\frac{\partial\Psi_r}{\partial r} + \frac{1}{r^2}\left[2n\Psi_\theta - n^2\Psi_r - \Psi_r\right] + q^2\Psi_r = 0 \qquad [4.173]$$

and a second equation, in which the index θ appears instead of the index r and vice versa:

$$\frac{\partial^2\Psi_\theta}{\partial r^2} + \frac{1}{r}\frac{\partial\Psi_\theta}{\partial r} + \frac{1}{r^2}\left[2n\Psi_r - n^2\Psi_\theta - \Psi_\theta\right] + q^2\Psi_\theta = 0 \qquad [4.174]$$

The subtraction of the two equations leads to a Bessel equation of order $(n + 1)$, whose solution is:

$$\Psi_r(r) - \Psi_\theta(r) = 2CJ_{n+1}(qr) \qquad [4.175]$$

Adding the two equations leads to a Bessel equation of order $(n - 1)$, whose solution is:

$$\Psi_r(r) + \Psi_\theta(r) = 2DJ_{n-1}(qr) \qquad [4.176]$$

hence:

$$\begin{cases} \Psi_r(r) = CJ_{n+1}(qr) + DJ_{n-1}(qr) \\ \Psi_\theta(r) = -CJ_{n+1}(qr) + DJ_{n-1}(qr) \end{cases} \quad [4.177]$$

Four unknown constants A, B, C, D appear in the solutions, while there are only three boundary conditions, expressing the fact that the cylinder surface $r = a$ is free: $\sigma_{rr} = 0$, $\sigma_{rz} = 0$ and $\sigma_{r\theta} = 0$. It is, therefore, possible to impose a condition between the components ψ_r, ψ_θ, ψ_z. The choice of $\Psi_r = -\Psi_\theta$ is convenient since it gives $D = 0$ (Meeker and Meitzler 1964). Consequently, the scalar potential ϕ and the components of the vector $\underline{\psi}$ are:

$$\phi = AJ_n(pr) \begin{Bmatrix} \cos n\theta \\ \sin n\theta \end{Bmatrix} e^{i(kz-\omega t)}, \quad \psi_r = CJ_{n+1}(qr) \begin{Bmatrix} \sin n\theta \\ \cos n\theta \end{Bmatrix} e^{i(kz-\omega t)},$$

$$\psi_\theta = -CJ_{n+1}(qr) \begin{Bmatrix} \cos n\theta \\ \sin n\theta \end{Bmatrix} e^{i(kz-\omega t)} \quad \psi_z = BJ_n(qr) \begin{Bmatrix} \sin n\theta \\ \cos n\theta \end{Bmatrix} e^{i(kz-\omega t)}$$

[4.178]

BOUNDARY CONDITIONS.– The expressions for the displacement components: $\underline{u} = \text{grad } \phi + \text{rot } \underline{\psi}$, in terms of these potentials, are deduced from equations [A1.2] and [A1.6] in Appendix 1:

$$\begin{cases} u_r = \dfrac{\partial \phi}{\partial r} + \dfrac{1}{r}\dfrac{\partial \psi_z}{\partial \theta} - \dfrac{\partial \psi_\theta}{\partial z} \\ u_\theta = \dfrac{1}{r}\dfrac{\partial \phi}{\partial \theta} + \dfrac{\partial \psi_r}{\partial z} - \dfrac{\partial \psi_z}{\partial r} \\ u_z = \dfrac{\partial \phi}{\partial z} + \dfrac{1}{r}\dfrac{\partial (r\psi_\theta)}{\partial r} - \dfrac{1}{r}\dfrac{\partial \psi_r}{\partial \theta} \end{cases} \quad [4.179]$$

The expressions for the stresses [A1.9] involve the Lamé constants λ and μ and the local dilatation:

$$\text{div } \underline{u} = \Delta \phi = -k_L^2 \phi = -\left(p^2 + k^2\right)\phi \quad [4.180]$$

Canceling the three normal stresses:

$$\begin{cases} \sigma_{rr} = -\lambda(p^2 + k^2)\phi + 2\mu \dfrac{\partial u_r}{\partial r} \\ \sigma_{r\theta}^* = \mu \left(\dfrac{\partial u_\theta}{\partial r} + \dfrac{1}{r}\dfrac{\partial u_r}{\partial \theta} - \dfrac{u_\theta}{r} \right) \\ \sigma_{rz} = \mu \left(iku_r + \dfrac{\partial u_z}{\partial r} \right) \end{cases} \quad [4.181]$$

at the free surface $r = a$ yields a linear and homogeneous system of three equations with three unknowns A, B, C, which admits non-zero solutions if the determinant of these coefficients is zero. Expressing this explicitly, we obtain the dispersion equation, which relates ω and k for a given circumferential order n. Numerical resolution shows that for any integer n and real wave number k, there is an infinite number of real roots, which are the angular frequencies $\omega_{n,m}$ of modes propagating along z. These modes, labeled by two indices (n and m) are grouped into three families:

– *compressional* waves $L(0, m)$, axially symmetric ($n = 0$) with displacements u_r and u_z independent of θ;

– *flexural* waves $F(n, m)$, with $n \neq 0$ whose displacement $\underline{u}\,(u_r, u_\theta, u_z)$ depends on r, θ and z;

– *torsional* waves $T(0, m)$, with $n = 0$ and a circumferential displacement u_θ independent of θ.

Equations [4.178] provide the essential features of these three families when $n = 0$ (compressional modes and torsional modes) and in the simple case $n = 1$ for flexural modes.

4.4.1. *Compressional modes*

The motion is described by the two displacement components u_r and u_z, which are independent of θ ($n = 0$). Thus, the potentials (equation [4.178]) are in the form:

$$\phi = A J_0(pr) e^{i(kz - \omega t)}, \quad \psi_\theta = -C J_1(qr) e^{i(kz - \omega t)} \quad \text{and} \quad \psi_r = \psi_z = 0 \quad [4.182]$$

Using the relations:

$$\frac{\mathrm{d}J_0(x)}{\mathrm{d}x} = -J_1(x), \quad \frac{\mathrm{d}[x J_1(x)]}{\mathrm{d}x} = x J_0(x) \quad \text{and} \quad \frac{\mathrm{d}J_1(x)}{\mathrm{d}x} = J_0(x) - \frac{J_1(x)}{x}$$

[4.183]

the radial and axial components of the displacement are deduced from equations [4.179]:

$$\begin{cases} u_r = [ikC J_1(qr) - pA J_1(pr)]\, e^{i(kz - \omega t)} \\ u_z = [ikA J_0(pr) - qC J_0(qr)]\, e^{i(kz - \omega t)} \end{cases} \quad [4.184]$$

These expressions are substituted into equations [4.181] of the stresses σ_{rr} and σ_{rz}. Setting these stresses to zero at the free surface $r = a$ leads to:

$$\begin{cases} \sigma_{rr} = \left\{ -\left[\lambda\left(p^2 + k^2\right) + 2\mu p^2\right] J_0(pa) + 2\mu \dfrac{p}{a} J_1(pa) \right\} A \\ \qquad + 2ik\mu \left[q J_0(qa) - \dfrac{1}{a} J_1(qa) \right] C = 0 \\ \sigma_{rz} = -2ikp\mu J_1(pa) A + \mu \left(q^2 - k^2\right) J_1(qa) C = 0 \end{cases} \quad [4.185]$$

Considering the equality $\lambda(p^2 + k^2) + 2\mu p^2 = \mu(q^2 - k^2)$, these two equations can be written in a matrix form:

$$\begin{pmatrix} -(q^2 - k^2) J_0(pa) + 2\dfrac{p}{a} J_1(pa) & 2ik \left[q J_0(qa) - \dfrac{1}{a} J_1(qa) \right] \\ -2ikp J_1(pa) & (q^2 - k^2) J_1(qa) \end{pmatrix} \begin{pmatrix} A \\ C \end{pmatrix} = \begin{pmatrix} 0 \\ 0 \end{pmatrix}$$

[4.186]

The dispersion equation, often called the *Pochhammer–Chree equation*, is obtained by setting the determinant of this matrix to zero:

$$\dfrac{2p}{a} (q^2 + k^2) J_1(pa) J_1(qa) - (q^2 - k^2)^2 J_0(pa) J_1(qa) - 4k^2 pq J_1(pa) J_0(qa) = 0$$

[4.187]

Given the expressions for p and q (equations [4.168] and [4.171]), this relation contains the angular frequency ω, the wave number k, the velocities of longitudinal and transverse bulk waves, V_L and V_T, and the cylinder radius a. Depending on the value of the phase velocity $V = \omega/k$ with respect to V_L and V_T, the parameters p and q are real or purely imaginary. In the latter case, the equations must be rewritten taking into account the identity $J_n(ix) = i^n I_n(x)$, $I_n(x)$ denoting the modified Bessel functions of the first kind. For a given wave number k, equation [4.187] admits an infinite number of roots $\omega_{0,m}$, corresponding to as many longitudinal modes L(0, m). Figure 4.31 represents the dispersion curves $\omega(k)$ for the first compressional modes propagating in a stainless steel cylinder ($\rho = 7\,930$ kg/m^3, $V_L = 5\,692$ m/s, $V_T = 3\,172$ m/s, i.e., $\nu = 0.27$). The ordinate and abscissa are the dimensionless quantities $\omega a/V_T$ and ka, respectively. All the modes are dispersive and the group velocity of the L(0,2) mode is canceled for $ka = 1.06$ (Meitzler 1965).

When k tends to zero, that is, for long wavelengths ($\lambda \gg a$) and for the first mode L(0,1), which has no cut-off frequency, ω, p and q tend to zero. Its phase velocity is obtained by replacing, in the dispersion equation, $J_0(pa)$ and $J_0(qa)$ by 1, and $J_1(pa)$ and $J_1(qa)$ by $pa/2$ and $qa/2$, respectively, the first terms of their development:

$$J_0(x) = 1 - \dfrac{x^2}{4} + \ldots \quad \text{and} \quad J_1(x) = \dfrac{x}{2} - \dfrac{x^3}{16} + \ldots$$

[4.188]

The result:

$$V = V_T \sqrt{2(1+\nu)} = \sqrt{\dfrac{E}{\rho}} = V_b$$

[4.189]

that is, the bar velocity ($V_b = 5\,072$ m/s for steel), can also be found through a direct calculation. Assuming that the radius a is small enough compared to the wavelength, so the stresses: σ_{rr}, σ_{rz}, $\sigma_{r\theta}$, $\sigma_{\theta z}$, $\sigma_{\theta\theta}$, which are zero on the lateral surface of the cylinder, are also zero in the interior. Hooke's law is then written as:

$$\varepsilon_{zz} = \frac{\partial u_z}{\partial z} = \frac{\sigma_{zz}}{E} \quad [4.190]$$

and the fundamental equation of dynamics [1.22]:

$$\rho \frac{\partial^2 u_z}{\partial t^2} = \frac{\partial \sigma_{zz}}{\partial z} = E \frac{\partial^2 u_z}{\partial z^2} \quad [4.191]$$

shows that the longitudinal wave velocity is $V_b = \sqrt{E/\rho}$. Considering the lateral motion of the particles, the formula established by Rayleigh:

$$V \approx \sqrt{\frac{E}{\rho}} \left[1 - \nu^2 \left(\frac{ka}{2}\right)^2 \right] \quad [4.192]$$

is found by retaining the x^2 term of $J_0(x)$.

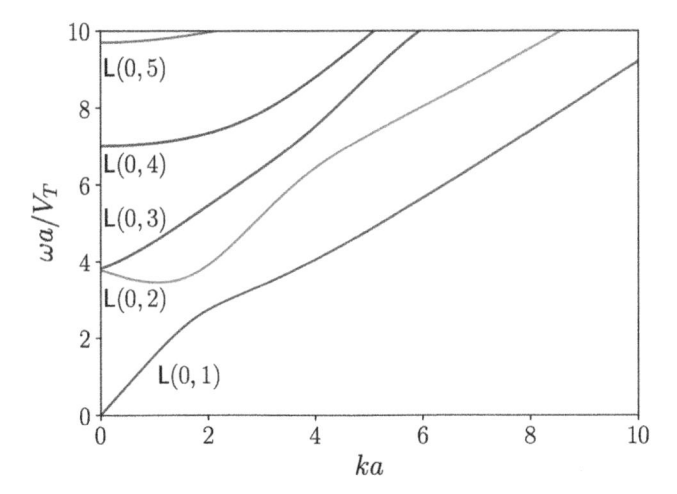

Figure 4.31. *Steel cylinder ($\nu = 0.27$). Dispersion curves for the first longitudinal modes with $\omega a/V_T$ on the ordinate and ka on the abscissa. The line with unit slope corresponds to a phase velocity equal to V_T. The first mode $L(0,1)$ is the only one with no cut-off frequency*

The other modes present a cut-off frequency: $\omega \to \omega_c$, $p \to \omega_c/V_L$ and $q \to \omega_c/V_T$, when ka is much smaller than unity. Since the last term in the Pochhammer–Chree equation [4.187] tends to zero, a first set of solutions is given by:

$$J_1\left(\frac{\omega_c a}{V_T}\right) = 0, \quad \text{hence} \quad \frac{\omega_c a}{V_T} = x_n \quad \longrightarrow \quad f_c d = \frac{x_n V_T}{\pi} \quad \text{with} \quad J_1(x_n) = 0$$

$$[4.193]$$

where $d = 2a$ is the diameter of the cylinder. The first zeros of $J_1(x)$ are $x_1 = 3.83$, $x_2 = 7.02$, $x_3 = 10.15$, etc., and in Figure 4.31 the cut-off frequencies correspond to the even compressional modes L(0,2), L(0,4) and L(0,6). The relative frequencies $\omega_c a/V_T$ are independent of the material of the guide. Considering the asymptotic behavior of $J_1(x)$ for a large x, the difference between two successive values of $f_c d$ is equal to V_T, as for the Lamb modes S_{2n}. A second set of solutions is given by the equation:

$$y J_0(y) = 2\kappa^2 J_1(y) = \frac{1-2\nu}{1-\nu} J_1(y) \quad \text{with} \quad y = \frac{\omega_c a}{V_L} = \kappa x \quad \text{hence} \quad f_c d = \frac{y_n}{\pi} V_L$$

[4.194]

The roots y_n depend on the material *via* the velocity ratio $\kappa = V_T/V_L$ or the Poisson's ratio ν. These are the cut-off frequencies of odd modes L(0,3), L(0,5), etc. The analogy between the compressional modes in a cylinder and symmetric plate modes persist at high frequencies ($\lambda << a$, i.e., $ka >> 1$). As for S_0, the phase velocity of the L(0,1) mode tends to the Rayleigh wave velocity: $V_R = 2\,940$ m/s, smaller than V_T. The velocity of the higher order modes tends toward V_T.

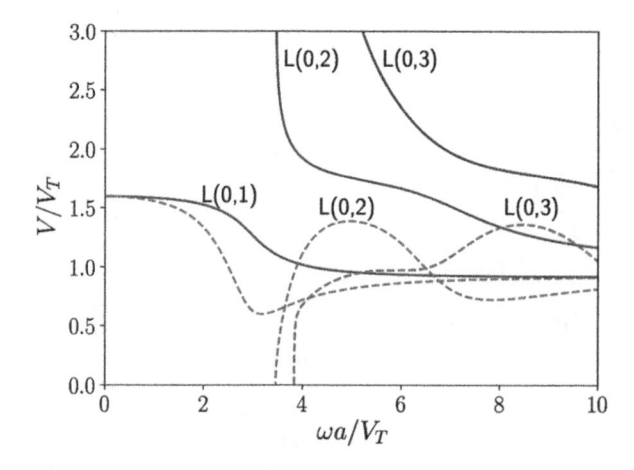

Figure 4.32. *First three longitudinal modes propagating in a steel cylinder. Variation of the phase velocity V_φ/V_T (solid lines) and group velocity V_g/V_T (dashed lines), as a function of the normalized angular frequency $\omega a/V_T$. For a color version of this figure, see www.iste.co.uk/royer/waves1.zip*

The variation of the phase velocity V_φ and group velocity V_g, normalized to the transverse bulk wave velocity, is plotted in Figure 4.32, as a function of the normalized angular frequency $\omega a/V_T$, for the first three longitudinal modes. Concerning the first mode, at low frequency ($\omega << V_T/a$), the lateral motions are not involved. The phase velocity is determined by the Young modulus

E: $V_\varphi \approx V_b = \sqrt{E/\rho}$. At high frequencies, the motion is concentrated near the surface; the velocity tends toward that of the Rayleigh wave V_R. In the intermediate range, the propagation is dispersive and the group velocity V_g passes through a minimum for an angular frequency $\omega a \approx 3.17 V_T$.

4.4.2. *Flexural modes*

The motion is described by the three displacement components u_r, u_θ, u_z, which vary with θ, as the functions $\sin n\theta$ and $\cos n\theta$. As for the compressional modes, the dispersion equation is obtained by substituting the expressions for these components in those for the stresses: σ_{rr}, σ_{rz} and $\sigma_{r\theta}$. Then the stress free boundary conditions at $r = a$ are written in a matrix form, so that the three equations in A, B, C thus established are only verified if the determinant of the matrix is zero. For a given circumferential order that is an integer n and a given real wave number k, the numerical solution of the dispersion equation provides the real angular frequencies $\omega_{n,m}$ of the different flexural modes F(n, m).

In order to get an idea of the vibration that propagates in the cylinder, let us consider the value $n = 1$. The mode is described by the potentials in equation [4.178] (the exponential propagation term is omitted here, as in the expressions of the displacement components that follow):

$$\phi = AJ_1(pr)\cos\theta, \qquad \psi_r = CJ_2(qr)\sin\theta$$
$$\psi_\theta = -CJ_2(qr)\cos\theta \quad \text{and} \quad \psi_z = BJ_1(qr)\sin\theta \qquad [4.195]$$

which give, according to [4.179]:

$$u_r = U_r(r)\cos\theta, \qquad u_\theta = U_\theta(r)\sin\theta \quad \text{and} \quad u_z = U_z(r)\cos\theta \qquad [4.196]$$

where $U_r(r)$, $U_\theta(r)$ and $U_z(r)$ only depend on r. The examination of the variations as a function of the angle θ of the displacement components u_r (along r), u_θ (along θ), and u_z (along z) explains the name "flexural waves". Indeed, for $\theta = 0$, $u_\theta = 0$, so that points in the xz plane are displaced in this plane. For $\theta = \pm\pi/2$, only u_θ is non-zero, so the points in the yz plane are displaced parallel to the xz plane.

The dispersion curves for the families F(n, m) are plotted in Figure 4.33 for a steel wire of diameter 0.755 mm, and a circumferential order n between 1 and 5. The curves for the compressional modes are reproduced on the left to highlight the similarity between the symmetric Lamb modes S$_n$ and the L$(0, m)$ modes, on the one hand, and between the antisymmetric modes A$_n$ and the flexural waves F$(1, m)$, on the other hand. Thus, the F$(1, 1)$ branch is similar to that of the A$_0$ mode. Major differences appear for the families F(n, m) of order $n \geq 2$. As for Lamb waves, the dispersion curves of the same family do not intersect and many ZGV modes exist (section 4.3.1.4); they are indicated by an arrow (Laurent *et al.* 2015).

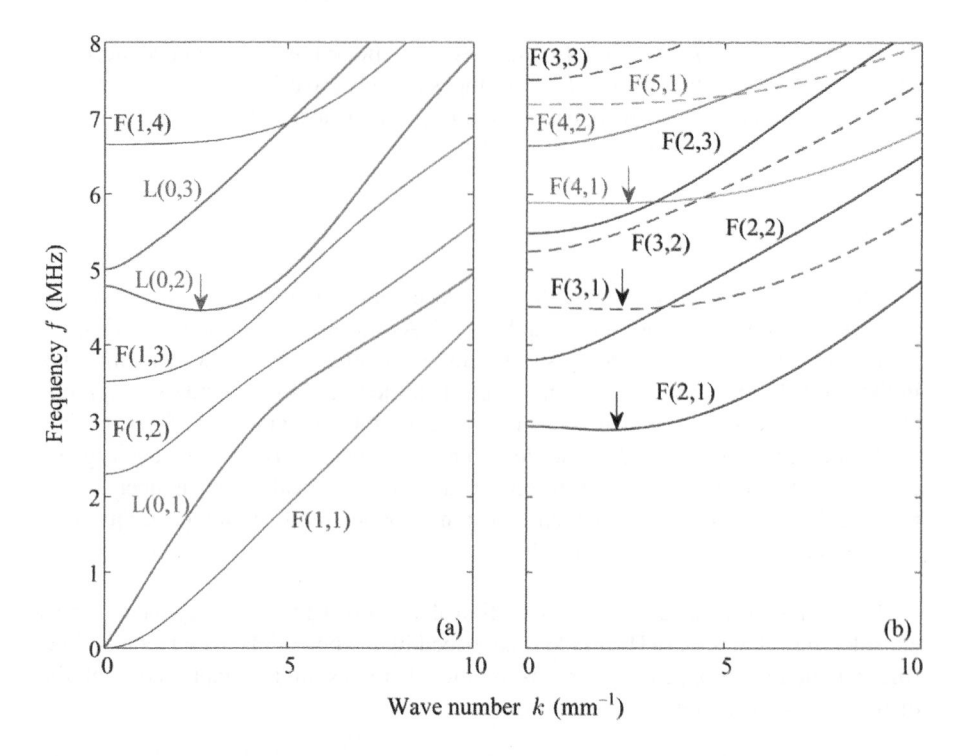

Figure 4.33. *Dispersion curves for the first flexural modes of a steel wire with diameter 0.775 mm ($V_L = 5\,650$ m/s, $V_T = 3\,010$ m/s). a) Modes $F(1, m)$ and $L(0, m)$ and b) modes $F(n, m)$ for $n = 2$ to 5. ZGV points are indicated by an arrow. For a color version of this figure, see www.iste.co.uk/royer/waves1.zip*

4.4.3. *Torsional modes*

The mechanical displacement, independent of the angle θ, has only one component: $u_\theta(r, z, t)$. It derives from the single potential ψ_z (equation [4.179]); from equation [4.183]:

$$\psi_z = B J_0(qr) e^{i(kz - \omega t)} \quad \text{hence} \quad u_\theta = -\frac{\partial \psi_z}{\partial r} = -q B J_1(qr) e^{i(kz - \omega t)}$$

[4.197]

The stress $\sigma_{r\theta}$ must be zero at the free surface $r = a$ of the cylinder:

$$\sigma_{r\theta} = \mu \left(\frac{\partial u_\theta}{\partial r} - \frac{u_\theta}{r} \right)_{r=a} = 0 \qquad [4.198]$$

that is, considering the relation [4.183]:

$$qa J_0(qa) - 2J_1(qa) = 0 \qquad [4.199]$$

The first solution of this equation is $qa = 0$. Defining $A = -Bq^2/2$, the expression for u_θ, when q tends to zero:

$$u_\theta = Are^{i(kz-\omega t)} \qquad [4.200]$$

is that of the lowest order torsional mode, for which each "section" of the cylinder rotates around its center, since the displacement is proportional to the radial coordinate. From the second equation [4.171], the value $q = 0$ also imposes that the phase velocity is constant and equal to V_T. This torsional mode is therefore not dispersive. Denoting by x_n the non-zero roots of the equation [4.199], the dispersion equation of the other modes is:

$$qa = x_n \quad \text{with} \quad q = \sqrt{k_T^2 - k^2} \quad \text{and} \quad x_1 = 5.14, \quad x_2 = 8.42, \quad x_3 = 11.62, \ldots \qquad [4.201]$$

that is:

$$\frac{\omega a}{V_T} = \sqrt{k^2 a^2 + x_n^2} \qquad [4.202]$$

These results have to be compared with those for the TH waves guided in an isotropic plate (section 4.2.1). Figure 4.34 shows the normalized dispersion curves for the first torsional modes. They are similar to those in Figure 4.4.

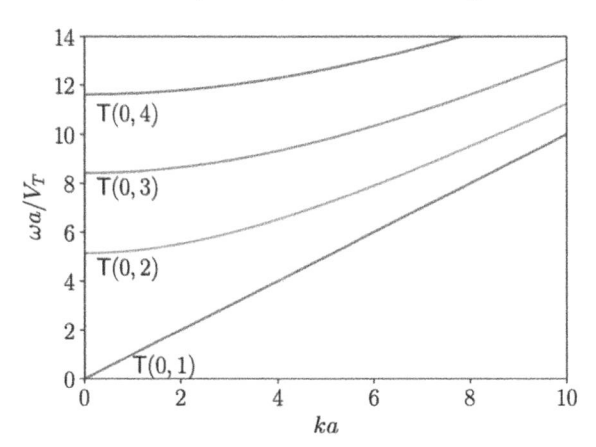

Figure 4.34. *Dispersion curves for the first torsional modes. The mode T(0,1) is non-dispersive*

4.4.4. Tubular waveguide

In principle, the propagation in a cylindrical tube can be analyzed by a method similar to that used for a cylinder. However, in order to express the boundary conditions (stresses σ_{rr}, σ_{rz} and $\sigma_{r\theta}$ being zero) on both the external and internal surfaces, it is necessary to introduce a Bessel function of the second kind into the expressions of the scalar potential and the three components of the potential vector. Six constants appear in the six equations that express the conditions to be satisfied on the two free surfaces. This system with six equations has solutions only if its determinant is zero, which leads to the dispersion equation. The results are only accessible by numerical techniques (Gazis 1959; Nicholson and McDicken 1991).

Figure 4.35 shows the variation, as a function of the normalized frequency $\omega a/V_T$, of the group velocities V_g/V_T of the first compressional and flexural modes, calculated using the Disperse software (Pavlakovic 1997), for a steel tube with mean radius $a = 3.75$ mm and shell thickness $h = 1.5$ mm. The group velocity of the L(0, 1) mode presents a deep minimum ($V_g = 583$ m/s) for $\omega a/V_T = 1.6$, that is, for $f = 222$ kHz. The attenuation due to radiation in a liquid, inversely proportional to V_g according to equation [4.131], is then very large (Royer et al. 1993).

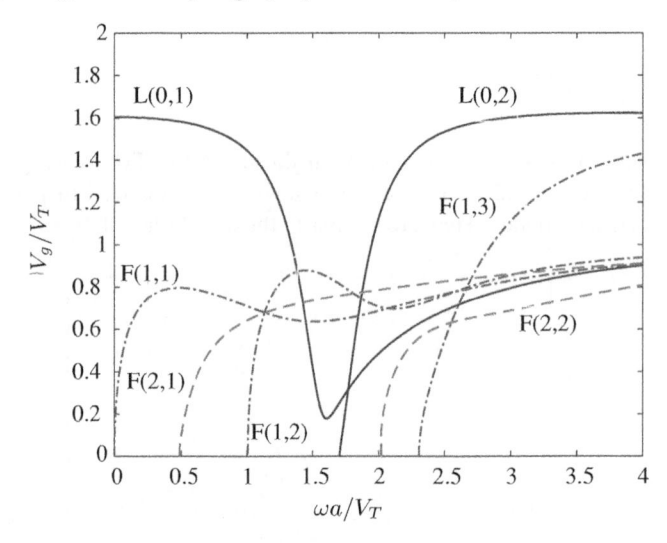

Figure 4.35. Steel tube (mean radius: 3.75 mm, thickness: 1.5 mm). Variation of the group velocity, as a function of normalized frequency, for the first compressional (L) and flexural (F) modes. For a color version of this figure, see www.iste.co.uk/royer/waves1.zip

Appendix 1

Differential Operators in Cylindrical and Spherical Coordinates

The use of cylindrical or spherical coordinates simplifies the solution of the equations governing the propagation in structures with axial or spherical symmetry. Thus, propagation in a homogeneous cylinder or radiation in an isotropic medium requires cylindrical or spherical coordinates. This appendix gives the formulae needed for calculations in these two curvilinear coordinate systems.

A1.1. Cylindrical coordinates

Let R_c be the reference frame of cylindrical coordinates (r, θ, z) and center O. The Cartesian coordinates are then defined as a function of the cylindrical coordinates by the relations:

$$x = r \cos \theta \quad \text{and} \quad y = r \sin \theta \tag{A1.1}$$

In this section, the usual differential operators are defined in cylindrical coordinates.

The gradient applied to a scalar f is a vector:

$$\underline{\text{grad}}\, f = \frac{\partial f}{\partial r} \underline{e}_r + \frac{1}{r} \frac{\partial f}{\partial \theta} \underline{e}_\theta + \frac{\partial f}{\partial z} \underline{e}_z \tag{A1.2}$$

and the gradient of a vector $\underline{A}$ is a tensor:

$$\underline{\underline{\text{grad}\,\underline{A}}} = \begin{pmatrix} \dfrac{\partial A_r}{\partial r} & \dfrac{1}{r}\left(\dfrac{\partial A_r}{\partial \theta} - A_\theta\right) & \dfrac{\partial A_r}{\partial z} \\ \dfrac{\partial A_\theta}{\partial r} & \dfrac{1}{r}\left(\dfrac{\partial A_\theta}{\partial \theta} + A_r\right) & \dfrac{\partial A_\theta}{\partial z} \\ \dfrac{\partial A_z}{\partial r} & \dfrac{1}{r}\dfrac{\partial A_z}{\partial \theta} & \dfrac{\partial A_z}{\partial z} \end{pmatrix} \qquad [A1.3]$$

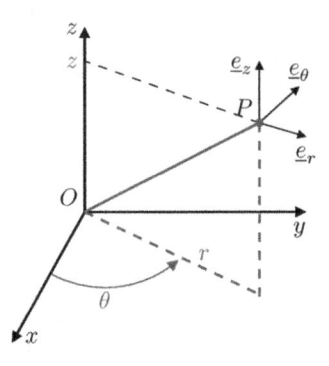

Figure A1.1. *Cylindrical coordinate system. For a color version of this figure, see www.iste.co.uk/royer/waves1.zip*

The divergence of a vector $\underline{A}$ is a scalar:

$$\text{div}\,\underline{A} = \dfrac{\partial A_r}{\partial r} + \dfrac{A_r}{r} + \dfrac{1}{r}\dfrac{\partial A_\theta}{\partial \theta} + \dfrac{\partial A_z}{\partial z} \qquad [A1.4]$$

and that of a tensor $\underline{\underline{\sigma}}$ is a vector:

$$\begin{aligned}\underline{\text{div}\,\underline{\underline{\sigma}}} &= \left(\dfrac{\partial \sigma_{rr}}{\partial r} + \dfrac{1}{r}\dfrac{\partial \sigma_{r\theta}}{\partial \theta} + \dfrac{\partial \sigma_{rz}}{\partial z} + \dfrac{\sigma_{rr} - \sigma_{\theta\theta}}{r}\right)\underline{e}_r \\ &+ \left(\dfrac{\partial \sigma_{\theta r}}{\partial r} + \dfrac{1}{r}\dfrac{\partial \sigma_{\theta\theta}}{\partial \theta} + \dfrac{\partial \sigma_{\theta z}}{\partial z} + \dfrac{\sigma_{r\theta} + \sigma_{\theta r}}{r}\right)\underline{e}_\theta \\ &+ \left(\dfrac{\partial \sigma_{zr}}{\partial r} + \dfrac{1}{r}\dfrac{\partial \sigma_{z\theta}}{\partial \theta} + \dfrac{\partial \sigma_{zz}}{\partial z} + \dfrac{\sigma_{zr}}{r}\right)\underline{e}_z \end{aligned} \qquad [A1.5]$$

The expression for the rotational of a vector $\underline{A}$ is:

$$\underline{\text{rot }} \underline{A} = \left(\frac{1}{r}\frac{\partial A_z}{\partial \theta} - \frac{\partial A_\theta}{\partial z}\right)\underline{e}_r + \left(\frac{\partial A_r}{\partial z} - \frac{\partial A_z}{\partial r}\right)\underline{e}_\theta + \frac{1}{r}\left(\frac{\partial(rA_\theta)}{\partial r} - \frac{\partial A_r}{\partial \theta}\right)\underline{e}_z \quad [A1.6]$$

Finally, the Laplacian applied to a scalar f or a vector $\underline{A}$ is given by:

$$\Delta f = \frac{\partial^2 f}{\partial r^2} + \frac{1}{r}\frac{\partial f}{\partial r} + \frac{1}{r^2}\frac{\partial^2 f}{\partial \theta^2} + \frac{\partial^2 f}{\partial z^2} \quad [A1.7]$$

and:

$$\Delta \underline{A} = \left(\Delta A_r - \frac{A_r}{r^2} - \frac{2}{r^2}\frac{\partial A_\theta}{\partial \theta}\right)\underline{e}_r + \left(\Delta A_\theta - \frac{A_\theta}{r^2} + \frac{2}{r^2}\frac{\partial A_r}{\partial \theta}\right)\underline{e}_\theta + (\Delta A_z)\underline{e}_z \quad [A1.8]$$

Considering these expressions, the components of the stress tensor of an isotropic solid are given by:

$$\begin{cases} \sigma_{rr} = \lambda \text{ div } \underline{u} + 2\mu \frac{\partial u_r}{\partial r} \\ \sigma_{\theta\theta} = \lambda \text{ div } \underline{u} + 2\mu \left(\frac{1}{r}\frac{\partial u_\theta}{\partial \theta} + \frac{u_r}{r}\right) \\ \sigma_{zz} = \lambda \text{ div } \underline{u} + 2\mu \frac{\partial u_z}{\partial z} \end{cases} \begin{cases} \sigma_{r\theta} = \mu\left(\frac{\partial u_\theta}{\partial r} - \frac{u_\theta}{r} + \frac{1}{r}\frac{\partial u_r}{\partial \theta}\right) \\ \sigma_{rz} = \mu\left(\frac{\partial u_r}{\partial z} + \frac{\partial u_z}{\partial r}\right) \\ \sigma_{\theta z} = \mu\left(\frac{1}{r}\frac{\partial u_z}{\partial \theta} + \frac{\partial u_\theta}{\partial z}\right) \end{cases} \quad [A1.9]$$

A1.2. Spherical coordinates

Let R_s be the reference frame with spherical coordinates (r, θ, φ) and center O. The Cartesian coordinates are then defined in terms of the spherical coordinates by the relations:

$$\begin{cases} x = r \sin \theta \cos \varphi \\ y = r \sin \theta \sin \varphi \\ z = r \cos \theta \end{cases} \quad [A1.10]$$

In this section, the usual differential operators are defined in spherical coordinates. The gradient applied to a scalar f and to a vector $\underline{A}$ is given by:

$$\underline{\text{grad }} f = \frac{\partial f}{\partial r}\underline{e}_r + \frac{1}{r}\frac{\partial f}{\partial \theta}\underline{e}_\theta + \frac{1}{r \sin \theta}\frac{\partial f}{\partial \varphi}\underline{e}_\varphi \quad [A1.11]$$

and:

$$\underline{\underline{\text{grad }\underline{A}}} = \begin{pmatrix} \dfrac{\partial A_r}{\partial r} & \dfrac{1}{r}\left(\dfrac{\partial A_r}{\partial \theta} - A_\theta\right) & \dfrac{1}{r}\left(\dfrac{1}{\sin\theta}\dfrac{\partial A_r}{\partial \varphi} - A_\varphi\right) \\ \dfrac{\partial A_\theta}{\partial r} & \dfrac{1}{r}\left(\dfrac{\partial A_\theta}{\partial \theta} + A_r\right) & \dfrac{1}{r}\left(\dfrac{1}{\sin\theta}\dfrac{\partial A_\theta}{\partial \varphi} - \dfrac{A_\varphi}{\tan\theta}\right) \\ \dfrac{\partial A_\varphi}{\partial r} & \dfrac{1}{r}\dfrac{\partial A_\varphi}{\partial \theta} & \dfrac{1}{r}\left(\dfrac{1}{\sin\theta}\dfrac{\partial A_\varphi}{\partial \varphi} + \dfrac{A_\theta}{\tan\theta} + A_r\right) \end{pmatrix} \quad [\text{A1.12}]$$

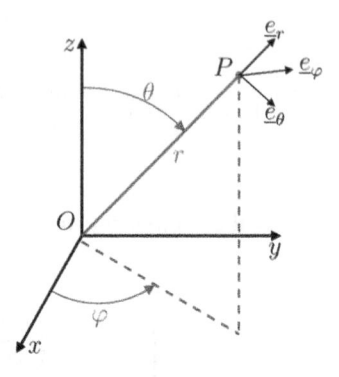

Figure A1.2. *Spherical coordinate system. For a color version of this figure, see www.iste.co.uk/royer/waves1.zip*

Similarly, the divergence applied to a vector $\underline{A}$ is given by:

$$\text{div }\underline{A} = \dfrac{\partial A_r}{\partial r} + \dfrac{2A_r}{r} + \dfrac{A_\theta}{r\tan\theta} + \dfrac{1}{r}\dfrac{\partial A_\theta}{\partial \theta} + \dfrac{1}{r\sin\theta}\dfrac{\partial A_\varphi}{\partial \varphi} \quad [\text{A1.13}]$$

and for a symmetric second-rank tensor $\underline{\underline{\sigma}}$:

$$\begin{aligned}\text{div }\underline{\underline{\sigma}} = &\left(\dfrac{\partial \sigma_{rr}}{\partial r} + \dfrac{1}{r}\dfrac{\partial \sigma_{r\theta}}{\partial \theta} + \dfrac{1}{r\sin\theta}\dfrac{\partial \sigma_{r\varphi}}{\partial \varphi} + \dfrac{2\sigma_{rr} - \sigma_{\theta\theta} - \sigma_{\varphi\varphi} + \sigma_{r\theta}\cot\theta}{r}\right)\underline{e}_r \\ &+ \left(\dfrac{\partial \sigma_{\theta r}}{\partial r} + \dfrac{1}{r}\dfrac{\partial \sigma_{\theta\theta}}{\partial \theta} + \dfrac{1}{r\sin\theta}\dfrac{\partial \sigma_{\theta\varphi}}{\partial \varphi} + \dfrac{(\sigma_{\theta\theta} - \sigma_{\varphi\varphi})\cot\theta + 3\sigma_{r\theta}}{r}\right)\underline{e}_\theta \\ &+ \left(\dfrac{\partial \sigma_{\varphi r}}{\partial r} + \dfrac{1}{r}\dfrac{\partial \sigma_{\varphi\theta}}{\partial \theta} + \dfrac{1}{r\sin\theta}\dfrac{\partial \sigma_{\varphi\varphi}}{\partial \varphi} + \dfrac{3\sigma_{r\varphi} + 2\sigma_{\theta\varphi}\cot\theta}{r}\right)\underline{e}_\varphi \quad [\text{A1.14}]\end{aligned}$$

The expression for the rotational of a vector $\underline{A}$ is:

$$\text{rot } \underline{A} = \frac{1}{r\sin\theta}\left[\frac{\partial}{\partial\theta}(A_\varphi \sin\theta) - \frac{\partial A_\theta}{\partial\varphi}\right]\underline{e}_r + \left[\frac{1}{r\sin\theta}\frac{\partial A_r}{\partial\varphi} - \frac{1}{r}\frac{\partial}{\partial r}(rA_\varphi)\right]\underline{e}_\theta$$

$$+ \frac{1}{r}\left[\frac{\partial}{\partial r}(rA_\theta) - \frac{\partial A_r}{\partial\theta}\right]\underline{e}_\varphi \qquad \text{[A1.15]}$$

Finally, the Laplacian applied to a scalar f is given by the relation:

$$\Delta f = \frac{1}{r^2}\left[r^2\frac{\partial^2 f}{\partial r^2} + 2r\frac{\partial f}{\partial r} + \frac{1}{\sin\theta}\frac{\partial}{\partial\theta}\left(\sin\theta\frac{\partial f}{\partial\theta}\right) + \frac{1}{\sin^2\theta}\frac{\partial^2 f}{\partial\varphi^2}\right] \qquad \text{[A1.16]}$$

Considering these expressions, the components of the stress tensor for an isotropic solid are written as:

$$\begin{cases} \sigma_{rr} = \lambda \text{ div }\underline{u} + 2\mu\dfrac{\partial u_r}{\partial r} \\[4pt] \sigma_{\theta\theta} = \lambda \text{ div }\underline{u} + 2\mu\left(\dfrac{1}{r}\dfrac{\partial u_\theta}{\partial\theta} + \dfrac{u_r}{r}\right) \\[4pt] \sigma_{\varphi\varphi} = \lambda \text{ div }\underline{u} + 2\mu\left(\dfrac{1}{r\sin\theta}\dfrac{\partial u_\varphi}{\partial\varphi} + \dfrac{u_r}{r} + \dfrac{u_\theta}{r\tan\theta}\right) \\[4pt] \sigma_{r\theta} = \mu\left(\dfrac{\partial u_\theta}{\partial r} + \dfrac{1}{r}\dfrac{\partial u_r}{\partial\theta} - \dfrac{u_\theta}{r}\right) \\[4pt] \sigma_{r\varphi} = \mu\left(\dfrac{1}{r\sin\theta}\dfrac{\partial u_r}{\partial\varphi} + \dfrac{\partial u_\varphi}{\partial r} - \dfrac{u_\varphi}{r}\right) \\[4pt] \sigma_{\theta\varphi} = \mu\left(\dfrac{1}{r}\dfrac{\partial u_\varphi}{\partial\theta} + \dfrac{1}{r\sin\theta}\dfrac{\partial u_\theta}{\partial\varphi} - \dfrac{u_\varphi}{r\tan\theta}\right) \end{cases} \qquad \text{[A1.17]}$$

Appendix 2

Symmetry and Tensors

This appendix deals with the symmetry of crystals and the representation of their physical properties by tensors. The anisotropy of crystals plays a very important role with respect to their macroscopic physical properties: crystals with the same point symmetry behave similarly under physical actions of the same orientation. Tensor analysis expresses this behavior well as it classifies physical quantities according to the laws for transforming their components when the reference axes are changed. When this change of axes corresponds to a symmetry operation, it results from the identity of the macroscopic properties of the crystal in both reference frames, relations between the components of the tensors representing these properties. Consequently, there is a reduction in the number of independent components. The non-existence of properties represented by tensors of a given rank in various classes of crystals is verified. The reduction of the number of independent elastic constants is treated in the last section of this appendix.

A2.1. Crystalline structure

Among solids, we must distinguish between amorphous bodies and crystals. There are also intermediate cases such as polycrystalline materials, consisting of microscopic single crystals, randomly oriented, whose behavior at a macroscopic scale is homogeneous and isotropic. Amorphous bodies, such as resin or glass, have no characteristic geometrical forms and are, in reality, extremely viscous liquids. When heated, they undergo pasty fusion: their fluidity increases continuously with temperature. During cooling, the only visible change is a gradual increase in viscosity and the temperature decrease *versus* time presents no plateau.

On the contrary, in the case of a crystal, during the cooling of the liquid, the temperature stabilizes as soon as solid seeds of polyhedral form appear: these are crystals. All these polyhedra are convex and similar, that is, the dihedral angles between the natural faces of crystals of the same variety are always constant and ordinary angles, whatever the appearance of the sample. The crystal faces can be developed in very different forms, but the orientation of the faces is constant: *the beam of normals to the crystal faces drawn from a fixed point forms an invariant geometric figure* (Romé de l'Isle's law of constant angles).

The examination of crystals of the same species shows that the crystalline medium is *anisotropic and homogeneous*. The orientation of the faces is one expression of the anisotropy, as is the appearance of shock figures, the different aspect of etch figures with respect to the attacked faces and the cleavage along preferred planes. These effects are at the origin of the hypothesis (formulated by R.J. Haüy in the 18th century) of an ordered, periodic structure of matter at the atomic scale. This hypothesis was verified in 1912 by X-ray diffraction (Von Laue, Friedrich) and, thereafter, confirmed by direct observations using instruments such as the electron microscope or the atomic force microscope.

The crystalline medium is *homogeneous*: the behavior of several macroscopic samples with the same dimensions and orientation, cut into a crystal, is identical. At the atomic scale, where the medium is discontinuous, the homogeneity persists in the sense given by Bravais: there exists in the crystal, along three distinct directions, a discrete infinity of points that are homologous to any given point, that is, having the same environment. The external appearance and the macroscopic properties of crystals suggest their classification according to the symmetry of the beam of normals to the natural faces, known as the *point symmetry*.

A2.1.1. *Lattice and atomic structure*

The crystalline medium is characterized by the presence of an infinite number of geometric points equivalent to any given point. All these homologous points have the same atomic environment, and they can be deduced from one another through a succession of elementary translations of the three vectors $\underline{a}$, $\underline{b}$, $\underline{c}$. Any point M homologous to the origin O has the position:

$$\underline{OM} = m\underline{a} + n\underline{b} + p\underline{c} \qquad [A2.1]$$

where m, n, p are relative integers. The set of these homologous points, called nodes, forms a three-dimensional lattice that expresses the periodicity of the crystal in all directions (Figure A2.1(a)). The atomic structure of a crystal is determined, unambiguously, once the lattice and the cluster of atoms arranged at the nodes of the lattice are known (Figure A2.1(b)).

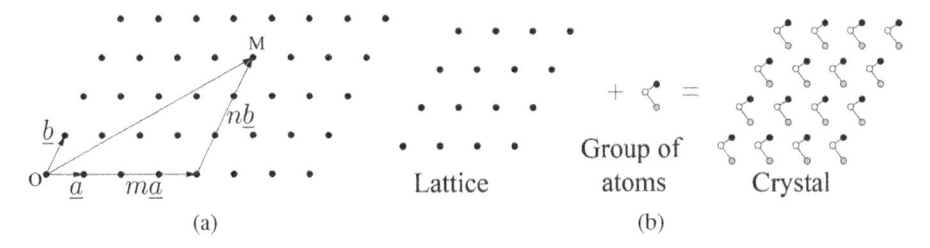

Figure A2.1. a) Two-dimensional crystalline lattice: all nodes are equivalent, and they have the same atomic environment as the origin node O. The position of these points is given by $OM = m\underline{a} + n\underline{b}$. b) Atomic structure: a crystal is formed by the adjunction of a group of atoms to each node of the lattice

A2.1.2. Rows, lattice planes and cells

The lattice can be decomposed into simple, one-dimensional units (rows), two-dimensional units (lattice planes) or three-dimensional units (cells).

A2.1.2.1. Rows

The lattice nodes are located at the intersections of three families of parallel lines (Figure A2.2). Along each of these lines (called rows), the nodes are equidistant. Three relative integers that are prime to each other, u, v, w, define the row $[u, v, w]$ by the vector:

$$\underline{R}_{u,v,w} = u\underline{a} + v\underline{b} + w\underline{c} \qquad [\text{A2.2}]$$

joining a node to its immediate neighbor. Thus, in Figure A2.2 ($w = 0$), the rows of family D_3 are denoted $[2, \bar{1}, 0]$, where $\bar{1}$ signifies -1.

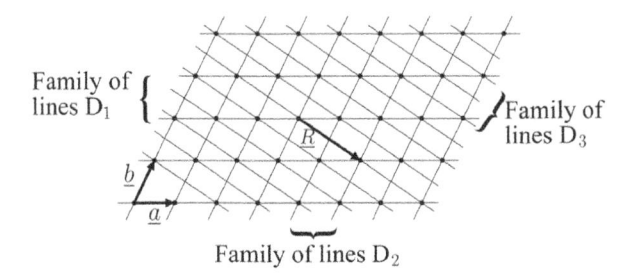

Figure A2.2. Rows. The nodes in a two-dimensional lattice are located at the intersections of any two of the families of lines, D_1, D_2, D_3, etc. For the family D_3, the node spacing is $\underline{R} = 2\underline{a} - \underline{b}$

A2.1.2.2. *Lattice planes and Miller indices*

It is possible to allocate all the nodes of a lattice to a family of parallel, equidistant planes, called lattice planes, which are derived from one another by the elementary translation of any row not contained in the planes of that family. Since there exists a node at the origin and another at the end of each vector $\underline{a}, \underline{b}, \underline{c}$, there is a plane that passes by each of these four points. In the general case, other planes are interposed between these, such that each basis vector is divided into equal segments (Figure A2.3). If the spacing between two adjacent planes is a/h along $\underline{a}$, b/k along $\underline{b}$, and c/l along $\underline{c}$, the numbers h, k, l are the indices of the family of reticular planes (Miller notation: (h, k, l)). A negative index has an over-bar; a zero index indicates that the planes are parallel to the corresponding axis.

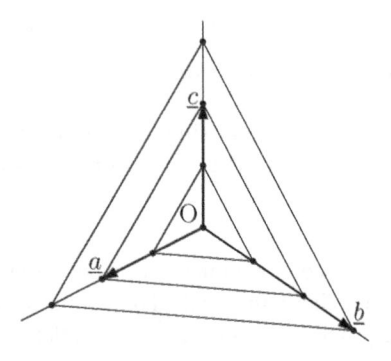

Figure A2.3. *Miller indices. The spacing between two neighboring lattice planes is: $a/2$ along $\underline{a}$, $b/3$ along $\underline{b}$, $c/2$ along $\underline{c}$. The Miller indices for this family of planes are: 2, 3, 2*

Since the number of nodes per unit volume is fixed, planes with small indices, such as (100), (010), (001), which are the most distant, have the largest node density per unit area. These planes with strong cohesion are potentially planes of cleavage, since the inter-reticular bonds are relatively weak; moreover, they reflect X-rays more strongly than other planes.

A2.1.2.3. *Cells*

The lattice can be also considered as a stack of parallelepipeds identical to that constructed on the three basis vectors $\underline{a}, \underline{b}, \underline{c}$. These parallelepipeds, which have nodes at their vertices, are called cells. A cell constructed with three basis vectors is a simple cell, because it contains only one node. Indeed, each of its eight nodes also belongs to the seven neighboring cells, and therefore it only counts as 1/8th. A given crystal has a unique lattice, but it is possible to generate it from other basis vectors and thus to define other cells.

A2.2. Point symmetry of crystals

Observations made by X-ray diffraction or by using a scanning tunneling microscope show that the dimensions of the crystalline cell are of a few angströms. From the macroscopic point of view, the elementary translations of the lattice are infinitesimal, so that the periodicity has no direct influence on the properties of the crystal, which appears to be continuous, and only the anisotropy is relevant. This *anisotropy* is not total: there exist directions equivalent to an arbitrary direction, along which the crystal properties are identical. The set of all the symmetry operations that bring the crystal in a position indistinguishable by macroscopic measurements, from the original position constitutes the *symmetry of orientation* of the crystal. This is also called *point symmetry* since, translations being excluded, the symmetry operations preserve at least one point fixed.

After having defined the elements of the point symmetry and established some useful relations, we examine the symmetry of the lattice and that of the crystal. The lattice periodicity reduces the number of elements of symmetry and imposes relations such that only seven classes of symmetry exist for lattices, each one determining a crystal system. The *symmetry of the crystal (lattice + group of atoms), at most equal to that of its lattice*, is obtained by removing certain elements from the symmetry class of the lattice. The 32 classes of point symmetry of crystals are thus naturally categorized into the seven crystal systems.

A2.2.1. *Point symmetry operations*

The elements of point symmetry are of two types:

– The direct symmetry elements, which are reduced to the axes of rotation. A crystal has a direct n-fold rotation axis (n being an integer), denoted by A_n, if a rotation by an angle $2\pi/n$ about this axis brings it to a position indistinguishable from the original position.

– The inverse symmetry elements, which can be divided into the center of symmetry and the inverse rotation axes. The center of symmetry C corresponds to the operation of symmetry about a point, also called an inversion. A crystal has an inverse n-fold axis of rotation, denoted by $\overline{A}_n$, if a rotation of $2\pi/n$ about this axis, followed by a symmetry with respect to a point on this axis, brings the crystal into a position indistinguishable from its original position. The product, symbolized by a dot, of these two operations, is commutative, so that $\overline{A}_n = A_n \cdot C = C \cdot A_n$.

Symmetry with respect to a plane is a particular case of an inverse rotation axis. Figure A2.4 shows that a twofold inverse axis is identical to a plane of symmetry or *mirror plane* M, perpendicular to the axis at the inversion point, so that $\overline{A}_2 \equiv M$. Similarly, a center of symmetry is a onefold inverse axis: $\overline{A}_1 \equiv C$.

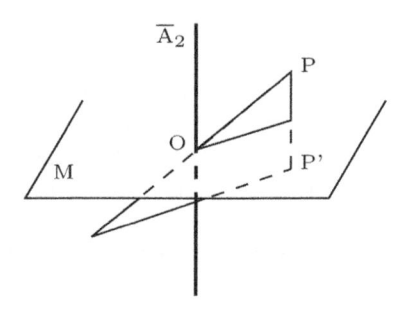

Figure A2.4. *A twofold inverse axis ($\overline{A}_2$) is equivalent to a perpendicular mirror plane M*

A2.2.1.1. *Stereographic projection*

To establish more complex relations between the elements of point symmetry, it is convenient to use a planar representation, for example the stereographic projection (Figure A2.5(a)), which maps, onto the equatorial plane E, the points on a sphere of radius R, using an inversion of power $2R^2$ and pole S (south) for points of the northern hemisphere and of pole N (north) for points of the southern hemisphere. Thanks to the change in the pole of inversion, all points are mapped inside the equatorial circle. Like any inversion, the stereographic projection conserves angles. The transforms of two curves drawn on the sphere, which intersect at a point C at an angle α, also intersect on the projection of C at the same angle α.

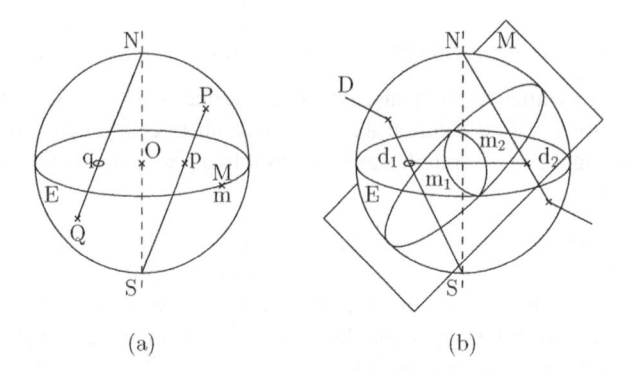

Figure A2.5. *Stereographic projection. a) Triangles SOp and SPN are similar, so that $Sp/2R = R/SP$. Since $Sp.SP = 2R^2$, the stereographic projection is an inversion of power $2R^2$. b) A line D is represented by the projections, d_1 and d_2, of its intersections with the sphere. A plane M is represented by the two circular arcs m_1 and m_2*

The projections of points in the northern hemisphere or on the equator are represented by a cross (×), while those of points in the southern hemisphere are represented by an encircled dot (⊙). A line is represented by the projection of its intersections with the sphere. A plane passing through the center is projected along two arcs of a circle, symmetrical with respect to its intersection with the equatorial plane (Figure A2.5(b)).

In a stereogram reflecting the properties of point symmetry, the north-south axis is generally carried by the symmetry axis with the highest order, and the center of the sphere is located at the center of symmetry, if any. Geometrical constructions are in dotted lines; the projections alone are in solid lines.

A2.2.1.2. *Equivalence relations*

Let us start with relations involving a single element of inverse symmetry.

a) *An inverse axis of odd order n is equivalent to a direct axis of order n and a center of symmetry.* Indeed, the repetition of n operations of inverse symmetry: $(\overline{A}_n)^n = (A_n)^n \cdot C^n = C^n$ because $(A_n)^n = I$ (identity operation). If n is odd ($n = 2p+1$), this gives $C^n = C^{2p} \cdot C = C$. On the other hand, it comes $\overline{A}_n \cdot C = A_n \cdot C^2 = A_n$. Thus, an inverse symmetry axis with odd order (for example, $n = 3$) is equivalent to the association of a direct axis and a center of symmetry, so that $\overline{A}_{2p+1} \equiv A_{2p+1}C$.

REMARK.– In the notation $A_{2p+1}C$, the two elements are independent, unlike the notation of the definition $\overline{A}_n = A_n \cdot C$, in which the dot indicates the succession of the two operations.

b) *An inverse axis of even order n implies a collinear, direct axis of order $n/2$.* Indeed: $(\overline{A}_n)^2 = (A_n)^2$. If $n = 2p$, $(A_n)^2$ corresponds to a rotation by an angle $2\pi/p$, so that $(\overline{A}_{2p})^2 = A_p$.

c) *If, in addition, $p = n/2$ is odd (for example, $n = 6$), there exists a mirror plane perpendicular to the axis.* Since p is odd, $C^p = C$ and the operation $(\overline{A}_{2p})^p = (A_{2p})^p \cdot C^p = A_2 \cdot C = \overline{A}_2$ is identical to a perpendicular mirror plane M. The property (b) is always true, so that:

$$\overline{A}_{2p} \equiv \frac{A_p}{M} \quad \text{if } p \text{ is odd} \qquad [A2.3]$$

The bar of the fraction indicates that the mirror is perpendicular to the axis. In the figures, the symmetry axes are represented by symbols that take into account the previous results:

– direct axis of order 2(○), 3(△), 4(□), 6(○);
– inverse axis of order $\overline{1}$(○), $\overline{3}$(◮), $\overline{4}$(▱), $\overline{6}$(◎).

If we add the inverse binary axis, represented by the equivalent perpendicular mirror, these symbols are those of the symmetry elements of crystals (section A2.2.2).

The simultaneous presence of several symmetry elements often implies the existence of other elements. A few examples are given as follows:

d) *A direct axis of even order $n = 2p$, and a center of symmetry imply the existence of a mirror plane perpendicular to the axis.* This property results from the operation: $(A_{2p})^p \cdot C = A_2 \cdot C = \overline{A}_2 \equiv M$, that is:

$$A_{2p}C \longrightarrow \frac{A_{2p}}{M}C \qquad [A2.4]$$

e) *An n-fold direct axis and a perpendicular binary axis imply n perpendicular binary axes forming angles π/n between them.* If the order of the axis A_n is odd ($n = 2p + 1$), two successive binary axes correspond to each other in the rotation of angle $p(2\pi/n) = \pi - (\pi/n)$. All these equivalent axes are denoted by the same symbol A'_2:

$$A_{2p+1}A'_2 \longrightarrow A_{2p+1}(2p+1)A'_2 \qquad [A2.5]$$

Conversely, if n is even ($n = 2p$), the angle $\pi - (2\pi/n) = (p-1)(2\pi/n)$ is a multiple of $2\pi/n$, so that the binary axes are only equivalent alternately. They form two groups with p elements each, denoted A'_2 and A''_2:

$$A_{2p}A_2 \longrightarrow A_{2p}\,pA'_2\,pA''_2 \qquad [A2.6]$$

Figure A2.6 highlights this distinction for the cases $n = 3$ and $n = 4$.

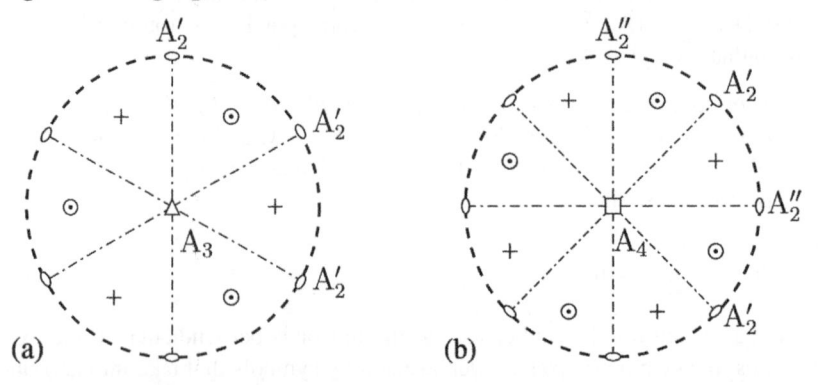

Figure A2.6. *Association of an n-fold direct axis with n perpendicular binary axes. a) If n is odd, all the binary axes are equivalent. b) If n is even, the axes are equivalent alternately*

f) *A direct n-fold axis and a mirror passing through this axis imply n mirror planes with an angle equal to π/n between them*. If n is odd, all the mirror planes are equivalent:

$$A_{2p+1}M' \longrightarrow A_{2p+1}(2p+1)M' \qquad [A2.7]$$

If n is even, the mirror planes are of two types, M' and M'':

$$A_{2p}M' \longrightarrow A_{2p}\,pM'pM'' \qquad [A2.8]$$

g) *An inverse n-fold axis (n being odd) and a binary axis lead to n equivalent binary axes and n mirror planes perpendicular to them*. From properties a, e and d, we have:

$$\overline{A}_{2p+1}A'_2 \equiv A_{2p+1}C(2p+1)A'_2 \longrightarrow A_{2p+1}\frac{(2p+1)A'_2}{(2p+1)M'}C \qquad [A2.9]$$

A2.2.2. Point symmetry of lattices: the seven crystal systems

The periodicity of the lattice implies the following symmetry properties:

1) any straight line parallel to an n-fold symmetry axis of the lattice, containing a node, is a symmetry axis of the same order;

2) any symmetry axis passing through a node is a row in the lattice;

3) any node of the lattice is a center of symmetry;

4) a symmetry axis of order n greater than two implies the presence of n perpendicular binary axes;

5) a lattice has only direct and inverse axes of order 1, 2, 3, 4 and 6.

Let us demonstrate this last property: any rotation of angle $\varphi = 2\pi/n$ must match a translation $\underline{T}$ of the lattice with another translation $\underline{T}'$. Since, in a reference frame formed by the lattice basis vectors, the components of the vectors $\underline{T}$ and $\underline{T}'$ are integers, the components of the rotation matrix α are also integers. Since the trace of a matrix is invariant in a change of axes, the trace $(2\cos\varphi + 1)$ of α must be equal to an integer p, that is:

$$2\cos\left(\frac{2\pi}{n}\right) + 1 = p \qquad [A2.10]$$

The integer p can only take the values -1, 0, 1, 2, 3, corresponding to $n = 2, 3, 4, 6, 1$.

Each class of point symmetry compatible with the periodicity of lattices defines a crystal system:

– the simplest class has only a center of symmetry:

$$C \qquad \textit{Triclinic system}$$

– the addition of a binary axis to this center implies a perpendicular mirror plane M, hence the symmetry:

$$\frac{A_2}{M}C \qquad \textit{Monoclinic system}$$

– if there is more than one binary axis, there are three binary axes at right angle: A_2, A_2', A_2'', and the center of symmetry implies three mirror planes M, M', M'' perpendicular to these axes:

$$\frac{A_2}{M}\frac{A_2'}{M'}\frac{A_2''}{M''}C \qquad \textit{Orthorhombic system}$$

– a lattice with an A_3 axis necessarily contains three binary axes A_2' and three mirror planes M', due to the center of symmetry:

$$A_3\frac{3A_2'}{3M'}C \qquad \textit{Trigonal system}$$

– if the lattice has an A_4 axis, the binary axes are only alternately equivalent:

$$\frac{A_4}{M}\frac{2A_2'}{2M'}\frac{2A_2''}{2M''}C \qquad \textit{Tetragonal system}$$

– with an A_6 axis, the binary axes are grouped three by three:

$$\frac{A_6}{M}\frac{3A_2'}{3M'}\frac{3A_2''}{3M''}C \qquad \textit{Hexagonal system}$$

The six classes listed above are the only symmetry classes for lattices with at most one axis of order n greater than two.

The only class containing more than one axis with order greater than two corresponds to the complete symmetry of the cube (Figure A2.7). This includes:

– four A_3 axes directed along the diagonals of the cube;

– three A_4 axes perpendicular to the faces;

– six A'_2 axes connecting the midpoints of opposite edges;

– a center of symmetry that implies mirror planes perpendicular to the A_4 and A'_2 axes.

Hence the symmetry is:

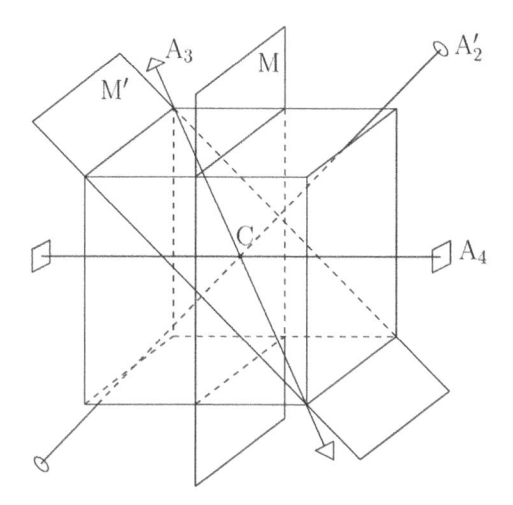

$$\frac{3A_4}{3M} 4A_3 \frac{6A'_2}{6M'} C \qquad \text{Cubic system}$$

Figure A2.7. *The symmetry elements of a cube are* $\frac{3A_4}{3M} 4A_3 \frac{6A'_2}{6M'} C$

A2.2.3. The 32-point symmetry classes of crystals

Seven crystal systems have been defined by arranging crystals according to the point symmetry of the lattice. The addition of the group of atoms to each node of the lattice can only decrease (at most maintain) the symmetry of the crystal structure, so that the crystal symmetry is, at most, equal to that of its lattice. Thus, unlike lattices, a crystal does not necessarily have a center of symmetry, nor the n binary axes imposed by the presence of a rotation axis of order n greater than two. Consequently, the point symmetry classes of crystals can be deduced from the corresponding lattice class by successively removing these optional elements. These classes are arranged, by crystal system, in Figure A2.8, where we have also indicated the symmetry elements (at the top), their stereographic projection, and the Hermannn–Mauguin notation (at the bottom).

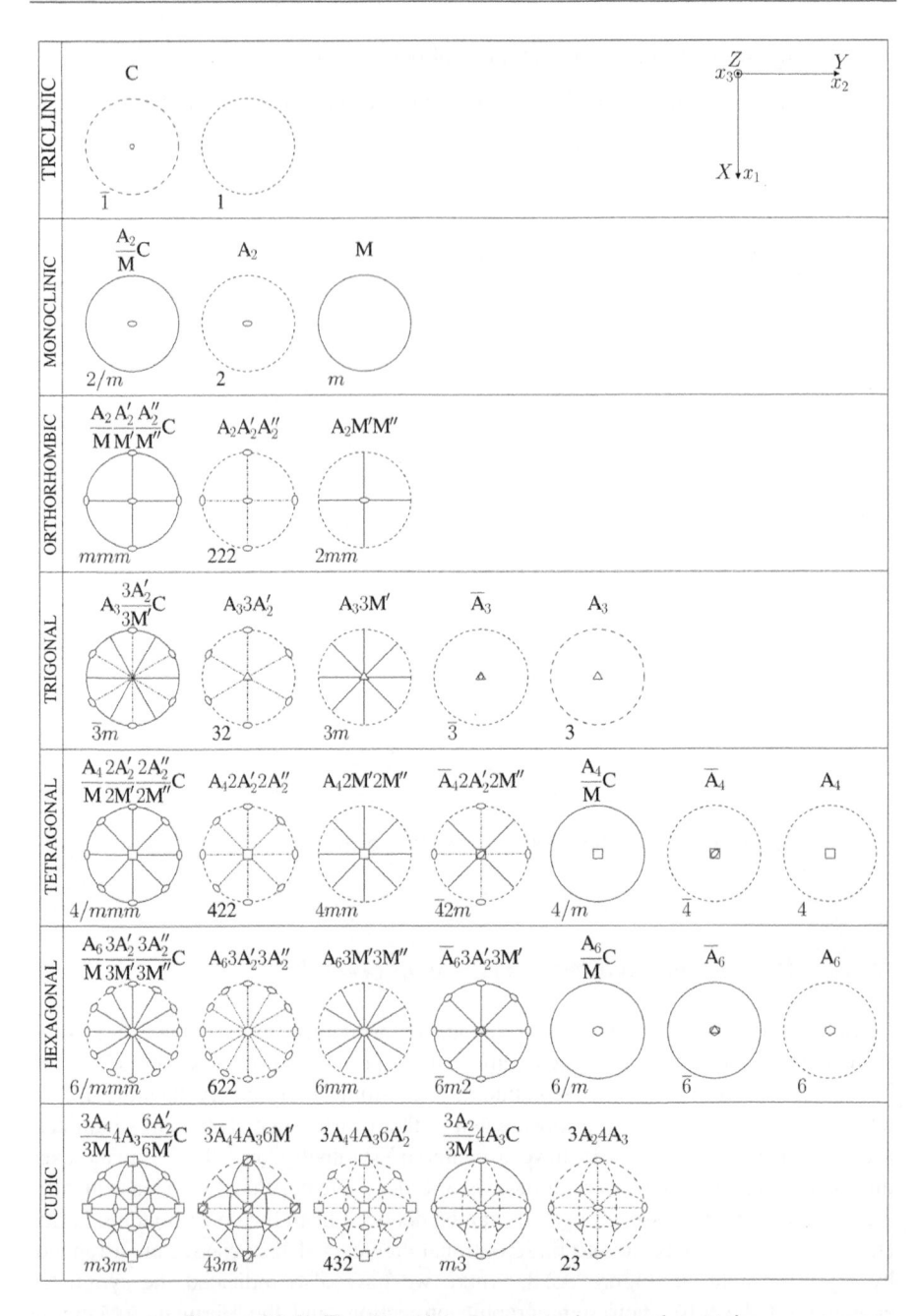

Figure A2.8. *The 32-point symmetry classes of crystals*

The first column groups the seven lattice classes determined in the previous section. Crystals with the same point symmetry as their lattice are called *holohedral*, the others are *merohedral*:

– the class with no symmetry element, obtained by removing the center of symmetry from the *triclinic* holohedral $\bar{1}$, is denoted 1;

– the removal of the symmetry center of the class of the *monoclinic* system, denoted $2/m$, conserves only the binary axis (class 2) or the mirror plane (class m);

– similarly, starting from the *orthorhombic* holohedral (denoted mmm), we get the class 222 by retaining the three binary axes $A_2 A'_2 A''_2$ or the class $2mm$ by retaining the A_2 axis and the two mirror planes M' and M'' passing through this axis;

– the symmetry of the *trigonal* system leads to the classes $A_3 3A'_2$ (denoted 32) and $A_3 3M'$ (denoted $3m$) by removing the center. It also leads to $A_3 C = \bar{A}_3$ (denoted $\bar{3}$) by removing the three binary axes A'_2 and, consequently, the three mirror planes M', and then leads to the class A_3 (denoted 3) by removing the center;

– the removal of the symmetry center from the class of the *tetragonal* system yields three new classes, by conserving:

  - all the binary axes: $A_4 2A'_2 2A''_2$ (class 422),

  - all the mirror planes: $A_4 2M'2M''$ (class $4mm$),

  - alternately, a binary axis and a mirror: $\bar{A}_4 2A'_2 2M''$ (class $\bar{4}2m$).

Then, we obtain the classes A_4 (4) and $\bar{A}_4$ ($\bar{4}$) by removing all these elements. Only the class $\frac{A_4}{M}C$, denoted $4/m$, has a center of symmetry.

– the seven classes of the hexagonal system are obtained in the same way;

– by removing the center, the *cubic* symmetry leads to the two classes $3A_4 4A_3 6A'_2$ (denoted 432) and $3\bar{A}_4 4A_3 6M'$ (denoted $\bar{4}3m$). The lowest symmetry compatible with the four ternary axes that are characteristic of the cubic system is $3A_2 4A_3$ (class 23). The addition of the center C gives rise to the mirror planes M perpendicular to the binary axes; this class of symmetry $\frac{3A_2}{3M}4A_3 C$ is denoted $m3$.

Each crystal system is characterized by the presence of particular symmetry elements:

– triclinic, with one onefold axis;

– monoclinic, with one binary (or dyad) axis;

– orthorhombic, with three binary axes;

– trigonal, with one ternary (or triad) axis;

– tetragonal, with one fourfold (or tetrad) axis;

– hexagonal, with one sixfold (or hexad) axis;

– cubic, with three direct binary axes and four direct ternary axes.

REMARK.–

– When the nature of the axis is not specified, it is either direct or inverse.

– All varieties of crystals are sorted into one of the 32 classes in Figure A2.8 and there are specimens of each class among natural or artificial crystals.

– Since crystals are anisotropic with respect to most physical properties, giving characteristic constants only makes sense if the position of the reference axes is specified with respect to the symmetry elements. The orthonormal reference frame is shown at the top right of Figure A2.8. The notation XYZ is often replaced by the notation $x_1 x_2 x_3$, which is more convenient for tensor calculations.

A2.3. Representation of physical properties of crystals by tensors

The notion of a tensor emerges as soon as we try to establish linear relations between effects and causes in an anisotropic medium. In a crystal, a cause applied along one direction usually gives rise to an effect oriented along another direction. For example, an electric field creates a polarization in a direction different from that of the field.

If the cause $\underline{c}$ and the effect $\underline{g}$ are vectorial quantities, the most general linear relation between the components of the effect, g_1, g_2, g_3, and those of the cause c_1, c_2, c_3, in the same system of axes, involves nine coefficients A_{ij}:

$$\begin{cases} g_1 = A_{11}c_1 + A_{12}c_2 + A_{13}c_3 \\ g_2 = A_{21}c_1 + A_{22}c_2 + A_{23}c_3 \\ g_3 = A_{31}c_1 + A_{32}c_2 + A_{33}c_3 \end{cases} \quad \text{or} \quad g_i = A_{ij}c_j \quad \text{with} \quad i, j = 1, 2, 3$$

[A2.11]

by summing from 1 to 3 over the repeated index j, called a dummy index (Einstein convention). The nine components A_{ij} form a second-rank tensor; a vector like $\underline{g}$ or $\underline{c}$ is called by generalizing *a first-rank tensor* and a scalar, like temperature, is a *zero-rank tensor*.

The vectors $\underline{c}$ or $\underline{g}$ represent physical quantities, while the tensor A_{ij} characterizes the material: it describes the vectorial response g_i of the crystal to the vectorial cause c_i. A scalar quantity such as temperature or energy is, thus, represented by a zero-rank tensor, a vectorial quantity (electric field, force) by a first-rank tensor, and other more complex quantities (strains, stresses) are represented by second-rank tensors (section 1.1). Physical properties of crystals are expressed by zero-rank tensors (specific heat), first-rank tensors (pyroelectricity), second-rank tensors (permittivity, electrical conductivity, thermal dilatation), third-rank tensors (piezoelectricity) and fourth-rank tensors (elasticity). For example, the relation between electric induction and electric field, $D_i = \epsilon_{ij} E_j$, defines the permittivity tensor ϵ_{ij}.

In practice, the nine numbers A_{ij} are not sufficient to characterize the physical property considered; we must specify the system of axes chosen, because in another reference frame, the tensor components take different values A'_{ij}. The two sets of values represent the same property, independent of the reference axes; consequently, a relationship exists between these components that involves the change of axes. These laws for transforming the components are used to define the rank of the tensors.

A2.3.1. Change of orthonormal reference axes

A change in the axes of reference is determined by the nine coefficients α_i^k of the relations expressing the new basis vectors $\underline{e}'_1, \underline{e}'_2, \underline{e}'_3$ in terms of the old basis vectors $\underline{e}_1, \underline{e}_2, \underline{e}_3$, that is, using the summation convention over the dummy index k:

$$\underline{e}'_i = \alpha_i^k \underline{e}_k \qquad [A2.12]$$

These coefficients can be arranged in a (3×3) array, called the matrix $\boldsymbol{\alpha}$, in which the row of index i consists of the components α_i^k expressing the vector $\underline{e}'_i$ on the basis $\underline{e}_k$. Conversely, the original basis vectors $\underline{e}_k$ are written in terms of the new basis vectors e'_j, using the elements β_k^j of the inverse matrix $\boldsymbol{\beta}$:

$$\underline{e}_k = \beta_k^j \underline{e}'_j \qquad [A2.13]$$

The relation between the coefficients β_k^j and α_i^k is obtained by substituting this expression into the relation [A2.12]:

$$\underline{e}'_i = \alpha_i^k \beta_k^j \underline{e}'_j \qquad [A2.14]$$

or, since $\underline{e}'_i = \delta_{ij} \underline{e}'_j$:

$$\left(\alpha_i^k \beta_k^j - \delta_{ij} \right) \underline{e}'_j = 0 \qquad [A2.15]$$

Since they form a basis, there is no linear relationship between the vectors $\underline{e}'_j$. Therefore, the coefficients in equation [A2.15] are all zero:

$$\alpha_i^k \beta_k^j = \delta_{ij} \qquad [A2.16]$$

The reference frames used in the following are all orthonormal, so that: $\underline{e}'_i \cdot \underline{e}'_j = \delta_{ij}$ and $\underline{e}_k \cdot \underline{e}_l = \delta_{kl}$. In this case, the inverse matrix β can be simply deduced from α. Indeed, using equation [A2.12], the first scalar product is written as:

$$\alpha_i^k \alpha_j^l (\underline{e}_k \cdot \underline{e}_l) = \delta_{ij} \qquad [\text{A2.17}]$$

and, taking into account the orthogonality of the old basis vectors: $\alpha_i^k \alpha_j^k = \delta_{ij}$. The comparison between this relation and equation [A2.16] shows that $\beta_k^j = \alpha_j^k$. The matrix β, inverse of the orthonormal axis change matrix α, is obtained by interchanging the rows and columns of α (transpose matrix α^t). The matrix α, whose inverse is its transpose matrix, is said to be orthogonal.

Let x_k be the coordinates of a point M, that is, the components of the vector $\underline{x} = \underline{OM}$ in the basis $\underline{e}_k$, and let x'_i be the coordinates of the same point in the basis $\underline{e}'_i$:

$$\underline{x} = x_k \underline{e}_k = x'_i \underline{e}'_i \qquad [\text{A2.18}]$$

and replacing $\underline{e}_k$ by equation [A2.13] gives:

$$\underline{x} = x_k \beta_k^j \underline{e}'_j = x'_i \underline{e}'_i \qquad [\text{A2.19}]$$

Since the decomposition of a vector in a basis is unique, the new coordinates x'_i are expressed in terms of the old coordinates by the relation $x'_i = x_k \beta_k^i$, which is similar to equation [A2.12], if the reference frames are orthonormal ($\beta_k^i = \alpha_i^k$):

$$x'_i = \alpha_i^k x_k \qquad [\text{A2.20}]$$

Therefore, the laws for transforming the coordinates of a point and the basis vectors [A2.12] are identical.

A2.3.2. *Definition of a tensor*

Physical quantities can be distinguished according to their behavior during a change of axes:

1) a *scalar* is a quantity independent of the chosen reference frame; it is an invariant or zero-rank tensor. The invariance of a scalar function of the coordinates $f(x_1, x_2, x_3)$ is expressed by the equality:

$$f(x_1, x_2, x_3) = f(x'_1, x'_2, x'_3) \qquad [\text{A2.21}]$$

Examples of scalars are as follows: temperature, energy and electric potential;

2) any set of three quantities A_i which transforms in a change of axis, like the basis vectors, is a first-rank tensor or vector:

$$A'_i = \alpha_i^k A_k \qquad [\text{A2.22}]$$

For example, the components of any vector in the vectorial space subtended by the three basis vectors form a first-rank tensor (equation [A2.20]). The three derivatives $\partial f/\partial x_i$ constitute a vector called the *gradient* of the scalar function $f(x_i)$. Indeed, the inversion of relation [A2.20]:

$$x_k = \beta_k^i x_i' = \alpha_i^k x_i' \qquad [A2.23]$$

shows that the components of the gradient are transformed like the basis vectors:

$$\frac{\partial f}{\partial x_i'} = \frac{\partial f}{\partial x_k}\frac{\partial x_k}{\partial x_i'} = \alpha_i^k \frac{\partial f}{\partial x_k} \qquad [A2.24]$$

3) any set of nine quantities A_{ij}, which transforms like the product of the components of two vectors, is a second-rank tensor:

$$A_{ij}' = \alpha_i^k \alpha_j^l A_{kl} \qquad [A2.25]$$

These nine quantities are arranged in a table with three rows and three columns, similar to that of the axis change matrix α. Nonetheless, matrices and tensors have different meanings: the elements α_i^k establish a correspondence between two systems of axes, while a tensor A_{ij} is a physical (or mathematical) quantity represented in one system of axes by nine numbers. However, in an orthonormal reference frame, from a mathematical point of view, there is no difference between matrices and tensors, for example, it is possible to attribute eigenvectors and eigenvalues to a tensor.

An example of second-rank tensor is the derivatives $\partial A_i/\partial x_k$ of the components of a vector (the demonstration is identical to that used for the gradient).

From a tensor A_{ij}, it is possible to form an invariant by adding all the diagonal terms. Using the convention on repeated indices, this quantity $A_{11} + A_{22} + A_{33}$, called the trace of the tensor, is written as A_{ii}. In a change of orthonormal axes:

$$A_{ii}' = \alpha_i^k \alpha_i^l A_{kl} = A_{ll} \qquad [A2.26]$$

because $\alpha_i^k \alpha_i^l = \delta_{kl}$. In particular, the trace of the tensor $A_{ij} = A_i B_j$, that is, the invariant $A_i B_i$, is the scalar product of the vectors A_i and B_j. The contracted product:

$$A_i A_i = A_1^2 + A_2^2 + A_3^2 \qquad [A2.27]$$

denoted $(A_i)^2$, represents the square of the length of the vector A_i;

4) the definition is easily generalized: a tensor of rank r is a set of 3^r components identified by r indices which, in a change of axes, transform as follows:

$$A'_{...ijk...} = ...\alpha_i^l \alpha_j^m \alpha_k^n ... A_{...lmn...} \qquad [A2.28]$$

The formation of the trace of a second-rank tensor is a particular case of the rule for contracting indices: the quantity $A_{...ijjl...}$ (with summation over the repeated index j)

coming from the tensor $A_{...ijkl...}$ of rank r is a tensor of rank $r-2$. Indeed, in a transformation of axes, the new components:

$$A'_{...ijjl...} = ...\alpha_i^m \alpha_j^n \alpha_j^p \alpha_l^q ... A_{...mnpq...} \qquad [A2.29]$$

are written, using $\alpha_j^n \alpha_j^p = \delta_{np}$, as:

$$A'_{...ijjl...} = ...\alpha_i^m \alpha_l^q ... A_{...mnnq...} \qquad [A2.30]$$

5) the following rule is often useful to recognize the tensor character of a quantity: the linear relation between an m-rank tensor $A_{...ij...}$ and an n-rank tensor $A_{...kl...}$ defines a tensor $C_{...ijkl...}$ of rank $m+n$:

$$A_{...ij...} = C_{...ijkl...} B_{...kl...} \qquad [A2.31]$$

DEMONSTRATION.– In a change of axes:

$$A'_{...pq...} = ...\alpha_p^i \alpha_q^j A_{...ij...} \qquad [A2.32]$$

and:

$$B_{...kl...} = ...\beta_k^r \beta_l^s B'_{...rs...} \qquad [A2.33]$$

The matrix β expresses the old components in terms of the new ones. Since $\beta_k^r = \alpha_r^k$, by substituting [A2.33] into [A2.31], and then [A2.31] into [A2.32] we get:

$$A'_{...pq...} = (...\alpha_p^i \alpha_q^j \alpha_r^k \alpha_s^l ... C_{...ijkl...}) B'_{...rs...} \qquad [A2.34]$$

The factor in brackets is none other than $C'_{...pqrs...}$. Consequently, the quantities $C_{...ijkl...}$ are transformed like the components of a tensor of rank $m+n$.

APPLICATION.– *The Kronecker symbol δ_{ij} is a second-rank tensor, because $A_i = \delta_{ij} A_j$. In elasticity, the stiffness constants that relate the stress tensor σ_{ij} to the strain tensor ε_{kl} form a fourth-rank tensor (section 1.1.4): $\sigma_{ij} = c_{ijkl} \varepsilon_{kl}$.*

A2.4. Reduction of the number of independent components imposed by the symmetry elements

Let $A_{...ijk...}$ be the components of a tensor, in the reference frame $Ox_1 x_2 x_3$, expressing a physical property of a crystal, and let $A'_{...ijk...}$ be the components of this tensor in the same reference frame for a new orientation of the crystal, obtained by applying the operation S. To express the $A'_{...ijk...}$ as a function of the $A_{...pqr...}$, from the point of view of the relative orientation of the crystal with respect to the reference frame, it is the same as applying the inverse operation S^{-1} to the system of axes, maintaining the crystal fixed. Consequently, with $\boldsymbol{\alpha}$ being the axis change matrix corresponding to the operation S^{-1}, and according to relation [A2.28]:

$$A'_{...ijk...} = ...\alpha_i^p \alpha_j^q \alpha_k^r ... A_{...pqr...} \qquad [A2.35]$$

If S is one of the operations of the crystal point symmetry class, the new orientation of the crystal with respect to the reference frame $Ox_1x_2x_3$ is indistinguishable, from the initial orientation, by macroscopic physical measurements. This is expressed by the equality: $A'_{...ijk...} = A_{...ijk...}$. Therefore, the invariance of properties for particular operations of symmetry imposes relations:

$$A_{...ijk...} = ...\alpha_i^p \alpha_j^q \alpha_k^r ... A_{...pqr...} \qquad [A2.36]$$

which reduce the number of independent components of tensors. Since the inverse of a symmetry operation of the crystal is a symmetry operation, $(S^{-1} = S^{n-1})$, $\boldsymbol{\alpha}$ is the basis change matrix associated with one of the operations of the point symmetry class of the crystal.

A2.4.1. *Matrices of point symmetry operations*

The symmetry elements of crystals are the center of symmetry, the direct or inverse axis of rotation and the mirror plane (section A2.2.1). The basis vector change matrices are for:

– *a symmetry with respect to a point taken as the origin:* $\boldsymbol{\alpha}_c = -\underline{\underline{1}}$;

– *a rotation of angle* $\varphi = 2\pi/n$ *about the axis* Ox_3:

$$\boldsymbol{\alpha}_{x_3} = \begin{pmatrix} \cos\varphi & \sin\varphi & 0 \\ -\sin\varphi & \cos\varphi & 0 \\ 0 & 0 & 1 \end{pmatrix} \qquad [A2.37]$$

– *a rotation-inversion of the order* n $(\overline{A}_n = A_n \cdot C)$: $\overline{\boldsymbol{\alpha}}_{x_3, 2\pi/n} = \boldsymbol{\alpha}_{x_3, 2\pi/n} \cdot \boldsymbol{\alpha}_c$;

– *a symmetry with respect to a plane* perpendicular to Ox_3:

$$\boldsymbol{\alpha}_{M \perp x_3} = \begin{pmatrix} 1 & 0 & 0 \\ 0 & 1 & 0 \\ 0 & 0 & -1 \end{pmatrix} \qquad [A2.38]$$

A2.4.2. *Effect of a center of symmetry*

Since the matrix $\boldsymbol{\alpha}$ is diagonal ($\alpha_i^j = -\delta_{ij}$) and for a tensor of rank r, the invariance condition [A2.36] reduces to:

$$A_{...ijk...} = ...\alpha_i^i \alpha_j^j \alpha_k^k ... A_{...ijk...} = (-1)^r A_{...ijk...} \qquad [A2.39]$$

When r is odd, we have $(-1)^r = -1$ and this relation implies that all components are zero. The physical properties represented by the tensors of odd rank do not manifest in crystals belonging to one of the 11 centro-symmetric classes. In particular, this is the case for piezoelectricity ($r = 3$).

If the rank r is even, the relation [A2.39] brings in nothing new. Therefore, a center of symmetry has no influence on the physical properties represented by even-rank tensors. Moreover, there is no need to distinguish between the nature of the axis (direct or inverse), so that the classes of the same system have a minimum restriction to the number of independent components, because they all have in common one or more direct or inverse axes (section A2.2.3). These observations are illustrated by the example of the elastic tensor.

A2.5. Reduction of the number of independent elastic constants

The general condition for invariance [A2.36] is written as:

$$c_{ijkl} = \alpha_i^p \alpha_j^q \alpha_k^r \alpha_l^s c_{pqrs} \qquad [A2.40]$$

Since the rank of tensor c_{ijkl} is even, the center of symmetry of crystals in the *triclinic system* does not impose any reduction in the number (21) of independent elastic constants.

A2.5.1. *Crystals with at least one binary axis*

The binary axis of crystals of the *monoclinic system* has the same effect, whether it is direct or inverse (mirror plane). Since the matrix [A2.37] is diagonal ($\varphi = \pi$), the condition [A2.40], which becomes:

$$c_{ijkl} = \alpha_i^i \alpha_j^j \alpha_k^k \alpha_l^l c_{ijkl} \qquad [A2.41]$$

implies the cancelation of constants with an odd number of index equal to 3 (one or three), for which the product $\alpha_i^i \alpha_j^j \alpha_k^k \alpha_l^l$ has the value -1. The number of independent components is reduced to 13, that is, the components for which the group of indices $ijkl$ includes zero times, twice or four times the index 3:

$$\underline{\underline{C}} = \begin{pmatrix} C_{11} & C_{12} & C_{13} & 0 & 0 & C_{16} \\ C_{12} & C_{22} & C_{23} & 0 & 0 & C_{26} \\ C_{13} & C_{23} & C_{33} & 0 & 0 & C_{36} \\ 0 & 0 & 0 & C_{44} & C_{45} & 0 \\ 0 & 0 & 0 & C_{45} & C_{55} & 0 \\ C_{16} & C_{26} & C_{36} & 0 & 0 & C_{66} \end{pmatrix} \quad \begin{array}{c} \text{monoclinic} \\ A_2 \text{ axis } //Ox_3 \end{array} \qquad [A2.42]$$

Crystals in the *orthorhombic system* are characterized by the presence of three direct or inverse orthogonal binary axes. The previous reasoning applied to each of these axes, that is, to each index, leads to the matrix [A2.43], which does not contain more than nine independent constants: those whose indices, $ijkl$, are repeated an even number of times:

$$\underline{\underline{C}} = \begin{pmatrix} C_{11} & C_{12} & C_{13} & 0 & 0 & 0 \\ C_{12} & C_{22} & C_{23} & 0 & 0 & 0 \\ C_{13} & C_{23} & C_{33} & 0 & 0 & 0 \\ 0 & 0 & 0 & C_{44} & 0 & 0 \\ 0 & 0 & 0 & 0 & C_{55} & 0 \\ 0 & 0 & 0 & 0 & 0 & C_{66} \end{pmatrix} \quad orthorhombic \quad [A2.43]$$

Crystals in the *cubic system* have at least four triad axes and three direct binary axes, which are taken as the reference axes. The non-zero components are the same as those for the orthorhombic system. The rotation of $2\pi/3$ about the triad axis directed along the [111] axis transforms Ox_1 into Ox_2, Ox_2 into Ox_3, and Ox_3 into Ox_1. The c_{ijkl} are, therefore, invariant in a circular permutation (1,2,3) of the indices, which leads to the relations:

$$c_{1111} = c_{2222} = c_{3333}, \quad c_{1122} = c_{2233} = c_{3311} \quad \text{and} \quad c_{1212} = c_{2323} = c_{3131}$$

[A2.44]

The number of independent elastic moduli is reduced to three (C_{11}, C_{12}, C_{44}), and Table [A2.43] becomes:

$$\underline{\underline{C}} = \begin{pmatrix} C_{11} & C_{12} & C_{12} & 0 & 0 & 0 \\ C_{12} & C_{11} & C_{12} & 0 & 0 & 0 \\ C_{12} & C_{12} & C_{11} & 0 & 0 & 0 \\ 0 & 0 & 0 & C_{44} & 0 & 0 \\ 0 & 0 & 0 & 0 & C_{44} & 0 \\ 0 & 0 & 0 & 0 & 0 & C_{44} \end{pmatrix} \quad cubic \quad [A2.45]$$

A2.5.2. Crystals with a principal axis A_n (n > 2)

Crystals in trigonal, tetragonal or hexagonal systems have a single direct or inverse axis of order n greater than 2. With $\varphi = 2\pi/n \neq \pi$, the rotation matrix α about this principal axis, carried by Ox_3, is no longer diagonal [A2.37]. The invariance relation [A2.40] is difficult to exploit, because it involves many components. Thus, we return to a diagonal rotation matrix by reasoning in the basis of the eigenvectors $\underline{\xi}^{(1)}$, $\underline{\xi}^{(2)}$, $\underline{\xi}^{(3)}$ of α. In order to diagonalize the matrix α, we must solve the system of equations:

$$\left(\alpha_i^k - \lambda \delta_{ik}\right) \xi_k = 0 \quad [A2.46]$$

The eigenvalues λ are determined by the compatibility condition:

$$\det\left(\alpha_i^k - \lambda \delta_{ik}\right) = 0 \quad [A2.47]$$

The development of this determinant:

$$[(\lambda - \cos\varphi)^2 + \sin^2\varphi](1-\lambda) = 0 \qquad [\text{A2.48}]$$

provides:

$$\lambda^{(1)} = e^{i\varphi}, \quad \lambda^{(2)} = e^{-i\varphi} \quad \text{and} \quad \lambda^{(3)} = 1 \qquad [\text{A2.49}]$$

Each of these values corresponds to an eigenvector $\underline{\xi}^{(i)}$, whose components are obtained by solving the system [A2.46]; for $\lambda^{(1)}$: $\xi_2 = i\xi_1$ and $\xi_3 = 0$, for $\lambda^{(2)}$: $\xi_2 = -i\xi_1$ and $\xi_3 = 0$, for $\lambda^{(3)}$: $\xi_2 = \xi_1 = 0$ and any ξ_3.

Let us call γ_{ijkl} the elastic moduli in the orthonormal basis $\underline{\xi}^{(1)}, \underline{\xi}^{(2)}, \underline{\xi}^{(3)}$. Since the axis change matrix is diagonal in this basis, the invariance relation [A2.41] is written as:

$$\gamma_{ijkl} = \lambda^{(i)}\lambda^{(j)}\lambda^{(k)}\lambda^{(l)}\gamma_{ijkl} \qquad [\text{A2.50}]$$

If ν_1 and ν_2 are the number of indices equal to 1 and 2 in the permutation $ijkl$, and given the relations [A2.49], it comes:

$$\lambda^{(i)}\lambda^{(j)}\lambda^{(k)}\lambda^{(l)} = \exp\left[i(\nu_1 - \nu_2)\frac{2\pi}{n}\right] \qquad [\text{A2.51}]$$

Consequently, γ_{ijkl} is non-zero when $\nu_1 - \nu_2$ is a multiple of the order n of the axis, because the product $\lambda^{(i)}\lambda^{(j)}\lambda^{(k)}\lambda^{(l)}$ is then equal to unity. This is always the case for the five moduli:

$$\gamma_{1122}, \quad \gamma_{1212}, \quad \gamma_{3312}, \quad \gamma_{2313} \quad \text{and} \quad \gamma_{3333} \qquad [\text{A2.52}]$$

for which $\nu_1 = \nu_2$. There is no other modulus for the crystals in the hexagonal system ($n = 6$), which thus have five independent moduli. Crystals in the trigonal system have seven independent elastic constants, since γ_{1113} and γ_{2223} are also non-zero ($\nu_1 - \nu_2 = \pm 3$). This is the same for crystals in the tetragonal system ($n = 4$), for which γ_{1111} and γ_{2222} are different from zero ($\nu_1 - \nu_2 = \pm 4$).

The return to the constants c_{ijkl} is done by the axis change relation:

$$c_{ijkl} = a_i^p a_j^q a_k^r a_l^s \gamma_{pqrs} \qquad [\text{A2.53}]$$

in which the basis change matrix $\boldsymbol{a}$ from the system $Ox_1x_2x_3$ (constants c_{ijkl}) to the basis $\underline{\xi}^{(1)}, \underline{\xi}^{(2)}, \underline{\xi}^{(3)}$ (constants γ_{ijkl}) is formed by the juxtaposition of eigenvectors $\underline{\xi}^{(1)}, \underline{\xi}^{(2)}, \underline{\xi}^{(3)}$, normalized so that the matrix $\boldsymbol{a}$ is orthogonal:

$$\boldsymbol{a} = \begin{bmatrix} \frac{1}{\sqrt{2}} & \frac{i}{\sqrt{2}} & 0 \\ \frac{i}{\sqrt{2}} & \frac{1}{\sqrt{2}} & 0 \\ 0 & 0 & 1 \end{bmatrix} \qquad [\text{A2.54}]$$

The reduced number of constants γ_{pqrs} simplifies the expansion of [A2.53]. Moreover, since $a_3^3 = 1$ is the only non-zero coefficient with an index equal to 3, the only constants appearing in this development are the constants γ_{pqrs} that have the same distribution of indices equal to 3 as the modulus c_{ijkl}.

Let us start with the *trigonal system*, for which γ_{3333}, γ_{3312}, γ_{2313}, γ_{1113}, γ_{2223}, γ_{1122} and γ_{1212} are non-zero. Depending on the number of indices equal to 3:

– $c_{3333} = \gamma_{3333}$, consequently $C_{33} \neq 0$;

– c_{3313} and c_{3323} are zero because $\gamma_{3313} = \gamma_{3323} = 0$, hence $C_{35} = C_{34} = 0$;

– the moduli with two indices equal to 3 (c_{ij33} and c_{i3k3}) are expressed in terms of $\gamma_{1233} = \gamma_{2133}$ and $\gamma_{1323} = \gamma_{2313}$:

$$c_{ij33} = \left(a_i^1 a_j^2 + a_i^2 a_j^1\right)\gamma_{1233} \qquad [A2.55]$$

On the one hand, $a_1^1 a_1^2 + a_1^2 a_1^1 = a_2^1 a_2^2 + a_2^2 a_2^1$: $c_{1133} = c_{2233}$, that is, $C_{13} = C_{23}$, and, on the other hand, $a_1^1 a_2^2 + a_1^2 a_2^1 = 0$: $c_{1233} = 0$, that is, $C_{36} = 0$. Similarly, the expansion:

$$c_{i3k3} = \left(a_i^1 a_k^2 + a_i^2 a_k^1\right)\gamma_{1323} \qquad [A2.56]$$

leads to $c_{1313} = c_{2323}$ and $c_{2313} = 0$, giving $C_{55} = C_{44}$ and $C_{45} = 0$;

– for the moduli with a single index equal to 3, relation [A2.53] is written as:

$$c_{ijk3} = a_i^1 a_j^1 a_k^1 \gamma_{1113} + a_i^2 a_j^2 a_k^2 \gamma_{2223} \qquad [A2.57]$$

hence:

$$c_{22k3} = -\frac{1}{2} a_k^1 \gamma_{1113} + \frac{1}{2} a_k^2 \gamma_{2223} = -c_{11k3} \qquad [A2.58]$$

and:

$$c_{i223} = -\frac{1}{2} a_i^1 \gamma_{1113} + \frac{1}{2} a_i^2 \gamma_{2223} = -c_{i113} \qquad [A2.59]$$

or, in matrix notation:

$$k = 2 \;\Rightarrow\; C_{24} = -C_{14}, \qquad k = 1 \;\Rightarrow\; C_{25} = -C_{15} \qquad [A2.60]$$
$$i = 1 \;\Rightarrow\; C_{46} = -C_{15}, \qquad i = 2 \;\Rightarrow\; C_{24} = -C_{56} \qquad [A2.61]$$

or again:

$$C_{14} = -C_{24} = C_{56} \quad \text{and} \quad C_{25} = -C_{15} = C_{46} \qquad [A2.62]$$

– the moduli without any index equal to 3: c_{1111}, c_{2222}, c_{1112}, c_{2221}, c_{1122} and c_{1212}, are expressed in terms of γ_{1122} and γ_{1212}:

$$c_{ijkl} = \left(a_i^1 a_j^1 a_k^2 a_l^2 + a_i^2 a_j^2 a_k^1 a_l^1\right) \gamma_{1122} + \left(a_i^1 a_j^2 + a_i^2 a_j^1\right)\left(a_k^1 a_l^2 + a_k^2 a_l^1\right) \gamma_{1212} \quad [\text{A2.63}]$$

hence:

$$c_{iiii} = 2\left(a_i^1 a_i^2\right)^2 (\gamma_{1122} + 2\gamma_{1212}) = -\frac{1}{2}(\gamma_{1122} + 2\gamma_{1212}) \quad [\text{A2.64}]$$

that is:

$$C_{11} = C_{22} = -\frac{1}{2}(\gamma_{1122} + 2\gamma_{1212}) \quad [\text{A2.65}]$$

Let us examine the modulus c_{1112}:

$$c_{1112} = a_1^1 a_1^2 \left(a_1^1 a_2^2 + a_1^2 a_2^1\right)(\gamma_{1122} + 2\gamma_{1212}) = 0 \quad [\text{A2.66}]$$

because $a_1^1 a_2^2 + a_1^2 a_2^1 = 0$. In the same way, it can be shown that $c_{2221} = 0$. Consequently: $C_{16} = C_{26} = 0$.

Since there are only two independent constants with no index equal to 3, there exists a relation between c_{1111}, c_{1122} and c_{1212}. Indeed, from the development [A2.63] it follows:

$$c_{1122} = \frac{1}{2}(\gamma_{1122} - 2\gamma_{1212}) = C_{12} \quad \text{and} \quad c_{1212} = -\frac{1}{2}\gamma_{1122} = C_{66} \quad [\text{A2.67}]$$

Considering relation [A2.65], we get:

$$C_{66} = \frac{C_{11} - C_{12}}{2} \quad [\text{A2.68}]$$

The table below summarizes these results:

$$\underline{\underline{C}} = \begin{pmatrix} C_{11} & C_{12} & C_{13} & C_{14} & -C_{25} & 0 \\ C_{12} & C_{22} & C_{13} & -C_{14} & C_{25} & 0 \\ C_{13} & C_{13} & C_{33} & 0 & 0 & 0 \\ C_{14} & -C_{14} & 0 & C_{44} & 0 & C_{25} \\ -C_{25} & C_{25} & 0 & 0 & C_{44} & C_{14} \\ 0 & 0 & 0 & C_{25} & C_{14} & C_{66} \end{pmatrix} \quad \text{trigonal} \quad [\text{A2.69}]$$

In classes 32, 3m and $\bar{3}m$, the binary axes perpendicular to the principal axis impose additional conditions. Taking Ox_1 to be along one of these binary axes, constants c_{ijkl} having an odd number of indices equal to 1 are zero (see the monoclinic system). This reduces the number of independent elastic moduli to 6 for these classes, since $C_{15} = c_{1113} = 0$.

For crystals in the *tetragonal system*, the relations between the moduli with four, three and two indices equal to 3 are the same as those established above, because of the identity of the non-zero components γ_{ijkl} involved in both cases. Constants with a single index equal to 3 are zero, since $\gamma_{1113} = \gamma_{2223} = 0$:

$$C_{14} = C_{24} = C_{15} = C_{25} = C_{46} = C_{56} = 0 \qquad [A2.70]$$

The remaining moduli with no index equal to 3 are expressed as functions of γ_{1111}, γ_{2222}, γ_{1122} and γ_{1212} as:

$$c_{ijkl} = a_i^1 a_j^1 a_k^1 a_l^1 \gamma_{1111} + a_i^2 a_j^2 a_k^2 a_l^2 \gamma_{2222} + c_{ijkl}^{(3)} = 0 \qquad [A2.71]$$

where $c_{ijkl}^{(3)}$ represents the development [A2.63] of the corresponding constant in the case $n = 3$. Taking into account relation [A2.65], it comes:

$$c_{iiii} = \frac{1}{4}(\gamma_{1111} + \gamma_{2222}) - \frac{1}{2}(\gamma_{1122} + 2\gamma_{1212}), \text{ or } C_{11} = C_{22} \qquad [A2.72]$$

Since $c_{1112}^{(3)}$ and $c_{2221}^{(3)}$ are zero:

$$c_{1112} = \frac{i}{4}(\gamma_{1111} - \gamma_{2222}) = -c_{2221}, \qquad \text{hence} \qquad C_{26} = -C_{16} \qquad [A2.73]$$

The four independent moduli without an index equal to 3, C_{11}, C_{12}, C_{16} and C_{66} are added to C_{13}, C_{33} and C_{44}, to form Table [A2.74]. For the classes 422, 4mm, $\bar{4}2m$, $4/mmm$, with the Ox_1 axis parallel to one of the binary axes perpendicular to the principal axis, the constant $c_{1112} = C_{16}$ is zero because it has an odd number of indices equal to 1:

$$\underline{\underline{C}} = \begin{pmatrix} C_{11} & C_{12} & C_{13} & 0 & 0 & C_{16} \\ C_{12} & C_{11} & C_{13} & 0 & 0 & -C_{16} \\ C_{13} & C_{13} & C_{33} & 0 & 0 & 0 \\ 0 & 0 & 0 & C_{44} & 0 & 0 \\ 0 & 0 & 0 & 0 & C_{44} & 0 \\ C_{16} & -C_{16} & 0 & 0 & 0 & C_{66} \end{pmatrix} \qquad \text{tetragonal} \qquad [A2.74]$$

In elasticity, the principal axis of crystals of the *hexagonal system* behaves like a sixfold direct axis, that is, as a binary and a ternary axis combined. Table [A2.75] thus results from the combination of Tables [A2.42] and [A2.69], characteristic of monoclinic and trigonal systems:

$$\underline{\underline{C}} = \begin{pmatrix} C_{11} & C_{12} & C_{13} & 0 & 0 & 0 \\ C_{12} & C_{11} & C_{13} & 0 & 0 & 0 \\ C_{13} & C_{13} & C_{33} & 0 & 0 & 0 \\ 0 & 0 & 0 & C_{44} & 0 & 0 \\ 0 & 0 & 0 & 0 & C_{44} & 0 \\ 0 & 0 & 0 & 0 & 0 & C_{66} \end{pmatrix} \qquad \text{hexagonal} \qquad [A2.75]$$

with $C_{66} = (C_{11} - C_{12})/2$. This tensor is identical to that of a transverse isotropic material. This property is expressed in several ways:

– the planes perpendicular to the principal axis are isotropic with respect to elastic properties;

– all planes passing through the principal axis are equivalent, as are all directions equally inclined on this axis.

The results of the reduction in the number of independent moduli are given in Table 1.5. The relations, indicated by the symbols due to K.S. Van Dyke, apply directly to the stiffness constants C_{IJ}.

Appendix 3

Transport of Energy

The considerations that follow apply to surface and interface waves as well as guided waves. The only condition is the absence of losses, whether by internal dissipation or by radiation toward the core of a material (if it is semi-infinite), or, in the case of a waveguide, across the lateral surfaces that delimit it. However, only guided waves will be discussed in this appendix. Figure A3.1 represents the general structure of a waveguide, infinite in the x_1 direction, laterally limited by a boundary Σ, which carries electric charge with a density σ_e and upon which external mechanical forces, with a density p_i per unit area, are exerted.

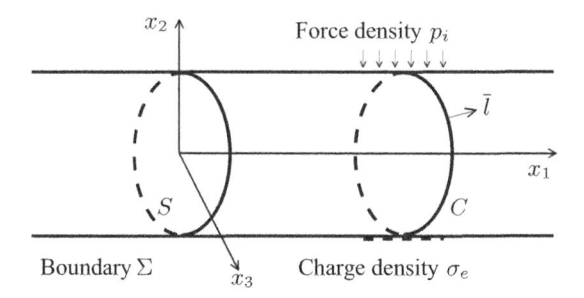

Figure A3.1. *The waves propagate along the x_1 axis of the waveguide. Quantities outside the guide are marked by a "prime"*

A3.1. Energy balance

In sections 1.1.3 and 1.4.1, it was shown that the expression for the power supplied, per unit volume, by mechanical and electrical sources is:

$$p_s(\underline{x},t) = \frac{d(e_k + e_p)}{dt} + \frac{\partial J_k}{\partial x_k} \qquad [A3.1]$$

where e_k and e_p are the kinetic and potential energy densities per unit volume, and where J_k is the Poynting vector [1.221]:

$$J_k = -\sigma_{ik}\dot{u}_i + \Phi\dot{D}_k \qquad [A3.2]$$

In the case of a waveguide, the power $p_i\dot{u}_i$ supplied by the mechanical sources and the power $\Phi\dot{\sigma}_e$ supplied by electrical sources at the surface are involved in the energy balance. With S denoting the normal section of the guide, the kinetic energy density (E_k) and the potential energy density (E_p) per unity length in the x_1 direction are given as:

$$E_k(x_1,t) = \iint_S e_k(\underline{x},t)\,dx_2\,dx_3 \quad \text{and} \quad E_p(x_1,t) = \iint_S e_p(\underline{x},t)\,dx_2\,dx_3 \qquad [A3.3]$$

The integration of [A3.1] over S leads to the expression for the power supplied P_s per unit length of the guide:

$$P_s(x_1,t) = \frac{d(E_k+E_p)}{dt} + \int_S \frac{\partial J_k}{\partial x_k}\,dx_2\,dx_3 + \int_C (p_i\dot{u}_i + \Phi\dot{\sigma}_e)\,dC \qquad [A3.4]$$

The last integral represents the power supplied by the mechanical and electrical surface sources, and dC is the length element along the contour C of the normal section S of the guide. By introducing the power P crossing the section of the guide:

$$P(x_1,t) = \int_S J_1(\underline{x},t)\,dx_2\,dx_3 \qquad [A3.5]$$

and by applying the two-dimensional divergence theorem, the first integral in the right-hand side of [A3.4] becomes:

$$\int_S \frac{\partial J_k}{\partial x_k}\,dx_2\,dx_3 = \frac{\partial P}{\partial x_1} + \iint_S \left[\frac{\partial J_2}{\partial x_2} + \frac{\partial J_3}{\partial x_3}\right]\,dx_2\,dx_3 = \frac{\partial P}{\partial x_1} + \int_C J_n\,dC \qquad [A3.6]$$

J_n is the component of the Poynting vector normal to the contour C. Taking into account relation [A3.2], it is expressed by:

$$J_n = J_k l_k = -T_i \dot{u}_i + \Phi \dot{D}_n \qquad [A3.7]$$

where $T_i = \sigma_{ik} l_k$ is the mechanical traction and $D_n = D_k l_k$ the normal component of the electric induction. By substituting into equation [A3.4], we get:

$$P_s = \frac{d(E_k+E_p)}{dt} + \frac{\partial P}{\partial x_1} + \int_C \left[(p_i - T_i)\dot{u}_i + \Phi(\dot{\sigma}_e + \dot{D}_n)\right]\,dC \qquad [A3.8]$$

According to the mechanical boundary conditions ($T_i - T'_i = p_i$ and $u'_i = u_i$) and electrical boundary conditions ($\Phi' = \Phi$ and $D'_n - D_n = \sigma_e$), the integral on the contour C is expressed in terms of quantities related to the external medium (marked by a prime):

$$P_s = \frac{d(E_k + E_p)}{dt} + \frac{\partial P}{\partial x_1} - \int_C T'_i \dot{u}'_i \, dC + \int_C \Phi' \dot{D}'_n \, dC \qquad [A3.9]$$

The first integral, which expresses the radiation of mechanical energy into the external medium, is zero if one of the two following conditions is satisfied:

– the surface Σ is free, so that $T'_i = 0$;

– the external medium is infinitely rigid, so that $u'_i = 0$.

The second integral, which expresses the radiation of electrical energy toward the exterior, is also zero if either:

– the surface Σ is metallized (short-circuit condition), so that: $\Phi' = 0$;

– the external medium has zero permittivity (open-circuit condition), so that $D'_n = 0$.

The law of conservation of energy is thus expressed for any section of abscissa x_1, and at any time t, by:

$$P_s(x_1, t) = \frac{d(E_k + E_p)}{dt} + \frac{\partial P}{\partial x_1} \qquad [A3.10]$$

These *non-dissipative boundary conditions* are very important in practice, because in a periodic regime they ensure the conservation of the average power transported by the wave in regions of the guide free of source.

In the following, we will assume that the sources are only of electrical origin. In the case of a waveguide made up of an insulating material, the only charges are on the surface. Taking into account the conservation of electric charge:

$$j_n - j'_n = \dot{\sigma}_e \qquad [A3.11]$$

where j_n and j'_n are the normal current densities in each medium. The power P_s supplied by the surface electric sources per unit length along x_1 is:

$$P_s = \int_C \Phi \dot{\sigma}_e \, dC = \int_C \Phi(j_n - j'_n) \, dC \qquad [A3.12]$$

Since the potential Φ on the surface Σ is constant on the contour C at position x_1, and since the conduction current density j_n is zero inside the guide, assumed to be insulating:

$$P_s = \int_C \Phi(-j'_n) \, dC = \Phi j \qquad \text{where} \qquad j(x_1, t) = -\int_C j'_n \, dC \qquad [A3.13]$$

is the intensity of the input current per unit length (the minus sign comes from the orientation of the normal toward the exterior).

A3.2. Harmonic case

The behavior of the medium being linear, by hypothesis, all the quantities vary sinusoidally with the same angular frequency ω, it is convenient to use complex notations. The power supplied by the electrical sources (per unit volume) is the real part of the complex power $\overline{p}_s$ and the complex Poynting vector $\overline{J}_k$ is written as:

$$\overline{J}_k = \frac{1}{2}\left(-\sigma_{ik}\dot{u}_i^\star + \Phi \dot{D}_k^\star\right) \qquad [\text{A3.14}]$$

Knowing that in the harmonic case ($e^{i\omega t}$):

$$\ddot{u}_i = i\omega \dot{u}_i, \qquad \dot{S}_{ik}^\star = -i\omega^\star S_{ik}^\star \quad \text{and} \quad \dot{D}_j^\star = -i\omega^\star D_j^\star \qquad [\text{A3.15}]$$

after simplification by the factor $e^{i(\omega-\omega^\star)t}$, the relation [1.220] is written as:

$$\overline{p}_s = \frac{i\omega}{2}\rho \dot{u}_i \dot{u}_i^\star - \frac{i\omega^\star}{2}\left(\sigma_{ik}S_{ik}^\star + E_j D_j^\star\right) + \frac{\partial \overline{J}_k}{\partial x_k} \qquad [\text{A3.16}]$$

The quantity $\rho \dot{u}_i \dot{u}_i^\star$ is equal to four times the average value of the kinetic energy density:

$$\langle e_k \rangle = \frac{1}{2}\operatorname{Re}\left[\frac{1}{2}\rho \dot{u}_i \dot{u}_i^\star\right] = \frac{1}{4}\rho \dot{u}_i \dot{u}_i^\star \qquad [\text{A3.17}]$$

If the medium is linear and non-dissipative, its constants, and consequently the quantity $\sigma_{ik}S_{ik}^\star + E_j D_j^\star$, are real. The latter is equal to four times the average potential energy density:

$$\langle e_p \rangle = \frac{1}{2}\operatorname{Re}\left[\frac{1}{2}\left(\sigma_{ik}S_{ik}^\star + E_j D_j^\star\right)\right] = \frac{1}{4}\left(\sigma_{ik}S_{ik}^\star + E_j D_j^\star\right) \qquad [\text{A3.18}]$$

The complex Poynting theorem takes the form:

$$\overline{p}_s = 2i\left[\omega \langle e_k \rangle - \omega^\star \langle e_p \rangle\right] + \frac{\partial \overline{J}_k}{\partial x_k} \qquad [\text{A3.19}]$$

or again, writing $\omega = \omega' + i\omega''$:

$$\overline{p}_s = 2i\omega'\langle e_k - e_p \rangle - 2\omega''\langle e_k + e_p \rangle + \frac{\partial \overline{J}_k}{\partial x_k} \qquad [\text{A3.20}]$$

By integrating over the normal section S, let us express the complex power $\overline{P}_s$, supplied by the electrical sources per unit length of the guide. Since the electric charges are distributed on the surface, according to [A3.13], the integral on the left-hand side is equal to $\frac{1}{2}\Phi j^\star$. If the lateral surface Σ does not give rise to any dissipation, the flux of the complex Poynting vector is expressed, as in the previous section, as a function of the complex power $\overline{P}$ transported along the axis of the guide, so that:

$$\overline{P}_s = \frac{1}{2}\Phi j^\star = 2i\omega'\langle E_k - E_p\rangle - 2\omega''\langle E_k + E_p\rangle + \frac{\partial \overline{P}}{\partial x_1} \qquad [\text{A3.21}]$$

where E_k and E_p are, respectively, the kinetic and potential energy densities per unit length.

All quantities associated with the eigenmodes of a homogeneous guide vary as $e^{i(\omega t - kx_1)}$, so that if we take $k = k' + ik''$, the products of the type $ab^\star$ are written as:

$$ab^\star = a_0 b_0^\star e^{i(k^\star - k)x_1} = a_0 b_0^\star e^{2k'' x_1} \qquad [\text{A3.22}]$$

thus:

$$\Phi j^\star = \Phi_0 j_0^\star e^{2k'' x_1} \quad \text{and} \quad \overline{P} = \overline{P}_0 e^{2k'' x_1} \qquad [\text{A3.23}]$$

After simplification by the factor $e^{2k'' x_1}$, relation [A3.21] becomes:

$$\frac{1}{2}\Phi_0 j_0^\star = 2i\omega'\langle E_k - E_p\rangle - 2\omega''\langle E_k + E_p\rangle + 2k''\overline{P}_0 \qquad [\text{A3.24}]$$

This is the expression of the complex power supplied by the electrical sources depending on the sum and the difference of the kinetic and potential energy densities per unit length of the guide and on the complex Poynting vector.

A3.3. Susceptance and free modes

In a steady-state regime and in the absence of dissipation, ω and k are real. The previous expression shows that the complex power is purely imaginary, because $\omega'' = k'' = 0$. The potential Φ_0 and the current density j_0 are thus in phase quadrature. Let:

$$j_0 = iB(\omega, k)\Phi_0 \qquad [\text{A3.25}]$$

be the most general linear relation between these quantities. $B(\omega, k)$ is called the susceptance per unit length of the guide. This is a function with a real value when ω and k are real. If ω'' and k'' are no longer zero, but infinitely small: $\omega'' = \delta\omega$ and $k'' = \delta k$, equation [A3.24] implies:

$$\left[iB(\omega, k) - \delta\omega\frac{\partial B}{\partial \omega} - \delta k\frac{\partial B}{\partial k}\right]^\star |\Phi_0|^2 = 4i\omega\langle E_k - E_p\rangle - 4\delta\omega\langle E_k + E_p\rangle$$
$$+ 4\delta k\left(\langle P\rangle + i\,\text{Im}[\overline{P}_0]\right)$$

$$[\text{A3.26}]$$

where the quantity $\langle P \rangle = \text{Re}[\overline{P_0}]$ represents the average values of the power crossing the section of the guide. We deduce from the equality of the real parts:

$$\langle E_k + E_p \rangle = \frac{1}{4}\frac{\partial B}{\partial \omega}|\Phi_0|^2 \quad \text{and} \quad \langle P \rangle = -\frac{1}{4}\frac{\partial B}{\partial k}|\Phi_0|^2 \qquad [A3.27]$$

and from the equality of the imaginary parts, with δk being infinitely small:

$$\langle E_k - E_p \rangle = -\frac{1}{4}\frac{B}{\omega}|\Phi_0|^2 \qquad [A3.28]$$

Finally, in the harmonic case, the average values of the kinetic and potential energies are given by:

$$\langle E_k \rangle = \frac{1}{8}\left(\frac{\partial B}{\partial \omega} - \frac{B}{\omega}\right)|\Phi_0|^2 \quad \text{and} \quad \langle E_p \rangle = \frac{1}{8}\left(\frac{\partial B}{\partial \omega} + \frac{B}{\omega}\right)|\Phi_0|^2 \qquad [A3.29]$$

Since the kinetic and potential energies, E_k and E_p, are positive, the inequality:

$$\frac{\partial B}{\partial \omega} \geq \left|\frac{B}{\omega}\right| \qquad [A3.30]$$

equivalent to the Foster's inequality for electrical impedances, shows that the susceptance of the guide is always an increasing function of frequency (Ramo et al. 1984).

In the absence of any source ($j = 0$), the eigenmodes of the guide, also called *free modes*, satisfy the equation:

$$B(\omega, k) = 0 \qquad [A3.31]$$

which expresses the dispersion relation between ω and k. The group velocity $d\omega/dk$ is obtained by differentiating this relation:

$$dB = \frac{\partial B}{\partial \omega}d\omega + \frac{\partial B}{\partial k}dk = 0 \quad \text{hence} \quad V_g = \frac{d\omega}{dk} = -\frac{\partial B}{\partial k}\bigg/\frac{\partial B}{\partial \omega} \qquad [A3.32]$$

According to relations [A3.27], it is equal to the energy velocity:

$$V_e = \frac{\langle P \rangle}{\langle E_k + E_p \rangle} = -\frac{\partial B}{\partial k}\bigg/\frac{\partial B}{\partial \omega} = V_g \qquad [A3.33]$$

For these free modes, the average values of kinetic and potential energies per unit length are equal (relation [A3.29] with $B = 0$):

$$\langle E_k \rangle = \langle E_p \rangle = \frac{1}{8}\frac{\partial B}{\partial \omega}|\Phi_0|^2 \qquad [A3.34]$$

Let us note that this property is not true for the volume energy densities. The equality $e_k = e_p$ is only verified for plane waves (section 1.3.4). In a waveguide, these densities vary from one point to another in the section S; the equality is found once again after integration over the cross-section of the guide. It follows that the average power transported along the guide axis:

$$\langle P \rangle = V_g \langle E_k + E_p \rangle = 2V_g \langle E_k \rangle \qquad [\text{A3.35}]$$

can be expressed in terms of only the mechanical displacement u_i of the wave:

$$\langle P \rangle = V_g \int_S \frac{1}{2} \rho |\dot{u}_i|^2 \, \mathrm{d}x_2 \, \mathrm{d}x_3 = \frac{1}{2} \rho \omega^2 V_g \int_S |u_i|^2 \, \mathrm{d}x_2 \, \mathrm{d}x_3 \qquad [\text{A3.36}]$$

In summary, in the absence of dissipation:

– the energy transport velocity is equal to the group velocity;

– the average kinetic energy, per unit length, is equal to the average potential energy;

– the calculation of the average energy transported by the guided wave does not require the computation of mechanical stresses.

REMARK.– In the case of non-dispersive surface or interface waves, the susceptance $B(\omega, k)$ is only a function of the ratio ω/k and relation [A3.31] is a characteristic equation that determines the phase velocity $V = \omega/k$ of the wave. Formula [A3.36] still applies by replacing V_g with V.

References

Chapter 1

Abramowitz, M. and Stegun, I.A. (1965). *Handbook of Mathematical Functions, with Formulas, Graphs, and Mathematical Tables*. Dover Publications Inc., Mineola.

Arlt, G. and Quadflieg, P. (1968). Piezoelectricity in iii-v compounds with a phenomenological analysis of the piezoelectric effect. *Phys. Status Solidi B*, 25(1), 323–330.

Bateman, T.B. (1962). Elastic moduli of single-crystal zinc oxide. *J. Appl. Phys.*, 33(11), 3309–3312.

Bateman, T.B., McSkimin, H.J., Whelan, J.M. (1959). Elastic moduli of single-crystal gallium arsenide. *J. Appl. Phys.*, 30(4), 544–545.

Bechmann, R. (1958). Elastic and piezoelectric constants of alpha-quartz. *Phys. Rev.*, 110, 1060–1061.

Briggs, A. (1995). *Advances in Acoustic Microscopy*, Volume 1. Springer, Berlin.

Bruneau, M. and Scelo, T. (2006). *Fundamentals of Acoustics*. ISTE Ltd, London.

Chang, Y.A. and Himmel, L. (1966). Temperature dependence of the elastic constants of Cu, Ag, and Au above room temperature. *J. Appl. Phys.*, 37(9), 3567–3572.

Christensen, R.M. (2003). *Theory of Viscoelasticity*, 2nd edition. Dover Publications Inc., Mineola.

Clorennec, D., Prada, C., Royer, D. (2007). Local and noncontact measurements of bulk acoustic wave velocities in thin isotropic plates and shells using zero group velocity Lamb modes. *J. Appl. Phys.*, 101(3), 034908.

Coquin, G.A., Pinnow, D.A., Warner, A.W. (1971). Physical properties of lead molybdate relevant to acousto-optic device applications. *J. Appl. Phys.*, 42(6), 2162–2168.

Fisher, E.S. and Renken, C.J. (1964). Single-crystal elastic moduli and the hcp → bcc transformation in Ti, Zr, and Hf. *Phys. Rev.*, 135, A482–A494.

Hall, J.J. (1967). Electronic effects in the elastic constants of n-type silicon. *Phys. Rev.*, 161, 756–761.

Higuet, J., Valier-Brasier, T., Dehoux, T., Audoin, B. (2011). Beam distortion detection and deflectometry measurements of gigahertz surface acoustic waves. *Rev. Sci. Instrum.*, 82(11), 114905.

Jaffe, H. and Berlincourt, D.A. (1965). Piezoelectric transducer materials. *Proceedings of the IEEE*, 53(10), 1372–1386.

Kinsler, L.E., Frey, A.R., Coppens, A.B., Sanders, J.V. (2009). *Fundamentals of Acoustics*, 4th editions. John Wiley & Sons, New York.

Lanceleur, P., Ribeiro, H., De Belleval, J.-F. (1993). The use of inhomogeneous waves in the reflection-transmission problem at a plane interface between two anisotropic media. *J. Acoust. Soc. Am.*, 93(4), 1882–1892.

Nagy, P.B. and Nayfeh, A.H. (1996). Viscosity-induced attenuation of longitudinal guided waves in fluid-loaded rods. *J. Acoust. Soc. Am.*, 100(3), 1501–1508.

Norris, A.N. (2017). An inequality for longitudinal and transverse wave attenuation coefficients. *J. Acoust. Soc. Am.*, 141(1), 475–479.

Ohmachi, Y. and Uchida, N. (1970). Temperature dependence of elastic, dielectric, and piezoelectric constants in TeO_2 single crystals. *J. Appl. Phys.*, 41(6), 2307–2311.

Raetz, S., Laurent, J., Dehoux, T., Royer, D., Audoin, B., Prada, C. (2015). Effect of refracted light distribution on the photoelastic generation of zero-group velocity Lamb modes in optically low-absorbing plates. *J. Acoust. Soc. Am.*, 138(6), 3522–3530.

Royer, D. and Dieulesaint, E. (1996). *Ondes élastiques dans les solides : propagation libre et guidée*, Volume 1. Masson, Paris.

Salençon, J. (1988). *Mècanique des milieux continus II – élasticité-milieux curvilignes*. Ellipses, Paris.

Slobodnik, A.J. and Sethares, J.C. (1972). Elastic, piezoelectric, and dielectric constants of $Bi_{12}GeO_{20}$. *J. Appl. Phys.*, 43(1), 247–248.

Szabo, T.L. and Wu, J. (2000). A model for longitudinal and shear wave propagation in viscoelastic media. *J. Acoust. Soc. Am.*, 107(5), 2437–2446.

Verma, R.K. (1960). Elasticity of some high-density crystals. *J. Geophys. Res.*, 65(2), 757–766.

Warner, A.W., Onoe, M., Coquin, G.A. (1967). Determination of elastic and piezoelectric constants for crystals in class (3m). *J. Acoust. Soc. Am.*, 42(6), 1223–1231.

Williams, M.L., Landel, R.F., Ferry, J.D. (1955). The temperature dependence of relaxation mechanisms in amorphous polymers and other glass-forming liquids. *J. Am. Chem. Soc.*, 77(14), 3701–3707.

Chapter 2

Auld, B.A. (1990). *Acoustic Fields and Waves in Solids*, Volume 1. Krieger Publishing Company, Malabar.

Breazeale, M.A., Adler, L., Scott, G.W. (1977). Interaction of ultrasonic waves incident at the Rayleigh angle onto a liquid-solid interface. *J. Appl. Phys.*, 48(2), 530–537.

Briggs, A. (1995). *Advances in Acoustic Microscopy*, Volume 1. Springer, Berlin.

Castagnède, B., Jenkins, J.T., Sachse, W., Baste, S. (1990). Optimal determination of the elastic constants of composite materials from ultrasonic wave-speed measurements. *J. Appl. Phys.*, 67(6), 2753–2761.

Lefebvre, G., Wunenburger, R., Valier-Brasier, T. (2018). Ultrasonic rheology of visco-elastic materials using shear and longitudinal waves. *Appl. Phys. Lett.*, 112, 241906.

Royer, D. and Dieulesaint, E. (1996). *Ondes élastiques dans les solides : propagation libre et guidée*, Volume 1. Masson, Paris.

Sachse, W. and Pao, Y.-H. (1978). On the determination of phase and group velocities of dispersive waves in solids. *J. Appl. Phys.*, 49(8), 4320–4327.

Schoch, A. (1950). Schallreflexion, Schallbrechung und Schallbeugung. *Ergeb. Exakt. Naturw.*, 23, 127–234.

Simon, A., Lefebvre, G., Valier-Brasier, T., Wunenburger, R. (2019). Viscoelastic shear modulus measurement of thin materials by interferometry at ultrasonic frequencies. *J. Acoust. Soc. Am.*, 146(5), 3131–3140.

Chapter 3

Adler, E.L. (1998). SAW, PSEUDOSAW, and HVPSEUDOSAW in Langasite. *Proc. IEEE Ultrasonics Symp.*, 1, 307–310.

Barnett, D.M. and Lothe, J. (1974). Consideration of the existence of surface wave (Rayleigh wave) solutions in anisotropic elastic crystals. *J. Phys. F: Met. Phys.*, 4(5), 671–686.

Bleustein, J.L. (1968). A new surface wave in piezoelectric materials. *Appl. Phys. Lett.*, 13(12), 412–413.

Campbell, C. (1989). *Surface Acoustic Wave Devices and their Signal Processing Applications*. Academic Press, Cambridge.

Cegla, F.B., Cawley, P., Lowe, M.J.S. (2005). Material property measurement using the Quasi-Scholte mode – A waveguide sensor. *J. Acoust. Soc. Am.*, 117(3), 1098–1107.

Chadwick, P. and Borejko, P. (1994). Existence and uniqueness of Stoneley waves. *Geophys. J. Int.*, 118(2), 279–284.

Dransfeld, K. and Salzmann, E. (1970). Excitation, detection, and attenuation of high-frequency elastic surface waves. *Physical Acoustics*, 7, 219–272.

Farnell, G.W. (1970). Properties of elastic surface waves. *Physical Acoustics*, 6, 109–166.

Favretto-Cristini, N., Komatitsch, D., Carcione, J.M., Cavallini, F. (2011). Elastic surface waves in crystals. Part 1: Review of the physics. *Ultrasonics*, 51(6), 653–660.

Feldmann, M. and Hénaff, J. (1977). Propagation des ondes élastiques de surface. *Rev. Phys. Appl.*, 12, 1775–1788.

Feldmann, M. and Hénaff, J. (1986). *Traitement du signal par ondes èlastiques de surface*. Masson, Paris.

Fujishima, S. (1990). Piezoelectric devices for frequency control and selection in Japan. *IEEE Ultrasonics Symp. Proc.*, 87–94.

Gedge, M. and Hill, M. (2012). Acoustofluidics 17: Theory and applications of surface acoustic wave devices for particle manipulation. *Lab Chip*, 12, 2998–3007.

Glorieux, C., Van de Rostyne, K., Nelson, K., Gao, W., Lauriks, W., Thoen, J. (2001). On the character of acoustic waves at the interface between hard and soft solids and liquids. *J. Acoust. Soc. Am.*, 110(3), 1299–1306.

Gulyaev, Y.V. (1969). Electroacoustic surface waves in solids. *J.E.T.P. Lett.*, 9, 37–38.

Pratt, R.G., Simpson, G., Crossley, W.A. (1972). Acoustic-surface-wave properties of $Bi_{12}GeO_{20}$. *Electron. Lett.*, 8(5), 127–128.

Rayleigh, L. (1885). On waves propagated along the plane surface of an elastic solid. *Proc. Lond. Math. Soc.*, s1–s17(1), 4–11.

Royer, D. and Dieulesaint, E. (1984). Rayleigh wave velocity and displacement in orthorhombic, tetragonal, hexagonal, and cubic crystals. *J. Acoust. Soc. Am.*, 76(5), 1438–1444.

Slobodnik, A.J., Delmonico, R.T., Conway, E.D. (1973). *Microwave Acoustics Handbook*. Air Force Cambridge Research Laboratories, L.G. Hanscom Field, Massport, Bedford.

Stoneley, R. (1924). Elastic waves at the surface of separation of two solids. *Proc. Roy. Soc. A*, 106(738), 416–428.

Stoneley, R. (1955). The propagation of surface elastic waves in a cubic crystal. *Proc. Roy. Soc. A*, 232(1191), 447–458.

Valier-Brasier, T., Dehoux, T., Audoin, B. (2012). Scaled behavior of interface waves at an imperfect solid-solid interface. *J. Appl. Phys.*, 112(2), 024904.

Viktorov, I.A. (1967). *Rayleigh and Lamb Waves: Physical Theory and Applications.* Plenum Press, New York.

Chapter 4

Aubert, V., Wunenburger, R., Valier-Brasier, T., Rabaud, D., Kleman, J.-P., Poulain, C. (2016). A simple acoustofluidic chip for microscale manipulation using evanescent scholte waves. *Lab Chip*, 16(13), 2532–2539.

Cès, M., Clorennec, D., Royer, D., Prada, C. (2011). Edge resonance and zero group velocity Lamb modes in a free elastic plate. *J. Acoust. Soc. Am.*, 130(2), 689–694.

Fraser, W.B. (1976). Orthogonality relation for the Rayleigh–Lamb modes of vibration of a plate. *J. Acoust. Soc. Am.*, 59(1), 215–216.

Gazis, D.C. (1959). Three-dimensional investigation of the propagation of waves in hollow circular cylinders I. Analytical foundation. *J. Acoust. Soc. Am.*, 31(5), 568–573.

Lamb, H. (1917). On waves in an elastic plate. *Proc. Roy. Soc. A*, 93, 114–128.

Laurent, J., Royer, D., Hussain, T., Ahmad, F., Prada, C. (2015). Laser induced zero-group velocity resonances in transversely isotropic cylinder. *J. Acoust. Soc. Am.*, 137, 3325–3334.

Love, A.E.H. (1911). *Some Problems of Geodynamics.* Cambridge University Press, Cambridge.

Meeker, T.R. and Meitzler, A.H. (1964). Guided wave propagation in elongated cylinders and plates. *Physical Acoustics*, 1A, 111–167.

Meitzler, A.H. (1965). Backward-wave transmission of stress pulses in elastic cylinders and plates. *J. Acoust. Soc. Am.*, 38(5), 835–842.

Mindlin, R.D. and Medick, M.A. (1959). Extensional vibrations of plates. *J. Appl. Mech.*, 26, 561–569.

Nicholson, N.C. and McDicken, W.N. (1991). Mode propagation of ultrasound in hollow waveguides. *Ultrasonics*, 29(5), 411–416.

Pagneux, V. (2006). Revisiting the edge resonance for Lamb waves in a semi-infinite elastic plate. *J. Acoust. Soc. Am.*, 120(2), 649–656.

Pavlakovic, B., Lowe, M., Alleyne, D., Cawley, P. (1997). Disperse: A general purpose program for creating dispersion curves. *Rev. of Progress in Quantitative Nondestructive Evaluation*, 16A, 185–192.

Prada, C., Clorennec, D., Royer, D. (2008). Local vibration of an elastic plate and zero-group velocity Lamb modes. *J. Acoust. Soc. Am.*, 124(1), 203–212.

Rayleigh, L. (1888). On the free vibrations of an infinite plate of homogeneous isotropic elastic matter. *Proc. Lond. Math. Soc.*, s1–20(1), 225–237.

Royer, D., Levin, L., Legras, O. (1993). A liquid level sensor using the absorption of guided acoustic waves. *IEEE Trans. Ultrason. Ferr.*, 40(4), 418–421.

Royer, D., Clorennec, D., Prada, C. (2009). Lamb mode spectra versus the Poisson ratio in a free isotropic elastic plate. *J. Acoust. Soc. Am.*, 125(6), 3683–3687.

Tolstoy, I. and Usdin, E. (1957). Wave propagation in elastic plates: Low and high mode dispersion. *J. Acoust. Soc. Am.*, 29(1), 37–42.

Appendixes

Ramo, S., Whinnery, J.R., Van Duzer, T. (1984). *Fields and Waves in Communication Electronics.* John Wiley & Sons, New York.

General bibliography

Achenbach, J. (2012). *Wave Propagation in Elastic Solids.* North Holland, New York.

Aki, K. and Richards, P.G. (2002). *Quantitative Seismology.* University Science Books, Herndon.

Auld, B.A. (1990). *Acoustic Fields and Waves in Solids*, Volume 1. R.E. Krieger Publishing Compagny, Malabar.

Brekhovskikh, L.M. and Godin, O.A. (1998). *Acoustics of Layered Media I: Plane and Quasi-Plane Waves*, 2nd edition. Springer, Berlin.

Bruneau, M. and Potel, C. (2009). *Materials and Acoustics Handbook.* ISTE Ltd, London and John Wiley & Sons, New York.

Campbell, C. (1989). *Surface Acoustic Waves Devices and their Signal Processing Applications.* Academic Press, Boston.

Cheeke, J.D.N. (2002). *Fundamentals and Applications of Ultrasonic Waves.* CRC Press, Boca Raton.

Elmore, W.C. and Heald, M.A. (1985). *Physics of Waves.* Dover Books, New York.

Gusev, V. and Karabutov, A. (1993). *Laser Optoacoustics.* AIP Press, Maryland.

Hutchins, D.A. and Hayward, G. (1990). Radiated fields of ultrasonic transducers. *Physical Acoustics*, 19, 1–80.

Kennett, B. (1983). *Seismic Wave Propagation in Stratified Media.* Cambridge University Press, Cambridge.

Kino, G.S. (1987). *Acoustic Waves: Devices, Imaging, and Analog Signal Processing.* Prentice-Hall, Upper Saddle River.

Martin, P.A. (2006). *Multiple Scattering–Interaction of Time-Harmonic Waves with N Obstacles.* Cambridge University Press, Cambridge.

Miklowitz, J. (2012). *The Theory of Elastic Waves and Waveguides*. North-Holland, New York.

Mindlin, R.D. and Yang, J. (2006). *An Introduction to the Mathematical Theory of Vibrations of Elastic Plates*. World Scientific, Singapore.

Morgan, D.P. (1991). *Surface-Wave Devices for Signal Processing*. Elsevier, New York.

Nayfeh, A.H. (1995). *Wave Propagation in Layered Anisotropic Media*. North-Holland, New York.

Ristic, V.M. (1983). *Principles of Acoustic Devices*. John Wiley & Sons, New York.

Rose, J.L. (2004). *Ultrasonic Waves in Solid Media*. Cambridge University Press, Cambridge.

Rose, J.L. (2014). *Ultrasonic Guided Waves in Solid Media*. Cambridge University Press, Cambridge.

Royer, D. and Dieulesaint, E. (1996). *Ondes élastiques dans les solides : propagation libre et guidée*, Volume 1. Masson, Paris.

Royer, D. and Dieulesaint, E. (1999). *Ondes élastiques dans les solides : génération, interaction acousto-optique, applications*, Volume 2. Masson, Paris.

Salençon, J. (1988). *Mécanique des milieux continus II – élasticité-milieux curvilignes*. Ellipses, Paris.

Index

A

acoustic
    impedance, 27, 94
    instantaneous power, 10
    intensity, 26, 47
anisotropy factor, 52, 146
attenuation by
    radiation, 150
        of Lamb wave, 224
        of Rayleigh wave, 168
    viscoelastic dissipation, 74, 114

B

Bessel functions, 237
Bleustein–Gulyaev wave, 173
bulk
    modulus, 16, 20, 36, 76
    wave
        polarization, 25, 29
        velocity, 20

C

Cauchy stress tensor, 5
centro-symmetric classes, 59
Christoffel
    equation, 41, 117
    tensor, 41, 117, 122
compliance
    constants, 38
    tensor, 39

composite material, 37, 40, 232
compressibility coefficient, 16
compressional
    modes (*see also* dispersion curves), 239, 246
    wave, 21
conservation of
    angular momentum, 8
    energy, 11, 55
    linear momentum, 7
    matter, 4
constitutive equation
    elasticity (anisotropic solid), 41
    elasticity (isotropic solid), 14
    piezoelectricity, 54
    viscoelasticity, 71
    viscous fluid, 80
continuity equation, 79
convergent wave, 29, 31
critical angle, 89, 94, 109, 127
crystallographic axes, 43, 52, 151, 259
cut-off frequency, 183, 206
cylindrical
    coordinates, 27, 235, 247
    waveguide, 235
    waves, 27

D

d'Alembert equation, 18
dielectric
    constants, 65

tensor, 61
dilatation, 9
dilatational wave, 31
dispersion curves
   compressional modes, 241, 242
   flexural modes, 243
   Lamb waves, 202, 213
   Love waves, 194
   TH guided waves, 191
   torsional modes, 245
divergent wave, 29–31

E

edge wave, 217
elastic constant, 12, 15, 36
electromechanical coupling coefficient
   bulk wave, 67
   surface wave, 157
energy
   flux, 91
   internal, 13, 55
   kinetic, 10, 46
   potential, 11, 14, 46, 55
   velocity, 46, 151, 161, 234, 284
enthalpy, 55, 56
entropy, 13, 55, 57
evanescent wave, 87, 89, 92, 104, 118, 122, 165, 190, 228

F, G

Fabry–Pérot interferences, 96
flexural modes, 243, 246
free energy, 13
Green's theorem, 4
group velocity, 45, 181, 186, 188, 203, 234

H, I

Hankel functions, 29
Helmholtz
   decomposition, 17, 27, 31
   equation, 19
Hermannn–Mauguin notation, 263
Hooke's law, 12, 33, 74
   generalized, 71
impedance matching, 96

inhomogeneous wave, 118, 122, 129, 228
interface
   solid-fluid, 105, 164
   solid-solid, 127, 169
   solid-vacuum, 100, 132
   stiffness, 84, 95, 100, 173
Internal energy, 13, 55

K, L

Kelvin–Voigt viscoelastic model, 72
kinetic energy, 10, 46
Lamb waves, 196
Lamé
   constants, 15
   modes, 211, 214
Lamé–Navier equation, 17
Levi–Civita tensor, 8
longitudinal wave, 19
Love wave, 190

M, N

Maxwell
   relations, 12
   viscoelastic model, 73
mechanical traction, 4
Miller indices, 255
Neper (unity), 76
Newtonian fluid, 78

P

phase velocity, 41, 203
piezoelectric
   constants, 65
   effect, 54
   tensor, 61
plane wave, 21
plate
   free, 190, 196
   immersed, 113, 221
Pochhammer-Chree equation, 240
Poisson's ratio, 16, 19, 36, 211
potential energy, 11, 14, 46, 55
Poynting
   theorem (complex), 282

theorem (real), 11
  vector, 11, 91, 121, 235, 280
pseudo-surface acoustic wave, 147

Q, R

quarter wavelength plate, 99
quasi-longitudinal wave, 41, 127, 234
quasi-transverse wave, 41, 127, 147, 234
Rayleigh
  equation, 135
  generalized wave, 141
  leaky wave, 113, 168, 169
  wave, 127, 132
Rayleigh–Lamb equation, 200
resonance
  edge, 217
  thickness-shear, 207
  thickness-stretch, 207

S

sagittal plane, 132, 141, 143, 146, 153
scalar potential, 18, 28, 31, 63, 133, 197, 236
Scholte wave, 166
slowness surface, 48, 67, 69, 129, 150, 151, 234, 235
Snell–Descartes law, 85, 97, 117
spherical
  coordinates, 31, 249
  wave, 31
stiffness
  constants, 17, 40
  tensor, 12, 38
Stoneley wave, 126, 169, 170
strain tensor, 9
stress
  definition, 12
  tensor, 12, 15
symmetry
  axes, 258
  cubic, 51, 66, 147, 150, 151
  monoclinic, 33, 121, 228
  orthorhombic, 34, 231
  orthotropic, 127, 143
  tetragonal, 276

transverse isotropic, 176, 233
trigonal, 70

T

tensor
  cubic system, 36, 273
  hexagonal system, 277
  isotropic material, 36
  monoclinic system, 34, 272
  orthorhombic system, 35, 273
  tetragonal system, 277
  transverse isotropic system, 35
TH guided wave, 190
torsional modes, 244
transverse wave, 19

V

vector potential, 18, 31, 133, 197, 237
velocity
  bar, 219
  bulk waves, 18, 19
  plate, 204
  Rayleigh wave, 135
Viktorov's formula, 135
viscosity, 78, 83
Voigt's notation, 32

W

wave
  number, 22
  packet, 185
  plane, 23
  surface, 48
  vector, 23
waveguide
  cylindrical, 239
  elementary, 181
  planar, 190, 196
  tubular, 246

Y, Z

Young's modulus, 16, 21, 36
ZGV (zero group velocity) modes, 209

Other titles from

iSTE

in

Waves

2021

DAHOO Pierre-Richard, LAKHLIFI Azzedine
Infrared Spectroscopy of Symmetric and Spherical Top Molecules for Space Observation 1 (Infared Spectroscopy Set – Volume 3)

Infrared Spectroscopy of Symmetric and Spherical Top Molecules for Space Observation 2 (Infrared Spectroscopy Set – Volume 4)

FAVENNEC Pierre-Noël
Electromagnetic Waves 1: Maxwell's Equations, Wave Propagation
Electromagnetic Waves 2: Antennas

LACAZE Pierre Camille, LACROIX Jean-Christophe
Nanotechnology and Nanomaterials for Energy

RÉVEILLAC Jean-Michel
Recording and Voice Processing 1: History and Generalities
Recording and Voice Processing 2: Working in the Studio

SAKHO Ibrahima
Nuclear Physics 1: Nuclear Deexcitations, Spontaneous Nuclear Reactions

2020

DANIELE Vito G., LOMBARDI Guido
Scattering and Diffraction by Wedges 1: The Wiener-Hopf Solution - Advanced Applications (Waves and Scattering Set – Volume 1)

Scattering and Diffraction by Wedges 2: The Wiener-Hopf Solution - Advanced Applications (Waves and Scattering Set – Volume 2)

SAKHO Ibrahima
Introduction to Quantum Mechanics 2: Wave-Corpuscle, Quantization & Schrödinger's Equation

2019

BERTRAND Pierre, DEL SARTO Daniele, GHIZZO Alain
The Vlasov Equation 1: History and General Properties

DAHOO Pierre-Richard, LAKHLIFI Azzedine
Infrared Spectroscopy of Triatomics for Space Observation (Infrared Spectroscopy Set – Volume 2)

RÉVEILLAC Jean-Michel
Electronic Music Machines: The New Musical Instruments

ROMERO-GARCIA Vicente, HLADKY-HENNION Anne-Christine
Fundamentals and Applications of Acoustic Metamaterials: From Seismic to Radio Frequency
(Metamaterials Applied to Waves Set – Volume 1)

SAKHO Ibrahima
Introduction to Quantum Mechanics 1: Thermal Radiation and Experimental Facts Regarding the Quantization of Matter

2018

SAKHO Ibrahima
Screening Constant by Unit Nuclear Charge Method: Description and Application to the Photoionization of Atomic Systems

2017

DAHOO Pierre-Richard, LAKHLIFI Azzedine
*Infrared Spectroscopy of Diatomics for Space Observation
(Infrared Spectroscopy Set – Volume 1)*

PARET Dominique, HUON Jean-Paul
Secure Connected Objects

PARET Dominque, SIBONY Serge
Musical Techniques: Frequencies and Harmony

RÉVEILLAC Jean-Michel
Analog and Digital Sound Processing

STAEBLER Patrick
Human Exposure to Electromagnetic Fields

2016

ANSELMET Fabien, MATTEI Pierre-Olivier
Acoustics, Aeroacoustics and Vibrations

BAUDRAND Henri, TITAOUINE Mohammed, RAVEU Nathalie
The Wave Concept in Electromagnetism and Circuits: Theory and Applications

PARET Dominique
Antennas Designs for NFC Devices

PARET Dominique
Design Constraints for NFC Devices

WIART Joe
Radio-Frequency Human Exposure Assessment

2015

PICART Pascal
New Techniques in Digital Holography

2014

APPRIOU Alain
Uncertainty Theories and Multisensor Data Fusion

JARRY Pierre, BENEAT Jacques N.
RF and Microwave Electromagnetism

LAHEURTE Jean-Marc
UHF RFID Technologies for Identification and Traceability

SAVAUX Vincent, LOUËT Yves
MMSE-based Algorithm for Joint Signal Detection, Channel and Noise Variance Estimation for OFDM Systems

THOMAS Jean-Hugh, YAAKOUBI Nourdin
New Sensors and Processing Chain

TING Michael
Molecular Imaging in Nano MRI

VALIÈRE Jean-Christophe
Acoustic Particle Velocity Measurements using Laser: Principles, Signal Processing and Applications

VANBÉSIEN Olivier, CENTENO Emmanuel
Dispersion Engineering for Integrated Nanophotonics

2013

BENMAMMAR Badr, AMRAOUI Asma
Radio Resource Allocation and Dynamic Spectrum Access

BOURLIER Christophe, PINEL Nicolas, KUBICKÉ Gildas
Method of Moments for 2D Scattering Problems: Basic Concepts and Applications

GOURE Jean-Pierre
Optics in Instruments: Applications in Biology and Medicine

LAZAROV Andon, KOSTADINOV Todor Pavlov
Bistatic SAR/GISAR/FISAR Theory Algorithms and Program Implementation

LHEURETTE Eric
Metamaterials and Wave Control

PINEL Nicolas, BOURLIER Christophe
Electromagnetic Wave Scattering from Random Rough Surfaces: Asymptotic Models

SHINOHARA Naoki
Wireless Power Transfer via Radiowaves

TERRE Michel, PISCHELLA Mylène, VIVIER Emmanuelle
Wireless Telecommunication Systems

2012

LALAUZE René
Chemical Sensors and Biosensors

LE MENN Marc
Instrumentation and Metrology in Oceanography

LI Jun-chang, PICART Pascal
Digital Holography

2011

BECHERRAWY Tamer
Mechanical and Electromagnetic Vibrations and Waves

BESNIER Philippe, DÉMOULIN Bernard
Electromagnetic Reverberation Chambers

GOURE Jean-Pierre
Optics in Instruments

GROUS Ammar
Applied Metrology for Manufacturing Engineering

LE CHEVALIER François, LESSELIER Dominique, STARAJ Robert
Non-standard Antennas

2010

BEGAUD Xavier
Ultra Wide Band Antennas

MARAGE Jean-Paul, MORI Yvon
Sonar and Underwater Acoustics

2009

BOUDRIOUA Azzedine
Photonic Waveguides

BRUNEAU Michel, POTEL Catherine
Materials and Acoustics Handbook

DE FORNEL Frédérique, FAVENNEC Pierre-Noël
Measurements using Optic and RF Waves

FRENCH COLLEGE OF METROLOGY
Transverse Disciplines in Metrology

2008

FILIPPI Paul J.T.
Vibrations and Acoustic Radiation of Thin Structures

LALAUZE René
Physical Chemistry of Solid-Gas Interfaces

2007

KUNDU Tribikram
Advanced Ultrasonic Methods for Material and Structure Inspection

PLACKO Dominique
Fundamentals of Instrumentation and Measurement

RIPKA Pavel, TIPEK Alois
Modern Sensors Handbook

2006

BALAGEAS Daniel *et al.*
Structural Health Monitoring

BOUCHET Olivier *et al.*
Free-Space Optics

BRUNEAU Michel, SCELO Thomas
Fundamentals of Acoustics

FRENCH COLLEGE OF METROLOGY
Metrology in Industry

GUILLAUME Philippe
Music and Acoustics

GUYADER Jean-Louis
Vibration in Continuous Media

Printed and bound by CPI Group (UK) Ltd, Croydon, CR0 4YY
09/05/2023
03217480-0002